D0124673

The Scientific Revolution

The Scientific Revolution

An Encyclopedia

William E. Burns

ABC-CLIO

Santa Barbara, California
Denver, Colorado
Oxford, England

Library of Congress Cataloging-in-Publication Data

Burns, William E., 1959–
The scientific revolution : an encyclopedia / William E. Burns
    p. cm.
Includes bibliographical references and index.
ISBN 0-87436-875-8 (hardcover : alk. paper)
1. Science—Europe—History—Encyclopedias.   I. Title.
Q127.E8 B87 2001
509.4′03—dc21   2001003382

07 06 05 04 03 02 01 10 9 8 7 6 5 4 3 2 1
This book is available on the World Wide Web as an e-book. Visit abc-clio.com for details.

ABC-CLIO, Inc.
130 Cremona Drive, P.O. Box 1911
Santa Barbara, California 93116-1911

This book is printed on acid-free paper ∞.

Manufactured in the United States of America

To my grandfather, Robert Sargent

Contents

The Scientific Revolution: An Encyclopedia

Foreword

Ever since historians first coined the term "scientific revolution" to describe the developments in natural knowledge in the sixteenth and seventeenth centuries, we have been alternately fascinated with and ambivalent about the idea that early modern physicians, astronomers, natural philosophers, naturalists, artisans, and engineers "revolutionized" our understanding of nature. Whether or not we accept the literal or metaphorical meaning of the scientific revolution is almost beside the point at this stage in the discussion. There is no doubt that early modern inquirers into the natural world fundamentally changed our understanding of the cosmos, both in the macrocosm of the heavens and in the microcosm of terrestrial nature. Nicolaus Copernicus's challenge to the geocentrism of Ptolemaic astronomy, Leonhard Fuchs's attempts to create a new kind of botany, and Andreas Vesalius's critique of traditional Galenic anatomy may not be as radical, upon closer inspection, as their rhetoric made them appear, but each helped to initiate a fundamental revision of an entire domain of knowledge. Galileo Galilei turns out to be much more indebted to Aristotelian physics than his noisy attacks on the ancients would have us believe, and Isaac Newton undeniably viewed alchemy as one of the important techniques for grasping the truth of nature, even as he helped to create a new kind of mathematics and physics. And yet none of these important revisions of individual episodes has obscured the crucial fact that the quantity and quality of scientific activity grew immensely in the sixteenth and seventeenth centuries. During this period, small changes in knowledge, as much as singularly dramatic episodes such as the collapse of geocentric cosmology, paved the way for a dramatic revision of the nature and purpose of scientific inquiry in the modern era.

The scientific revolution continues to be one of the central episodes in the history of science and an important aspect of the intellectual and cultural history of the early modern period. It offers us a portrait of traditional knowledge at the height of its explanatory powers—knowledge that had endured for centuries because it was largely based on commonsense observations about how the world appeared and was guided by metaphysical truths that harmonized well with the predominantly Christian outlook of early modern Europeans. At the same time, it is an account of the gradual erosion of tradition, of a nagging dissatisfaction with the status quo. Put a different way, we might describe the scientific revolution as an era that made the idea of questioning knowledge as important as the search for a good answer. As the French philosopher René Descartes famously reminded readers of his *Discourse on Method* (1637), it was a moment that taught us to inquire into ourselves as part of understanding and possessing knowledge of the world.

The past 20 years have witnessed a complete rethinking of the scientific revolution, from its most minute details to the very concept of the "scientific revolution" itself. Many of the initial questions of this reexamination dealt with the early formulation of the scientific revolution by Herbert Butterfield as a series of progressive revolutions in various scientific disciplines that occurred across several centuries—a disciplinary narrative that we have since rejected, becoming aware of the importance of the encyclopedic framework of knowledge that connected rather than separated different forms of inquiry. Historians of science were strongly influenced by Thomas Kuhn's detailed study of the Copernican "revolution" as the paradigmatic episode of the transformation of scientific thought, beginning with Copernicus's *On the Revolutions of the Celestial Spheres* (1543) and culminating with Isaac Newton's *Mathematical Principles of Natural Philosophy* (1687). It was Kuhn who famously offered the transformation of western European astronomy as a case study in shifting paradigms of knowledge. As rich and suggestive as Kuhn's analysis was, it offered a highly selective portrait of early modern science, limiting the scientific revolution exclusively to developments in the mathematical sciences without explaining why fields such as medicine, natural history, and alchemy followed a different path. Subsequent scholarship has made it clear that we cannot understand the transformation in scientific knowledge as a story of a single methodology or discipline.

No account of classic interpretations of the scientific revolution could be complete without mentioning the influential work of Frances Yates, who argued passionately for the importance of the occult sciences in developments in early modern science, giving figures such as John Dee, Giordano Bruno, and Athanasius Kircher a significant place in the struggle to understand the relationship between the natural and the supernatural. In Yates's interpretation, the fascination with original wisdom (*prisca sapientia*) was a driving force behind the quest to interpret nature. While subsequent scholarship has diminished the emphasis she placed on

Hermetic philosophy as a common point of reference for many early modern natural philosophers, her sense that the history of science must look backwards as well as forwards has been retained. If Kuhn could not imagine a revolutionary moment in science before the Copernican "revolution," making Copernicus a window into a modern way of shaping knowledge, Yates could not help but see the same episode as the culmination of everything that had preceded it.

While a great deal of ink was spilled in the early 1980s trying to prove or disprove any of these strongly argued theses about the "essential" nature of the scientific revolution, recent scholarship has taken a more ecumenical view of the subject. Stepping back from these debates, scholars have offered readers a wealth of new case studies that demonstrate how important it is to understand the scientific revolution from its sources, published and unpublished. More sociologically informed analyses of key episodes in the history of early modern science have made the scientific revolution a case study for the relations between knowledge and power. Steven Shapin and Simon Schaffer's important study of Thomas Hobbes's critique of Robert Boyle's experimental philosophy offers an excellent example, offering new sources and new methodologies that help us understand experimental philosophy as an activity that was defined in relationship with emerging ideas of scientific community in the mid-seventeenth century with the appearance of such organizations as the Royal Society of London. Similarly, Mario Biagioli's account of Galileo's career and trial underscores the importance of patronage in the lives of many early modern natural philosophers as they moved among the universities, academies, and courts. While such an approach has not been without its critics, it has nonetheless made the history of the scientific revolution a history of institutions, practices, and community as much as a history of ideas.

There is general agreement nowadays that early modern science was a broad and diverse enterprise, encompassing a wide range of approaches to the natural world, both empirical and mathematical, practical and theoretical, secular and sacred. The problem of teaching the scientific revolution is no longer what to include in relation to a tightly woven account of the subject, but what you inevitably leave out in choosing among the wealth of exciting research initiatives in this field. Should a course on the scientific revolution encompass the sixteenth and the seventeenth centuries? Should it connect this period to the age of Enlightenment? Should it include the growing body of literature on popular as well as learned understandings of nature? Should it encompass accounts of women natural philosophers and the role of gender in creating knowledge? While the emergence and diffusion of heliocentrism is still the central drama of the scientific revolution, it is surrounded by dozens of other equally important episodes that speak to such issues as the nature of empiricism, the emergence of scientific methodology, the formation of scientific communities, and the political, economic, and cultural contexts that gave meaning to the act of interpreting nature.

William Burns's guide to the scientific revolution offers readers a synthesis of recent research on the scientific revolution in an accessible format that reflects recent trends in scholarship and will guide readers through this literature. The *Dictionary of Scientific Biography*, the standard introductory reference work for the history of science, is now more than two decades old. More recent reference projects such as William Bynum, Janet Browne, and Roy Porter's *Dictionary of the History of Science* (1981) and R. C. Olby's *Companion to the History of Science* (1990) offer only limited entries on the scientific revolution. Burns's book can profitably be used with introductory surveys of the scientific revolution as a guide to further reading on some of the most interesting aspects of this subject.

Readers of this book will find biographical entries for well-known figures such as Tycho Brahe and Robert Boyle juxtaposed with equally important figures such as the Spanish physician Francisco Hernández, whose natural history of Mexico represents an early example of the relationship between science and empire, and Margaret Cavendish, the most prolific female natural philosopher of the early modern period. The religious aspects of the scientific revolution enjoy a new depth and significance in this volume. Only a decade ago, an entry on Jesuit science in such a book would have been very short indeed. But today there is a rich literature that reveals the important role of this religious order in promoting and organizing scientific activities, not only in Europe but throughout the world. Similarly, the inclusion of subjects such as prodigies and weather reflects a new interest in the more quotidian aspects of interpreting nature that engaged a much broader sector of the population than the handful of learned physicians and philosophers. And entries on such subjects as the circulation of the blood and the clitoris serve to remind readers that medical debates in the early modern period entailed a rethinking of the human body that was no less earthshaking or controversial than the realignment of the heavens.

Like any reference work, this guide reflects the interests and preferences of its author. But it also casts its net widely, defining "science" not just in a narrow sense but in relation to such subjects as art, music, and literature and introducing readers to an interdisciplinary literature on the scientific revolution that is not always included in standard accounts of this period. Open the book to any entry, and see why so many historians continue to think that the developments in natural knowledge in the early modern period represent a fundamental and interesting chapter in the history of knowledge.

Paula Findlen

Preface

One central aspect of the world we live in is the dominance of science. People look to science to cure diseases, feed the hungry, and provide vistas of wonder in the contemplation of the universe; people blame science for ecological collapse, racial, gender, and regional inequalities, and spiritual alienation. Millions of people are employed in science or science-based industries; scientists like Richard Feynman and Stephen Hawking have become folk heroes, while the "mad scientist" is one of our culture's most characteristic villains. We expect science to be a realm of constant innovation and change, sometimes benevolent and sometimes terrifying.

Yet it was not always thus. In European and other civilizations, theoretical science was a marginal activity before the seventeenth century, practiced by few and possessing little cultural authority. Science involved commentary on ancient texts as much as it did direct observation or experiment, and it was not marked by dynamism and progress. The early modern period was a time of great changes, both in the content of science and in its cultural role. This revolutionary process was marked by the overthrow of the authority of the ancient Greek natural philosophers, astronomers, and physicians. In content, astronomy moved from the Earth-centered universe of Ptolemy and Aristotle to the vast Sun-centered universe of Copernicus, Kepler, and Galileo. In physics, qualitative Aristotelianism was replaced by mathematical and mechanical Newtonianism.

Changes in astronomy and physics were the most dramatic and far-reaching during the period, and for that reason have perhaps had a disproportionate influence on subsequent conceptions of how scientific revolution occurs. Changes in other areas of the sciences were also of great importance. In medicine, the humoral medicine of Galen was challenged first by a plethora of new anatomical discoveries, the most dramatic being William Harvey's discovery of the circulation of the blood, and second by the rise of chemical and mechanical theories of bodily function. In natural history, Europe received a flood of new information from the newly encountered areas of the world, and eventually the understanding of plants, animals, and stones was dramatically transformed.

These changes in science came from a myriad of sources, both within the European tradition and beyond. Like many revolutionaries, early modern scientists often presented themselves as restorers of the past, as Pierre Gassendi presented atomism as the revival of the philosophy of Epicurus. The new science built on the old—on the Arab astronomers and mathematicians, the Aristotelianism of the medieval universities, and the scholarship of Renaissance humanists. The humanist revival of other ancient philosophical schools—Platonist, Epicurean, Skeptical, and Stoic—stimulated new thinking about the natural world. Physicians, mathematicians, astrologers, and alchemists all participated in scientific change, although magic itself, like

Aristotelian natural philosophy, was ultimately one of the losers in the revolution. The scientific revolutionaries themselves, for all their undoubted and awe-inspiring genius, were not abstract "great minds" standing outside the European society and culture of their day. Much scientific ingenuity was devoted to practical problems, such as the discovery of the longitude, and early modern science was shaped by the political, religious, and gendered positions of its makers. Like other revolutionaries, the makers of the scientific revolution quarreled fiercely among themselves, over everything from different claims to priority in a discovery to the role of the Bible in scientific questions—Galileo's enemies included astronomers as well as theologians.

With these dramatic changes, science assumed a much more central cultural role in Europe. In the seventeenth century, scientific institutions spread across Europe, developing from the temporary and informal Italian groupings, the Accademia dei Lincei and the Accademia del Cimento, into the permanent institutions of England and France founded in the 1660s, the English Royal Society and the French Royal Academy of Sciences. Networks of correspondence were supplemented by printed periodicals, including *Philosophical Transactions,* the *Journal des savants,* and others, that brought news of the new science to a broad public. The world of Copernicus, toiling in his lonely study and circulating manuscripts among a few peers, gave way to what is to us the far more recognizable world of Edmond Halley—promoter, editor, civil servant, sea captain, and scientist. New scientific instruments, such as the telescope, microscope, and air pump, were in the hands of more and more people. Experiments, which had played only a very minor role in science before the scientific revolution, now not only were essential to scientific work but were presented to curious audiences by lecturers and demonstrators as a form of intellectual entertainment. The heroic narrative of the rise of the new astronomy and physics from Copernicus's *On the Revolutions of the Celestial Spheres* in 1543 to Newton's *Mathematical Principles of Natural Philosophy* in 1687 was a founding myth of the eighteenth century Enlightenment, with its central scene being the trial of Galileo in 1633. Galileo and Newton, with a host of others, were exemplary figures, and the seventeenth century, the climax of the scientific revolution, was identified as the "century of genius."

Nor was this expansion of science's cultural presence restricted to Europe. The age of the scientific revolution was also the great age of European colonial expansion, and European science thereby acquired a global reach, from García d'Orta in India and Francisco Hernández in Mexico in the sixteenth century to Anna Maria Sibylla Merian in Surinam and Georg Eberhard Rumph in Indonesia at the close of the seventeenth. In China and Japan, the first faint stirrings of the process by which Western science—the science of the scientific revolution—was to overcome the other developed scientific traditions of the world were noticeable by 1700.

Like many revolutions, the scientific revolution overthrew one authority—that of the ancients—to lay the foundations for a new and more despotic authority, that of science itself. Science became an ideological justification both for conserving and changing the social order. Thomas Hobbes sought to ground political sovereignty in the mechanical philosophy, and political arithmeticians claimed to analyze human society quantitatively. Men used the latest scientific discoveries to support their power over women, and Europeans asserted on scientific grounds their superiorities over the world's other peoples. These potentials of science were hardly fully developed in the scientific revolution itself—the dominant set of intellectual categories for analyzing society was still religious at the close of the seventeenth century—but the seeds were there.

These transformations in both the content and the cultural role of science are the subject of this book. Chronologically, it covers the period from the remarkable generation whose contributions were made in the second quarter of the sixteenth century—the generation of Copernicus, Paracelsus, Vesalius, and Agricola—to the establishment of the Newtonian synthesis in England and the Netherlands in the early eighteenth century. Articles discuss the leading personalities, ideas, instruments, and institutions that created early modern science, as well as the scientific disciplines themselves. Attention is devoted to the losers—the Aristotelians, the Galenists, and the magicians such as Robert Fludd—as well as to the winners of the scientific revolution. Although no single volume could cover the entire breadth of early modern science, I hope this book gives a unified overview of that complex and fascinating phenomenon, the scientific revolution.

I would like to thank the Folger Shakespeare Library, the Library of Congress, the Franklin Library of the University of Pennsylvania, the McKeldin Library of the University of Maryland, and the Simpson Library of Mary Washington College, which provided many of the sources for this book. My editor at ABC-CLIO, Kevin Downing, was extremely helpful. I also thank Deborah Lynes and Patricia Heinicke Jr. for copyediting above and beyond the call of duty. I learned how to study the scientific revolution from the late Betty Jo Teeter Dobbs and Paula Findlen, who also generously provided the foreword for this book. I also thank Wilbur Applebaum, Nathan Sivin, and Pamela Griffith for their help.

William E. Burns

Academies and Scientific Societies

Scientific societies emerged in early modern Europe from the world of academies—structured gatherings of intellectuals and writers, sometimes linked to a court but independent of a university. The academic movement was closely associated with humanism and was originally strongest in Italy, the heartland of humanism. In the late fifteenth century, the famous Platonic Academy of Florence actively disseminated Neoplatonic philosophy and preached Hermeticism. Academies eventually gave rise to full-fledged scientific societies, of which by the late seventeenth century the most important were the Royal Society in England, chartered in 1662, and the Royal Academy of Sciences in France, founded in 1666.

The earliest academy known to devote itself to science was the Neapolitan Academy of the Secrets of Nature (Accademia dei Segreti), of which Giambattista della Porta was the leading spirit. That academy led an obscure existence for several years in the 1550s, until it was closed down by the religious authorities under suspicion of practicing magic. The next Italian scientific academy of importance was the Academy of Lynxes (Accademia dei Lincei), founded by the Roman nobleman Federico Cesi (1585–1630), of which the leading members were della Porta and Galileo Galilei. Like the Academy of the Secrets of Nature, it faced church opposition.

Organized natural philosophy in the seventeenth century was sometimes presented as inspired by Francis Bacon, whose *New Atlantis* (1627) set forth a vision of an idealized scientific society. Bacon's vision went beyond that of the early academies, in which members carried out their own projects but met for the purpose of discussion, to a more hierarchical vision in which scientific leaders would come up with plans and subordinates would carry them out. Bacon's influence was strongest in England, significant in France and to a lesser degree in Germany, and slight in Italy. In England, the English Civil War and Interregnum of the mid-seventeenth century saw the formation of several informal groups for the pursuit of natural knowledge, notably the "invisible college" associated with Samuel Hartlib (c. 1600–1662) and the group surrounding John Wilkins at Oxford. In France, there were informal gatherings around Théóphraste Renaudot and Marin Mersenne in Paris and other active groups in Montpellier and Caen. The most formal group (it had a written body of rules by 1657) to appear in mid-seventeenth-century France was the Parisian Montmor Academy (Académie Montmor), named after its patron, the wealthy nobleman and Cartesian Habert de Montmor (1600–1679), in whose home it met weekly. Emerging from an informal group presided over by Pierre Gassendi, who lived in Montmor's house, the Montmor Academy was a semipublic body by the late 1650s. Great occasions like the reading of Christiaan Huygens's paper on Saturn's rings in 1658 attracted leaders of the French state, church, and society in addition to academy members. In 1664, the Montmor Academy moved from Montmor's house to that of Melchisedech Thevenot (1620–1692). Germany had the College of the Curiosities of Nature (Collegium Naturae Curiosorum), a physician-dominated organization founded in 1651 that published collections of observations.

The differences between the sixteenth- and early-seventeenth-century academies and the scientific societies that emerged in the 1660s are that the new societies had a much stronger institutional and corporate existence no longer dependent on a particular patron or intellectual leader and that they were much more closely tied to the state. An important transitional institution was the Accademia del Cimento, which for the decade or so of its existence was situated between the private patronage of the Medici family and the public patronage of the Florentine state. The Royal Society and the Royal Academy of Sciences, from the beginning much more permanent institutions, were both located in national capitals of large kingdoms and flaunted their connection to monarchical patrons in their names. (The College of the Curiosities of Nature participated in this movement, in 1687 receiving an imperial charter and a new name, the Leopoldine Academy, after reigning Holy Roman Emperor Leopold I [r. 1658–1705].) Despite these similarities, the Royal Society and the Royal Academy presented fundamentally different models. The Royal Society, lacking state funding, was a gathering of natural philosophers and virtuosos, for which male gender and an interest in natural philosophy were sufficient membership qualifications.

The Royal Academy was a much more professional body, whose members were paid salaries and expected to produce scholarship that would redound to the glory and profit of the French state. The model of the Royal Academy proved the more influential, particularly given the high prestige of French culture generally in the age of the Sun King, Louis XIV (r. 1643–1715). Gottfried Wilhelm Leibniz, a great believer in scientific organization, promoted the establishment of academies on the French model in Berlin, the capital of Prussia, and in St. Petersburg, the capital of Russia. The Berlin Academy was founded in 1700, and the St. Petersburg Academy was founded in 1724, after Leibniz's death. These academies were mostly staffed by French and other western European scientists. The year 1700 also saw the foundation of the first Spanish scientific society, the Royal Society of Medicine and Other Sciences of Seville, a physician-dominated group that faced accusations of heresy.

Both the Royal Society and the Royal Academy dominated scientific culture in their respective territories, reducing provincial societies to adjunct status. The Royal Society had as affiliates the Dublin Philosophical Society founded by Sir William Petty (1623–1687) in 1684, the Oxford Philosophical Society (a revival of the Wilkins group), and a very short-lived Somerset Society, headed by the Reverend Joseph Glanvill (1636–1680). French provincial societies were squeezed by the French government's desire to centralize French intellectual life in Paris, and they were increasingly subordinate to the Royal Academy's agenda, carrying out astronomical observations for the academy, for example. They were more fortunate than Thevenot's Montmor Academy, which was driven out of existence entirely. The scientific academies of Italy were of the second rank after the breakup of the Accademia del Cimento. They included the physician-dominated Academy of Investigators in Naples and the Physical-Mathematical Academy in Rome, the latter patronized by the exiled Swedish monarch Queen Christina (1626–1689) and both more devoted to learned conversation than to advancing the frontiers of knowledge.

See also Accademia dei Lincei; Accademia del Cimento; Royal Academy of Sciences; Royal Society.
References
Brown, Harcourt. *Scientific Organization in Seventeenth-Century France (1620–1680).* Baltimore: Johns Hopkins University Press, 1934.
McClellan, James E. *Science Reorganized: Scientific Societies in the Eighteenth Century.* New York: Columbia University Press, 1985.
Moran, Bruce, ed. *Patronage and Institutions: Science, Technology, and Medicine at the European Court, 1500–1750.* Woodbridge, England: Boydell, 1991.

Accademia dei Lincei

The Academy of Lynxes (lynxes being thought to have keen vision), one of the earliest scientific academies, was created by the Roman aristocrat and botanist Federico Cesi (1585–1630), who hoped to be a great patron of learning. The academy went through a number of phases in its life from 1603 to 1629, its most illustrious member being Galileo Galilei. After its founding with elaborate rituals on Christmas Day, 1603, it entered its abortive first phase as an ascetic order composed of three of Cesi's personal followers. Cesi's father, the duke of Acquasparta, shut it down, fearing his son was becoming involved with men potentially as dangerous as the recently burned Giordano Bruno. On the old duke's death in 1610, Cesi reestablished the academy with a looser organization, its dominating intellectual presence being the natural magician Giambattista della Porta, head of the academy's Neapolitan branch. In 1611, Galileo was admitted and soon eclipsed della Porta, replacing a natural magic–based program with a mathematical one. The Accademia dei Lincei did not become officially Copernican, but it endorsed freedom of inquiry in natural philosophy and was somewhat hostile to the Jesuit Aristotelians of the

Collegio Romano. Elitist, the academy had only a few dozen members, and its infrequent meetings were principally devoted to discussion. The most active phase of the academy's history ended with the condemnation of Copernicanism by the church in 1616 and Cesi's retreat from Rome under acute financial embarrassment in 1618. However, it continued an ambitious publishing program, its central project being the publication of an Italian version of a Spanish work by Francisco Hernández on the natural history of Mexico. The first and only installment of this project appeared in 1628. The academy also published Galileo's *Assayer* (1623).

See also Galilei, Galileo; Porta, Giambattista della.
Reference
Biagioli, Mario. *Galileo, Courtier: The Practice of Science in the Culture of Absolutism.* Chicago: University of Chicago Press, 1993.

Accademia del Cimento

The first academy devoted to experimental science, the Academy of Experiment, was founded in Florence in 1657 by Ferdinand II (r. 1621–1670), the grand duke of Tuscany, and his brother Prince Leopold of Tuscany. The Tuscan court had been associated with science since the days of Galileo, and some of the academicians, such as the mathematician and engineer Vincenzo Viviani (1622–1703), were pupils of Galileo. The academy's natural philosophy was also in the Galilean tradition in being anti-Aristotelian.

The academy's exact membership is unknown, but it was quite small, possibly under a dozen. The most distinguished academician was the mathematician and physicist Giovanni Alfonso Borelli (1608–1679), professor of mathematics at the University of Pisa. The academy's only publication was the lavishly illustrated folio *Examples of Natural Experiments Made by the Academy of Experiment under the Protection of His Most Serene Highness Prince Leopold of Tuscany and Described by the Secretary of the Academy* (1667). It covered experiments made up to 1662, but it took five years to be published because of the need to get commentary from the academicians and the perfectionism of Lorenzo Magalotti (1637–1712), the academy's secretary, who was responsible for the work. The experiments mostly dealt with physics, including demonstrating the existence of a vacuum, against the Aristotelians, as well as investigating pneumatics, electricity, magnetism, and freezing (a particular interest of Leopold's.) The academy also engaged in astronomical research, including a close study of Saturn and its mysterious "handles" made in the summer of 1660 to test Christiaan Huygens's ring theory.

The academy had no charter and lacked the corporate existence of subsequent state-sponsored academies such as the Royal Society and the Royal Academy of Sciences. It was personally dependent on Prince Leopold, meeting not on a regular schedule but at Leopold's convenience. It did not carry on correspondence as a corporate entity, although Leopold and individual members corresponded on its behalf. The experiments in *Examples of Natural Experiments* were credited not to individual academicians but to the academy itself. The book circulated as a gift from Leopold rather than being sold in the market; this fact, along with the delay and the fact that many scientists outside Italy were ignorant of the language, limited the work's impact on scientific thought. By the time it was translated into English (1684), Latin (1731), and French (incomplete translations in 1754 and 1755), science had passed it by. The academy ended in 1667 when Leopold became a cardinal and the restless Borelli retired to Messina.

See also Courts.
References
Biagioli, Mario. "Scientific Revolution, Social Bricolage, and Etiquette." In Porter, Roy, and Mikulas Teich, eds. *The Scientific Revolution in National Context.* Cambridge and New York: Cambridge University Press, 1992.
Middleton, W. E. Knowles. *The Experimenters: A Study of the Accademia del Cimento.* Baltimore: Johns Hopkins University Press, 1971.

Acosta, José de (1540–1600)

The Jesuit José de Acosta was the greatest natural historian of the Spanish Empire in the Americas in the sixteenth century. From a merchant background, Acosta entered the Jesuits as a novice in 1552. He was educated at Jesuit schools and the University of Alcalá. Acosta arrived in Peru as a missionary in 1572 and left in 1586, rising to the post of provincial, or head of the Jesuits, in Peru. On returning to Spain, Acosta dabbled in politics, supporting the king of Spain in complex struggles involving Spain, various Jesuit factions, and the Spanish Inquisition. He ended his life as rector of the Jesuit College at Salamanca.

Acosta's books on his Peruvian experience, notably *Natural and Moral History of the Indies* (1590), promoted the evangelization of the natives of the Americas. Acosta covered the history, geography, weather, plants, animals, and native inhabitants of Peru and Mexico, arguing for the capacity of Native Americans to receive Christianity, provided it was presented in a way that Acosta deemed suitable for their understandings. Acosta argued that God had providentially prepared the Indians for Christian conversion, and that their customs, which shocked Europeans, did not render them less than human or beyond redemption. This argument led him to examine Native American civilization and place it in a hierarchy of human civilizations. The highest level was occupied by Europeans and other Old World peoples such as the Chinese and Japanese, who, it was generally agreed in Europe, were suitable for conversion. The Peruvian and Mexican Indians were on the second level, having cities and state organization but lacking written language and philosophy, and Acosta thought them also suitable for conversion. Societies could rise or decline in the hierarchy.

Like his anthropology, Acosta's natural science was basically Aristotelian. (Unlike many natural historians of the Spanish and Portuguese Empires, he was not a physician or primarily interested in medicinal uses.) Like other Jesuit Aristotelians, Acosta was not afraid to contradict specific Aristotelian assertions when they conflicted with reality, pointing out, for instance, that Aristotle's belief that the tropics would be too hot for human beings was clearly wrong. Although Acosta was struck by the strangeness of the flora, fauna, and inhabitants of the New World, he argued that they had ultimately originated in the Old World and that all were part of the same divine creation. Acosta's books on the Indies were written in Spanish rather than Latin, indicating that he was not primarily addressing a Europe-wide audience of the learned. However, they were shortly translated into several European languages as well as Latin, becoming a basic source of European knowledge of the New World.

See also Exploration, Discovery, and
   Colonization; Jesuits; Race.
Reference
Burgaleta, Claudio M., S.J. *José de Acosta, S.J.
   (1540–1600): His Life and Thought.* Chicago:
   Jesuit Way, 1999.

Acoustics
See Music.

Acta Eruditorum
See Periodicals.

Agricola, Georgius (1494–1555)

The German dyer's son Georg Bauer achieved distinction in the fields of humanism (Latinizing his name as was the humanist custom), medicine, geology, and technology. He received a B.A. from the University of Leipzig in 1515 and then studied medicine at Leipzig and at several Italian universities. As a medical humanist, he worked on the great Aldine edition of Galen's (A.D. 129–c. 199) work published at Venice. In Italy he became a friend and intellectual ally of the celebrated humanist scholar Desiderius Erasmus (1469–1536). Agricola returned to central Europe

in 1527 as town physician of Joachimstal in Bohemia, where the economy was dependent on mining. In 1533 he became the town physician of Chemnitz in Saxony, where he was one of the most wealthy and powerful citizens, serving as mayor four times and investing in mining stocks. As a physician, he studied the diseases of miners, introduced the practice of quarantine to Germany, and wrote *Of the Plague* in 1555, the result of his work during the plague epidemic of 1551–1552. He also studied the pharmacological uses of minerals. Agricola acquired Duke Moritz of Saxony (r. 1541–1553) as a patron, who subsidized him, employed him on diplomatic missions, and smoothed away any problems that arose for Agricola as a Catholic in Lutheran Saxony.

Agricola's chief publications in mineralogy were *On the Causes of Subterranean Things* and *On the Nature of Fossils,* both published in 1546. As was the common usage, Agricola referred to any natural object dug out of the earth as a fossil. *On the Causes* contains the first extensive discussions of the role of wind, water, and other forces in shaping the geological landscape. Agricola also was the first to extensively analyze the crust in terms of strata. *On the Nature of Fossils* contains extensive physical descriptions of over 80 minerals and a classification scheme for the contents of the Earth's crust, drawing a distinction between simple, homogeneously mixed, or "compound," and heterogeneously mixed substances. Agricola believed that metallic ores were carried through the Earth as solutions, or "juices," in water.

Agricola's first publication on mining and metallurgy was a dialogue, *Bermannus* (1530), for which Erasmus wrote the introduction. It was superseded by his posthumously published *Of Metallic Things,* the outstanding technological classic of the sixteenth century. Building on a mostly German tradition of studies in mining, Agricola treated mining in its legal, economic, administrative, and medical contexts as well as covering mining engineering, the identification, refinement, and

Georgius Agricola described this device for pumping fresh air into mines in Of Metallic Things. *(Burndy Library)*

processing of metallic ores, and various other chemical processes. The lavishly illustrated *Of Metallic Things* combines humanist learning—Agricola was intimately acquainted with classical literature on mining—with the results of Agricola's own observations and experiments. What theory the work contains is Aristotelian. It retained its leading status for over a century and was the starting point for Isaac Newton's investigations of metallurgy.

See also Mining.
Reference
Wolf, A. *A History of Science, Technology, and Philosophy in the Sixteenth and Seventeenth Centuries.* 2d ed., prepared by Douglas McKie. London: Allen and Unwin, 1950.

Agriculture

The economy of early modern Europe was principally based on agriculture, and many farmers and landowners were keenly interested in agricultural improvement. This interest overlapped with the practical application of natural philosophy.

The body of written agronomy available to farmers and landlords at the beginning of the early modern period was medieval and ancient Roman. Prized as sources of wisdom were not only the ancient Roman writers of agricultural manuals, such as the elder Cato, but also the poet Virgil, whose *Georgics* could be read as an agricultural textbook as well as a great Latin poem. All of these works were given good new humanist editions and were widely circulated in both Latin and vernacular translations. They inspired efforts to identify the more obscure crops grown by the ancients, many of which had been lost to Europeans since the barbarian invasions that destroyed Roman civilization. Plants could be referred to by several different names, even within the same country or region, and this difficulty alone is sufficient to explain why establishing a consistent terminology was a formidable task for early modern botanists and agriculturalists. The ancient works were joined by modern books, of which the most influential in the early phases of the scientific revolution was the *Theater of Agriculture* (1600) by the French Protestant landowner Oliver Des Serres (1539–1619). Like many works of agriculture, it was aimed at the wealthy and noble landowner rather than the working farmer. Writing at the end of the French Wars of Religion, Des Serres hoped that French nobles would devote themselves to cultivating their estates rather than causing trouble for the king. He encouraged the cultivation of domesticated grasses and of New World crops such as corn and potatoes, and he was an avid promoter of silkworm cultivation in France.

Early modern botanists concerned themselves principally with the medicinal rather than the agricultural uses of plants. However, there was interest in plants grown as crops for food and forage, and seeds for domesticated plants were exchanged among early modern botanists and natural historians such as Ulisse Aldrovandi. Aldrovandi received the seeds of many different crops from Spain, where the medieval Muslims had practiced a very productive and sophisticated agriculture, now in decline after the Christian conquest of Granada in 1492. The improvement of English agriculture was among the interests of the Hartlib circle in the mid-seventeenth century, and the Hartlib correspondence deals with the introduction of forage crops into England and the increasing domestication of clover, along with other ambitious projects to improve English agriculture. The Hartlib circle encouraged an experimental approach to agriculture, and others in the seventeenth-century followed suit. Captain Walter Blith, a veteran of Cromwell's army, conducted a series of agricultural experiments on a poor piece of land, manuring different sections with different manures and recording the results in his *English Improver Improv'd* (1652). A more sophisticated scientist, the German chemist Johann Rudolf Glauber (1604–1670) experimented with artificial fertilizers and tartar derived from wine, and he had several experimental plots. The late-seventeenth-century English landowner and agricultural writer John Worlidge built on Glauber's work, applying a basically Paracelsian salt-mercury-sulfur chemistry to agricultural questions. As befit an institution many of whose leading members were landowners, the Royal Society set up the Georgical Committee in 1664 to circulate a list of questions on land use to English farmers with the intention of improving English agriculture. One landowner, fellow of the Royal Society and member of the Georgical Committee John Evelyn (1620–1706), published a book on forestry, *Sylva* (1664), which was widely circulated. It included a treatise on growing fruit trees for cider, entitled "Pomona." As a result of the work of these and other English agricultural improvers, England had displaced the Netherlands as the home of Europe's most admired agriculture by the end of the seventeenth century. Most of the writings of natural philosophers and theoretical agriculturalists, however, had little impact on the farming practices of peasants and landowners, and the gap between the promises of agricultural improvement and the meager and halting reality was a frequent target for satirists.

See also Botany.

References

Ambrosoli, Mauro. *The Wild and the Sown: Botany and Agriculture in Western Europe: 1350–1850.* Translated by Mary McCann Salvatori. Cambridge and New York: Cambridge University Press, 1997.

Fussell, G. E. *Farms, Farmers and Society: Systems of Food Production and Population Numbers.* Lawrence, KS: Coronado, 1976.

Air Pumps

The air pump, which created a vacuum by removing air from a sealed container, was one of the most dramatic inventions of the seventeenth century. Its originator was Otto von Guericke (1602–1686), who publicly demonstrated an air pump in 1654, although he may have designed one much earlier. Von Guericke's most impressive demonstration, made before the Diet of the Holy Roman Empire at Ratisbon, was that of the Magdeburg Hemispheres, two hollow, bronze hemispheres joined together. The air was evacuated through a passage in one of them, which was then sealed. Two teams of eight horses could not pull the spheres apart. Von Guericke used pumps to demonstrate the existence of the vacuum and the weight of the air and to establish its density, eventually publishing a book called *New Experiments* (1673).

Earlier, von Guericke's work had inspired the Jesuit physics professor Kaspar Schott (1608–1666) to publish *Mechanica Hydraulica-Pneumatica* (1657). This book inspired the most notable user of the air pump, Robert Boyle. Boyle worked with an air pump that was built, after several tries (air pumps were notoriously difficult to build, maintain, and transport) by Robert Hooke. This air pump enabled Boyle to put things into the chamber before emptying it of air, and many of the experiments described in Boyle's *New Experiments Physico- Mechanicall Touching the Spring of the Air, and Its Effects* (1660) and *A Continuation of New Experiments Physico-Mechanicall Touching the Spring and Weight of the Air* (1669) record the effects of vacuum on various things. Animals died, candles went out, and the column of a barometer fell, demonstrating the existence of atmospheric pressure. Sound also died in the vacuum, but magnets were unaffected. The remainder of

This air pump belonged to Francis Hauksbee, a public lecturer on natural philosophy in the late seventeenth century. He required an air pump that was reliable, transportable, and somewhat ornamental. (Royal Society)

the seventeenth century saw further modifications making air pumps easier to use, and experiments by von Guericke, Huygens, and others. The air pump became a symbol of the "new philosophy," and as airpumps became more portable and reliable they became part of the standard equipment of the public scientific demonstrator and lecturer.

See also Boyle, Robert; Experiments; Hooke, Robert; Vacuum.
Reference
Wolf, A. *A History of Science, Technology, and Philosophy in the Sixteenth and Seventeenth Centuries.* 2d ed., prepared by Douglas McKie. London: Allen and Unwin, 1950.

Alchemy

Alchemy, at its most basic the art of refining and mixing various substances, had a long and complex history by the time of the scientific revolution, involving Chinese, Greek, Jewish, and Arab influences. Like modern chemistry, alchemy involved furnaces, beakers, and tubes, and the earliest laboratories were for alchemical work. Alchemists pioneered laboratory procedures such as distillation. Unlike modern chemistry, alchemy could also involve spells, prayers, incantations, and horoscopes taken at specific moments in the process. Alchemists also differed from modern chemists in viewing certain states of matter as more perfect than others, as gold represented the perfect state of metal. Alchemists could help this perfection emerge by purging impurities.

Alchemical practice was very diverse in the sixteenth century, from that of the philosophical alchemists, who interpreted alchemical substances allegorically and the alchemical process spiritually, to more practical workers who sought to transmute base metals into gold by preparing a substance known as the philosopher's stone. This was a goal for which early modern rulers were often willing to provide financial backing, and this form of alchemy was easily exploited by charlatans of the type satirized in Ben Jonson's (1572–1637) play *The Alchemist* (1610). Others applied alchemical principles to medicine, particularly the Paracelsian "chemical physicians." Alchemy was often associated with a closed system of communication, where alchemical adepts did not publicize the processes they had discovered but passed them down to their students, often using a cryptic notation or a pseudonym. The advent

of printing began to change this, and many alchemical texts were published, most notably the vast compilations *Theatrum Chemicum* (1602–1661) and *Theatrum Chemicum Britannicum* (1652). Alchemists such as Andreas Libavius promoted the teaching of alchemical knowledge in schools and universities.

Alchemists, unlike many natural philosophers, promoted an activist stance toward the natural world, striving not merely to understand it but to control natural processes, an aspect of the alchemical tradition that influenced Francis Bacon. In this, alchemical practice was compatible with a number of philosophical descriptions of the natural world, including Scholastic Aristotelianism as well as more radical approaches such as atomism. On the other hand, biblical alchemists such as Paracelsus promoted alchemical philosophy as an alternative to "pagan" Aristotelianism. Alchemists often employed a microcosm-macrocosm analogy, and many but not all alchemists saw natural substances as active, as opposed to the dead nature approach of some practitioners of the mechanical philosophy. Alchemical processes were referred to by terms that pictured chemical substances as organic, like "birth," "death," "resurrection," and "marriage." These terms were not always simply metaphors, but were considered accurate descriptions of processes. The changes undergone by chemical substances were also capable of a spiritual interpretation, and spiritual alchemists such as Robert Fludd promoted alchemy as a way of purifying and perfecting the soul of the adept as the alchemical substances were purified and perfected. The close connection between alchemy and religion extended as far as identifying the three material principles identified by Paracelsus—salt, sulfur, and mercury—with the Holy Trinity, or reading the story of the death and resurrection of Jesus Christ as a series of alchemical metaphors providing directions for the creation of the philosopher's stone, a move that aroused the particular anger of the antimagician Marin Mersenne. Mystical and millenarian alchemists such as the German

Protestant Jakob Böhme (1575–1624) also identified alchemical processes with the Second Coming of Jesus Christ. Alchemists fought bitter polemical wars among themselves as well as against the opponents of alchemy in general, such as the Jesuit scientist Athanasius Kircher or Galenic physicians who attacked alchemy along with Paracelsianism.

Eminent late-seventeenth-century scientists such as Robert Boyle, who believed that he had transmuted elements, and Isaac Newton, who left voluminous manuscript records of alchemical experiments, were alchemical practitioners, combining chemical and mechanical philosophy. Newton's willingness to consider action at a distance in the form of universal gravitation, forbidden by the strict mechanical philosophy, can be traced in part to his alchemical studies. The general tendency by this period was for practical and spiritual alchemy to drift further apart, the former being absorbed into chemistry, the latter into the magical tradition. Central Europe, where alchemy remained vital throughout the eighteenth century, was a partial exception where the spiritual and practical remained joined. Alchemical symbolism also played a major role in eighteenth-century freemasonry.

> *See also* Boyle, Robert; Chemistry; Croll,
>   Oswald; Libavius, Andreas; Newton, Isaac;
>   Paracelsianism; Starkey, George.

References
Dobbs, B. J. T. *The Foundations of Newton's
    Alchemy: or "The Hunting of the Greene Lyon."*
    Cambridge: Cambridge University Press,
    1975.
Smith, Pamela H. *The Business of Alchemy: Science
    and Culture in the Holy Roman Empire.*
    Princeton: Princeton University Press, 1994.
Taylor, F. Sherwood. *The Alchemists: Founders of
    Modern Chemistry.* London: W. Heinemann,
    1951.

Aldrovandi, Ulisse (1522–1605)

Renaissance Italy's foremost natural historian, Ulisse Aldrovandi was the son of a civil servant in Bologna in the Papal States.

Ulisse Aldrovandi's pictures and descriptions of domestic fowl are of interest to modern historians of poultry raising. This illustration entitled "The Female Paduan" is found in his Ornithologia *(1600).*

Educated at the Universities of Bologna and Padua, he received an M.D. and a Ph.D. from Bologna in 1553, joining the faculty the same year. Aldrovandi worked in many natural-historical fields, from botany to the study of monsters to embryology. He was the first to observe and describe the stages by which a chick develops in an egg. Aldrovandi participated in a network of botanists and natural historians, sending plants and other items across Italy and Europe, and he built the outstanding natural history collection of his time. His real stroke of good fortune occurred with the election of Pope Gregory XIII, a relation of his mother, in 1572. Gregory gave Aldrovandi money to publish his works and provided backing for his efforts to build the botanical garden at Bologna, founded in 1568. Aldrovandi was also allied with powerful cardinals, Italian princes, and members of the Senate of Bologna, and he was appointed *protomédico,* or city official in charge of regulating medical practice, in 1574, leading to a nasty but ultimately victorious struggle against the local college of physicians.

Aldrovandi's overall intellectual framework was Aristotelian. His lavishly illustrated folios were the ultimate expression of the Renaissance encyclopedic approach to nature, considering each creature in terms not only of its natural characteristics but also its literary, historical, and allegorical meanings. The time and energy he devoted to his collection, including over 20,000 items by the end of the century, not to mention his professional and civic duties, hindered him from publishing, and only four volumes were published in his lifetime. Aldrovandi willed his collection to the Senate of Bologna on condition that they see to the publication of his books. They hired two men in succession to supervise the collection, now a museum and a popular sight for visitors to Bologna, and to publish Aldrovandi's books, most of which appeared posthumously.

See also Museums and Collections; Natural
    History; Zoology.
Reference
Findlen, Paula. *Possessing Nature: Museums,
    Collecting, and Scientific Culture in Early Modern
    Italy.* Berkeley and Los Angeles: University of
    California Press, 1994.

Algebra

See Mathematics; Viète, Francois.

Anatomy

The scientific revolution was the heroic age of European anatomy, marked by radical changes and innovations in the understanding of the body's structure. Modern anatomy emerged in the sixteenth century from two sources. The first was the desire of painters and sculptors to more accurately represent the human form. Many artists studied anatomy, desiring to emulate the musculature of classical Greek sculpture. Although the most notable of these artists was Leonardo da Vinci (1452–1519), who dissected numerous bodies and made brilliant anatomical drawings that were never circulated widely, many other well-known Renaissance artists such as Michelangelo Buonarotti (1475–1564) also dissected and were interested in anatomy. There was a growing interest in pictorial representations of the body, as opposed to textual descriptions, and many distinguished artists worked on anatomical illustrations. The second and more important source of anatomical interest was among physicians. Dating from the fifteenth century, this interest really took off with the publication of a Latin translation of Galen's (A.D. 129–c. 199) recently discovered *On Anatomical Procedures* in 1531. The justification for anatomical study was not solely medical; it could also be presented as a religious exploration of the wonder of God's creation. This was a particularly effective argument for Protestant universities like the University of Wittenberg, a leader in instituting anatomy in Germany.

In the Middle Ages, anatomy had been a low-status discipline, often associated with surgeons rather than physicians and considered subordinate to physiology. The chairs of anatomy established at a number of universities in the sixteenth century initially had less prestige than the traditional medical chairs. The most notable anatomy professor in the sixteenth century was the Paduan Andreas Vesalius, whose lavishly illustrated *Of the Fabric of the Human Body* (1543) laid the foundation for subsequent anatomists. Like Galen's anatomical work, it was presented in a logical order, from the skeleton outward, as opposed to medieval anatomy texts, which worked in a practical order, from the viscera inward. However, Vesalius's study revealed many errors of Galen and the ancient anatomists, caused partly by the fact that they had been forbidden to dissect the human body and had to extrapolate from animals. Thus Galen, going by his experience of dogs, claimed that the human liver had five lobes, when in fact it has none.

Library Media Center
NEUQUA VALLEY HIGH SCHOOL
2360 95th STREET
NAPERVILLE, ILLINOIS 60564

In the late sixteenth century, the medical schools of northern Italy, particularly in Padua, were the centers of anatomical research. Although much of this work built on Vesalius, there was a shift of emphasis from Vesalius's Galenic preoccupation with structure to an Aristotelian investigation of function, of what various body parts actually do. Two of the innovative Italian anatomists of the late sixteenth century were Realdo Colombo (c. 1510–1559), known as a vivisector and a bitter critic and rival of Vesalius, and Bartolommeo Eustacchi (1520–1574), one of the first to write anatomical monographs on particular organs and the namesake of the eustachian tubes of the ear. Two who followed Vesalius as professors at Padua were Gabriele Fallopio (1523–1562), the first to describe what became known as the fallopian tubes, and Girolamo Fabrici, known as Fabricius of Acquapendente (1533–1619). Fabricius changed the focus of anatomical study away from the exclusive study of the human body. Inspired by Aristotle, he wanted to investigate not just the structure of the body, but how the different body parts functioned in different animals. He was the first to describe the valves of the veins and was a teacher and intellectual hero of William Harvey. Harvey's discovery of the circulation of the blood became the great triumph of early modern anatomy, settling the question of whether the moderns could surpass the ancients in the field.

The seventeenth century saw an explosion of anatomical study, no longer concentrated in Italy but spread throughout Europe. New techniques, such as the use of the microscope and the injection of colored wax to trace the different passages and channels of the body were applied to anatomical investigation. The innovations of this period can be evaluated simply by noticing how many names for body parts commemorate seventeenth-century scientists. Gasparo Aselli (1581–1625), a professor at Pavia, discovered the chyliferous vessels of the intestines and had the pancreas of Aselli named after him. Nicolaus Steno dis-

covered the duct of the parotid gland, thereafter called Steno's duct, and the English physician and dissector Edward Tyson (1650–1708) the mucilaginous glands of the human penis, now known as Tyson glands. The circlet of Willis, the ring of arteries at the base of the brain, was described by the distinguished brain anatomist Thomas Willis (1621–1675), one of the group of medical experimenters in the university town of Oxford. Willis correlated different mental functions with different areas of the brain. The greatest of the microscopic anatomists, Marcello Malpighi, was the discoverer and namesake of both the malpighian bodies in the kidney and the malpighian layer of the skin. Much study in the seventeenth century was given to the glands and to the lymphatic system, the discovery of which was the subject of a priority conflict between Olof Rudbeck and the Danish anatomist Thomas Bartholin (1616–1680), after whom the Bartholin duct and the Bartholin gland were named. Anatomy rose dramatically in prestige, becoming one of the most honored disciplines within medicine, and anatomy professors became leaders of medical faculties.

See also Dissection and Vivisection; Harvey, William; Malpighi, Marcello; Rudbeck, Olof; Vesalius, Andreas.

References
Conrad, Lawrence I., Michael Neve, Vivian Nutton, Roy Porter, and Andrew Wear. *The Western Medical Tradition: 800 B.C. to A.D. 1800.* Cambridge: Cambridge University Press, 1995.
Cunningham, Andrew. *The Anatomical Renaissance: The Resurrection of the Anatomical Projects of the Ancients.* Aldershot, England: Scolar Press, 1997.

Apothecaries and Pharmacology

The world of medical drugs was changing rapidly during the scientific revolution. Apothecaries, physicians, and surgeons, the three organized branches of medicine, were at the center of these changes. In addition to dispensing drugs, apothecaries in many areas

As pharmacology developed, apothecaries were required to have more scientific knowledge. This late-seventeenth-century engraving of an apothecary's shop features books as prominently as drugs. (National Library of Medicine)

provided primary medical care, which often led them into conflict with the better organized and more powerful physicians who wanted apothecaries to simply dispense the remedies physicians prescribed. Apothecaries were usually not educated in the medical theory taught in universities, and they learned their craft by apprenticeship in the apothecaries' guild or other organization. There were legendary quarrels between the powerful Society of Apothecaries of London, chartered in 1617, and the London College of Physicians over the right of apothecaries to practice medicine and the right of physicians to sell drugs. (Apothecaries gained the right to practice medicine in England in 1703 by a decision of the House of Lords.) On their other flank, apothecaries had to defend their monopoly over specifically medical substances from grocers and the wandering vendors of unlicensed medicines. Apothecaries were government regulated and often treated with suspicion, as it was believed that some dealt in poisons. Other herbal practitioners included housewives, expected to make up simple herbal recipes for their families, and cunning village men and women. Their practice was much less affected by scientific change than was that of the commercial apothecary, however.

Changes in the repertoire of medical drugs available in the apothecary's shop were driven by changes in medical theory, most notably Paracelsianism, and by the expansion of the range of drugs caused by the desire to identify "lost" drugs referred to in classical sources and by European exploration and expansion into new territories. Paracelsianism and the movement to chemical medicine encouraged the use of drugs that were the result of chemical preparations such as distillation rather than "simples"—parts of plants used in their more natural form. This use had a long history, and although some of these chemical procedures were quite controversial, the use of chemical medicines was approved by the medical establishment by the late sixteenth century. Some apothecaries

innovated in chemical medicine; the German alchemist and apothecary Johann Rudolf Glauber (1604–1670) devised a number of antimony-based medicines. The Society of Apothecaries of London engaged in large-scale medical manufacturing with a laboratory and a joint-stock company beginning in 1682.

The expansion of the pharmaceutical repertoire continued a trend from the Middle Ages. European apothecaries had long sold drugs from other parts of the world, with Venice being the great drug mart of Europe because of its leading role in Middle Eastern trade—particularly important because much of the literature in pharmacy was of Arab origin. Venice, which controlled the Greek-speaking islands of Crete and Cyprus, was also the center of the effort to recover ancient Greek medical drugs. The increased ease of transport caused by the Portuguese circumnavigation of Africa in 1498 added to the volume of exotic medicines. Although most of the influential writers on extra-European medical botany and pharmacology were physicians such as Garcia d'Orta and Christovão da Costa (c. 1515–1580) on India or Francisco Hernández and Nicolás Monardes (c. 1493–1588) on Mexico, apothecaries also played a role in this medical research. One Portuguese official in the Indies, Tomé Pires, a royal apothecary, sent back to Portugal a letter that gave one of the earliest European treatments of Indian drugs. Unfortunately, he was sent on an embassy to the Chinese court in 1516 and died in Chinese imprisonment. The Americas contributed new medicines that became very popular, notably the bark of the guaiac tree, thought to cure syphilis, often believed to be an American disease. Guaiac was being imported into Europe by the early sixteenth century. The other major American pharmaceutical innovation was the bark of the cinchona plant. Introduced by the Jesuits (hence the term "Jesuit's bark") in the early seventeenth century, it was used for the treatment of fevers and was particularly effective against malaria. Modern quinine is derived from it. In both cases, Europeans learned of the medical properties of the substance from the American natives—the usual pattern in pharmacological research.

This new medical knowledge, along with a great deal of old medical knowledge, was embodied in pharmacopeias. These pharmacopeias were usually the creation of physicians or organizations of physicians, who attempted to standardize medical recipes and set the rules for the appropriate use of drugs in a specific area. Although many apothecaries had close social or familial ties to physicians, others resented their claims to regulate all of medicine.

Like other medical professionals, apothecaries contributed to the volume of medical literature by writing on their own profession. Some also accumulated museums of pharmaceutical material, like the Italian apothecary Francesco Calzolari (1521–1600), whose museum included the unicorn's horns thought to cure poisoning. As scientific investigators, apothecaries were also found outside pharmacology in the related fields of botany, chemistry, and natural history. The French chemists Nicaise Le Febvre (c. 1610–1669) and Nicolas Lémery (1645–1715) were both royal apothecaries (Le Febvre to the king of England, Charles II.) An Italian apothecary, Giacinto Cestoni (1637–1718) was a significant natural historian and did microscopic observations.

See also Medicine; Orta, García d';
   Paracelsianism.
References
Barrett, C. R. B. *The History of the Society of Apothecaries of London.* London: E. Stock, 1905.
Conrad, Lawrence I., Michael Neve, Vivian Nutton, Roy Porter, and Andrew Wear. *The Western Medical Tradition: 800 B.C. to A.D. 1800.* Cambridge: Cambridge University Press, 1995.
Kremers, Edward, and George Urdang. *History of Pharmacy: A Guide and a Survey.* 2d ed., revised and enlarged. Philadelphia: Lippincott, 1951.

Arabic Science

The scientific revolution had inherited from the Arab world a rich body of thought in many scientific fields, but European attitudes toward Arabic science were ambiguous. Arab learning was considered particularly important in the fields of mathematics, astronomy, medicine, and alchemy, but Arab knowledge came under attack in the sixteenth century on two fronts. The most important was the humanist claim that the supreme masters in science were the ancient Greeks, which led to attacks on both the medieval Latin Scholastics and the Arabs as misinterpreters of Greek thought. The Arabs were also suspect due to their Islamic allegiance (although some scientists who used Arabic scholarship were Christians or members of other faiths) in a time when the Muslim Ottoman Empire was widely perceived as a military and political threat to Christian Europe.

Medieval Arabic thinkers, notably Avicenna (Ibn Sina, 980–1037), "the Prince of Physicians," and Averroës (Ibn Rushd, 1126–1198), "the Commentator" (on Aristotle), were weighty authorities in Renaissance universities. The Arabs had the most sophisticated astronomical theories and techniques, as well as the most accurate body of astronomical observations available in the West before Tycho Brahe. However, the Persian-language star catalog of the astronomer-prince of Samarkand, Ulugh Beg (1394–1449), the best available, was hard to find in the West. Edmond Halley's discovery of the long-term acceleration of the Moon was based on the observations of the Arab astronomer Al-Battani (c. 858–929), whose accuracy Halley and other early modern European astronomers greatly admired. The Arabs, notably Alhazen (Ibn al-Haytham, 965–1040), were admitted to have gone beyond the Greeks in optics and were obviously, given the vast extent of the lands of Islam, a potentially valuable source for geographical knowledge. Despite the revival of Greek mathematics, European mathematics continued to build on Arab advances, as can be seen from the Arab derivation of the words "algebra" and "algo-rithm." In medicine, the Arabs were less cre-ative and original, but the enormous *Canon* of Avicenna was the most complete and system-atic treatment of Galenic medicine available, and along with the work of his fellow-Muslim Rhazes (Al-Razi, c. 864–c. 930), it remained a central part of the medical curriculum at many universities through the seventeenth century. Jewish physicians in particular tend-ed to praise Arabic medicine and worked on translations of medical works from Arabic. Arabs also had a high reputation as magicians and were frequently invoked in astrological and alchemical literature.

Despite the excellence of Arabic scholar-ship, humanists attacked the Arabs as "barbar-ians" who had sullied the purity of Greek science. A common move in sixteenth- and seventeenth-century debate was to present oneself as purifying Greek truth from Roman, Arab, and medieval Scholastic corruption, and the Arabs, unlike the admired ancient Romans or even the Scholastics, could be insulted with impunity. There was an only partially success-ful drive to purify scientific language of Arab-derived words, as in the shift from "alchemy," incorporating the Arabic article *al,* to "chemia." However, because many classical Greek works survived only in Arabic transla-tions, even the humanists had to grudgingly admit the importance of Arabic knowledge. Interest in Arabic was expanding in the six-teenth and seventeenth centuries. Arabic presses and chairs of Arabic were established in a number of European centers, and European travelers and merchants in the Middle East collected Arabic manuscripts.

Early modern European interest in Arabic science focused on the past. Contemporary Arab and Islamic science, somewhat mori-bund in any case, aroused no interest, save in pharmacology, where Arabic-speakers were assumed to have knowledge of the medicinal uses of Middle Eastern plants. Nor did the contemporary Muslim world pay much attention to developments in European sci-ence other than some technological applica-tions. With the growing cultural arrogance of European science, interest in Arab science died out in the eighteenth century.

See also Jewish Culture; Medieval Science.
References
Russell, G. A., ed. *The 'Arabick' Interest of the Natural Philosophers in Seventeenth-Century England.* Leiden, the Netherlands: E. J. Brill, 1994.
Siraisi, Nancy G. *Avicenna in Renaissance Italy: The Canon and Medical Teaching in Italian Universities after 1500.* Princeton: Princeton University Press, 1987.

Aristotelianism

The texts of the ancient Greek philosopher Aristotle (384–322 *B.C.*) and the voluminous Greek, Arabic, medieval, and Renaissance commentaries on the man whom medieval scholars had called simply "the Philosopher" had a paradoxical effect on the scientific rev-olution. Its leaders often portrayed them-selves as rebels against Aristotle, but their rebellion itself was deeply marked by Aristo-telianism. Anyone with a university educa-tion, which the majority of early modern scientists had, was soaked in Aristotle, his interpreters, and his critics, and his influence shows in some surprising places. Without Aristotelianism, in fact, people might not have thought science worth doing at all.

Aristotle, whose texts were recovered for Latin Europe in the Middle Ages, dominated European thinking about the physical world through the sixteenth century. The thir-teenth-century rediscovery of Aristotle, through the Arab world, had had a radical impact on European culture and was mostly responsible for making the physical universe an object of scholarly interest in the first place. Along with Claudius Ptolemy (A.D. 90–168) and Galen (A.D. 129–c. 199) in the specialized fields of astronomy and medicine, Aristotle's work comprised the body of ancient authority that underlay natural phi-losophy. The Scholastic natural philosophy taught in Protestant as well as Catholic uni-versities expounded natural philosophy by commenting on Aristotle's scientific works (and some spurious texts alleged to be his). Although some humanists and Protestant reformers, most notably Martin Luther (1483–1546), attacked Aristotle, the early modern period was a golden age of Aristotelianism, partly because of the discovery of printing and partly because Aristotle still offered an intellectual system of unrivaled depth and range. Even Martin Luther's own university, Wittenberg, became an Aristotelian stronghold under Luther's disciple, Philip Melanchthon (1497–1560). More Latin Aristotle commentaries were produced in the period 1500–1650 than in the previous millennium. Aristotle and his commentators were not restricted to the classroom but were read in Greek, Latin, and the vernacular tongues in a variety of settings by a variety of people.

Early modern Aristotelianism was highly diverse and often original, rather than being just the sterile repetition of Aristotle's words. A direct legacy from the Middle Ages was Scholastic Aristotelianism, itself divided into a number of different schools such as Thomism (after the medieval philosopher St. Thomas Aquinas [1225–1274], himself undergoing a major revival in the sixteenth century) and Scotism (after another medieval philosopher, Duns Scotus [1265–1308]). These schools subordinated Aristotle to Christian doctrine, denying such Aristotelian claims as the eternal existence of the world. Very different were the commentaries of the Arab Muslim Averroës (1128–1198), called "the Commentator." These were retranslated from the Hebrew (rather than the Arabic original) and were widely influential in the early sixteenth century. Averroistic Aristotelianism, often associated with the University of Padua, asserted the autonomy of philosophy from Christian theology, attracting a great deal of suspicion from church authorities. (Padua and other northern Italian universities remained the most dynamic centers of Aristotelian natural philosophy in the early modern period.) Humanists, recovering the original Greek texts of Aristotle and many other ancient authors, rejected both Scholastic and Averroistic Aristotelianism, claiming to peel away the centuries of Arab and Latin commentary to recover the original Aristotle. The efforts of the humanists led to a revival of interest in Aristotelian texts that the Arab and medieval thinkers had largely ignored, notably his biological works, as well as in the ancient Greek commentaries on Aristotle that had been largely forgotten. Some Aristotelians modified Aristotle's teachings by incorporating approaches of other ancient philosophical schools or the discoveries of modern scientists. All of these trends influenced each other.

The basic procedure of Aristotelian natural philosophy takes generalizations about the physical world, originally based on knowledge acquired through the senses, and performs logical operations on them. The basis of the generalizations was not as important to medieval and Renaissance Aristotelians as was the logic applied to them. The emphasis on general statements meant that Aristotelians focused on the normal and the typical, viewing the study of the aberrant as not truly philosophical. Aristotelian philosophers aimed at certain rather than probable knowledge of the world. The empirical side of Aristotle received less emphasis than the rationalist side during the Middle Ages, but the humanistic revival of the Greek Aristotle brought Aristotelian empiricism to the fore. This influenced university teaching, as in the empiricism of the Padua logician Jacopo Zabarella (1533–1589), whose widely circulated treatises emphasized observation and reason and discouraged slavish adherence to the Aristotelian text.

Aristotelian natural philosophers usually saw the world as made of substances and qualities. In this view, things existing in the natural world are compounds of substance and qualities such as color. The material world itself is divided into four elements—earth, air, fire, and water—in various combinations. The universe, centered on the Earth, is divided between the corruptible "sublunary sphere," containing all beneath the Moon, and the perfect and unchanging heavens. This belief worked very well with the Christian contrast between earthly corruption and God's perfection. Aristotelian philosophy, in part originating with a critique of other ancient Greek philosophers who were mechanically minded, viewed natural things as endowed with purpose, the so-called Final Cause. Things fall, for example, because the Earth is the proper place for them. Other things, like the stars, don't fall because the heavens are the proper place for them. Motions were divided into natural motions— expressions of a body's true nature, and violent motions—those imposed on a body from outside. The hurling of a rock, for example, combines natural motion—the rock's eventual falling to the ground—and violent motion—the rock's going in the direction in which it was thrown. A body's natural motion includes things we would now call change and development as well as motion; for example, Aristotle viewed the growth of a tree as a form of motion. Aristotle believed that stars and planets move in circles because heavenly things are perfect and only circular motion is suitable to perfection. This distinction between earthly and heavenly things and the belief that the Earth is the proper place of earthly things explains why the Copernican replacement of the Earth by the Sun as the center of the universe was as much an assault on Aristotelian physics as on Ptolemaic astronomy—which in turn accounts for much of the opposition to Copernicanism.

A seemingly endless series of movements in Western thought, from late medieval nominalist philosophy to twentieth-century feminism, have proclaimed themselves rebels against Aristotle. In early modern thought, Ramism, Baconianism, Galilean mechanics, and Cartesianism were new intellectual movements opposing the Aristotelians of their time, while atomism, Platonism, Stoicism, and some types of magical thinking were revivals of ancient alternatives to Aristotle. The mechanical philosophy, whether atomist or Cartesian, was anti-Aristotelian in that it denied both the difference between a substance and its qualities and the endowment of natural things with purpose, reserving purpose for God. There was also no difference between natural and violent motion in the mechanical philosophy.

The growing emphasis on mathematical analysis in physics also worked against Aristotelianism, which did not grant mathematics a central intellectual role. The most systematic opponent of Aristotle was Descartes, who viewed himself as a new Aristotle; Cartesianism was designed to supplant Aristotelianism as the basis of the university curriculum, as it actually did in many parts of Europe beginning in the mid-seventeenth century. Like Aristotle, Descartes viewed his theory as applicable to all branches of knowledge, certainty as the goal of natural philosophy, and logical analysis as more important than empirical observation. Cartesians and Aristotelians also agreed on the nonexistence of the void.

Developments in astronomy and physics from Copernicus on gradually dismantled the Aristotelian world picture. The perfection of the heavens, the circular motion of the celestial bodies, and the centrality of the Earth were all rejected. Galileo, who completed Copernicus's task by inventing a mathematically based physics compatible with heliocentrism, opposed and mocked Aristotelian physics. Galileo's physics denied Aristotle's doctrine of natural place, separating a body's motion from its nature. (Despite his expressed scorn for Aristotelianism, Galileo's scientific method was deeply influenced by the Aristotelianism of Padua and the Jesuit Collegio Romano, and he often spoke respectfully of Aristotle himself.) In the course of the seventeenth century, those who stuck to Aristotle's authority, most notably the Jesuits, increasingly fell behind in the physical sciences.

Developments in the biological sciences were quite different in that there was no revolutionary rejection of Aristotle. On the contrary, Aristotle's biological works acted as an intellectual stimulus, particularly to the program followed by the anatomists Fabricius of Acquapendente (1533–1619), a professor at Padua, and his student William Harvey. Aristotle's emphasis on the function of organs and on the importance of dissecting and understanding animals as well as humans led to the discovery of the circulation of the blood by Harvey, who always acknowledged Aristotle as his master. Aristotle's ideas on biological processes continued to be influential throughout this period, although the growing weight of anatomical discovery made some of the specifics of Aristotle's biology outmoded.

As it entered the eighteenth century, Aristotelian natural philosophy retained some intellectual strength, still influential in university curricula in Catholic Europe and elsewhere. However, it clearly had lost its intellectual dynamism and had been defeated in the struggle against the new natural philosophies of Descartes and Newton.

See also Causation; Physics; Universities.
References
DiLiscia, Daniel, Eckhard Kessler, and Charlotte Methuen, eds. *Method and Order in Renaissance Philosophy of Nature: The Aristotle Commentary Tradition.* Aldershot, England: Ashgate, 1997.
Randall, John Herman. *The School of Padua and the Emergence of Modern Science.* Padua, Italy: Antenore, 1961.
Schmitt, Charles B. *Aristotle and the Renaissance.* Cambridge: Harvard University Press, 1983.

Art

Art and science dealt with many of the same issues in the early modern period, and there were many interchanges between the two. Both made greater use of mathematics than previously. The early fifteenth century saw Filippo Brunelleschi's (1377–1446) uniquely European invention of artistic perspective (originally for architectural rather than pictorial purposes), giving the illusion of three-dimensionality to a two-dimensional surface.

This meant many artists were interested in geometry and optical theory. As part of an overall trend toward realistic representation, there was greater interest in the proportion that different objects seen at different distances bear to each other. (Medieval artists assigned size on the basis of importance—thus Christ was often represented as the largest person in a scene, regardless of where he stood relative to the viewer and other people in the painting.) Renaissance artists were also interested in anatomy, as they desired to accurately represent the human form. This had not concerned medieval artists, who never painted nudes and concealed the contours of the human form behind layers of clothing. Art also intersected with alchemy and chemistry in the preparation and application of artistic materials. Artists themselves participated in the sciences, the most famous example being Leonardo da Vinci (1452–1519), who did important work in mechanics and anatomy as well as being one of the greatest painters of his era. Leonardo's anatomical drawings are among the finest of the early modern period, but since they were not published they had little influence. Michelangelo Buonarotti (1475–1564) collaborated in anatomy with the medical professor Realdo Colombo (c. 1510–1559). The German artist Albrecht Dürer (1471–1528) made independent contributions to geometry, as well as working on engineering problems.

Artists and scientists often moved in the same circles, particularly in Italy. Galileo Galilei's younger brother Michelangelo Galilei was a painter, and Galileo himself thought about being a painter in his youth. He associated with many painters with whom he discussed problems of perspectives and astronomy. Artists were interested in the new vistas opened by the telescope, and some, like Galileo's friend Ludovico Cigoli (1559–1613), incorporated the latest astronomical discoveries and theories into their paintings.

While the earliest treatises on perspective and proportion were written by practicing artists for other artists and some were

This drawing by Sébastien Leclerc shows the activities of two of Louis XIV's academies, the Royal Academy of Sciences and the Academy of Art. (Ecole de Beaux-Arts, Paris)

extremely primitive mathematically, by the middle of the sixteenth century the subject was attracting the attention of professional mathematicians, notably the mathematical humanist Federigo Commandino (1509–1575). This resulted in a much more mathematically sophisticated treatment of the problems involved, although not necessarily one of much use to practicing artists. Perspective problems helped geometry to extend from two dimensions, as in Euclid, to three. The mathematization of perspective culminated in the projective geometry of Girard Desargues (1591–1661), a member of the Mersenne circle. Desargues's innovative and hard to understand (even for other mathematicians) *Rough Draft on Conics* (1639) was the subject of a major dispute in the French Academy of Painting and Sculpture. The space created by perspective contributed to a cultural shift that affected science as well as other disciplines. Scholars could move from an idea of *place* to an idea of *space* as mathematically and geometrically defined and unbounded—eventually, the absolute space of Newtonian physics.

See also Illustration; Infinity.

References

Field, J. V. *The Invention of Infinity: Mathematics and Art in the Renaissance.* Oxford: Oxford University Press, 1997.

Reeves, Eileen. *Painting the Heavens: Art and Science in the Age of Galileo.* Princeton: Princeton University Press, 1997.

Astrology

By the beginning of the scientific revolution, astrology, a science of divination based on the idea of celestial influence from the stars and planets, went back thousands of years to the civilizations of the ancient Middle East. Although astrology and astronomy were recognized as separate disciplines, they were intertwined—Claudius Ptolemy (A.D. 90–168), the greatest ancient authority on astronomy, was also the greatest ancient authority on astrology. Early modern astrologers practiced everywhere from courts to villages, and astrological knowledge was embodied in a number of forms, from learned treatises to the almanacs that were one of the most commonly circulated forms of printed literature in the sixteenth and seventeenth centuries.

Astrology was a complex discipline, taking a number of different forms. Natal horoscopes interpreted the position of stars at the time of a person's birth, and buildings, cities, and countries were also considered to have dates of founding that could be analyzed astrologically. High astrology correlated the movements of the stars with large-scale events on Earth, such as plagues and wars. Astrologers took great interest in the relations between dramatic terrestrial events and dramatic celestial events, such as comets and the great conjunctions of Saturn and Jupiter that occurred about every 20 years. This type of astrology was particularly popular during times of crisis such as the English Civil War of the seventeenth century. Questions about marriage, lost or stolen goods, and many other topics could be resolved through horary astrology, the bread and butter of consulting astrologers. Horary astrologers took a chart of the positions of the stars and planets at the moment when someone asked them a question. Interpretation of the resulting "figure" revealed the answer. Astrological principles covered a number of natural domains. Herbs, colors, and gems were all thought to be ruled by different planets. Herbals published for use by physicians showed the planets that ruled different herbs, and how these matched up to the planets that governed various diseases.

Opposition to astrology also had a long and complex history. For most of the early modern period, religious and political objections to astrology, particularly those made by prestigious religious leaders such as John Calvin (1509–1564), were more significant than scientific ones. Most antiastrological arguments in this period were no different from those voiced in the ancient world. Astrological predictions were seen as leading to political instability because they inspire people with hopes of overthrowing the government. Some argued that by exalting the power of the stars, astrology diminishes God and human free will. The fact that many of the great names in the history of astrology were either ancient pagans or Muslims also rendered it suspect. In response, many astrologers were careful to emphasize that the stars do not determine but merely influence terrestrial events and human choices. Some clergymen practiced astrology, and others claimed that significant religious events such as the Creation or the end of the world could be dated astrologically. The three Magi who attended the infant Jesus guided by the star of Bethlehem provided an excellent Christian precedent for astrology.

Some scientists, most notably Johannes Kepler, practiced astrology, and Kepler claimed that "Mother Astronomy" depended on "Daughter Astrology" for her upkeep. Both he and Tycho Brahe performed astrological work for their royal patrons. However, as astronomy developed in the seventeenth century, it grew more distant from astrology. The needs of navigation supplanted astrology as the foremost practical application of astronomy. As astronomical tables and observations became more precise, astrologers lacking formal training were unable to keep up, and for that reason they attracted the ridicule of some scientists, such as John Flamsteed. The mechanical philosophy denied the possibility of occult influences from the planets, and mechanical philosophers such as Marin Mersenne opposed astrology along with other forms of magic. Astrology was based on a Ptolemaic and Earth-centered universe, and astrologers had difficulty translating their principles into a heliocentric universe, although this was attempted by the English clergyman Joshua Childrey (1623–1670). Some astrologers, notably Childrey, his countryman the almanac-maker John Gadbury (1627–1704), and the French mathematics professor Jean-Baptiste Morin (1583–1656), attempted to rebuild astrology on scientific and Baconian foundations, amassing numbers of horoscopes and analyzing common patterns rather than theorizing. These efforts failed to gain acceptance from the scientific community, and by the beginning of the eighteenth century, astrology, while flourishing at a popular level, was considered antiscientific superstition by Europe's educated elite.

See also Astronomy; Kepler, Johannes; Magic.
References
Curry, Patrick. *Prophecy and Power: Astrology in Early Modern England*. Princeton: Princeton University Press, 1989.
Tester, S. J. *A History of Western Astrology*. Woodbridge, England, and Wolfeboro, NH: Boydell, 1987.

Astronomy

Astronomy, the earliest precise science combining extensive observations with mathematical expression, underwent many important changes during the scientific revolution. The best known and most dramatic of these was the shift from the Earth-centered system of the ancient Greek Claudius Ptolemy (A.D. 90–168) to the Copernican Sun-centered system. Others include the opening of vast new areas of knowledge by the telescope, the bold intellectual claim of astronomers to describe the nature of the universe rather than merely "saving the phenomena," and the dissolution of the millennia-old connection of astronomy and astrology with the rise of the connection of astronomy with navigation.

The publication of Copernicus's *On the Revolutions of the Celestial Spheres* in 1543 (the same year as his death) was the culmination of the life's work of a man dissatisfied with the clumsiness of the Ptolemaic system. Copernicus's Sun-centered system did possess some advantages over the Ptolemaic system, notably in the elimination of equants—points other than the center of a planet's rotation relative to which the planet's motion is constant. These were useful mathematical devices, but difficult to represent physically. However, the principal task of working astronomers was the analysis and prediction of celestial motions—called saving the phenomena—rather than providing a physically accurate picture of the workings of the universe, the task of natural philosophers. Copernicus put forth his Sun-centered picture as physically accurate, but most astronomers for the next few decades followed the lead of his early editor Andreas Osiander (1498–1552) in treating the system as useful for making calculations rather than as physically accurate.

The most immediately successful new planetary system introduced in the sixteenth century was not Copernican, but that of Tycho Brahe, who put the Sun and Moon in orbit around the Earth and the planets around the Sun. Since these paths intersected, Tycho abolished the celestial orbs that Ptolemy and Copernicus had agreed carried the planets in their courses. Tycho's system combined the mathematical advantages of Copernicanism with a stable Earth, more reconcilable both to everyday experience and to Aristotelian physics. Tycho's most important contribution to astronomy, however, was his painstaking collection of accurate data on the positions and movements of the planets from his observatory, Uraniborg, on the Danish island of Hven. Tycho was the greatest of all observers before the advent of the telescope, working with the best instruments available at the time.

The two most important astronomers of the generation after Tycho were the Copernicans Johannes Kepler and Galileo Galilei. Galileo represents a major change in the field. He was not an astronomer at all in the classic sense, and the laborious continued observation and painstaking construction of elaborate charts and tables that was the work of the traditional astronomer had no appeal for him. But by pointing a telescope at the sky, Galileo fundamentally changed the character of astronomy. His spectacular discoveries—the moons of Jupiter and the phases of Venus, among others—both provided powerful evidence in favor of Copernicanism, or at least against the Ptolemaic

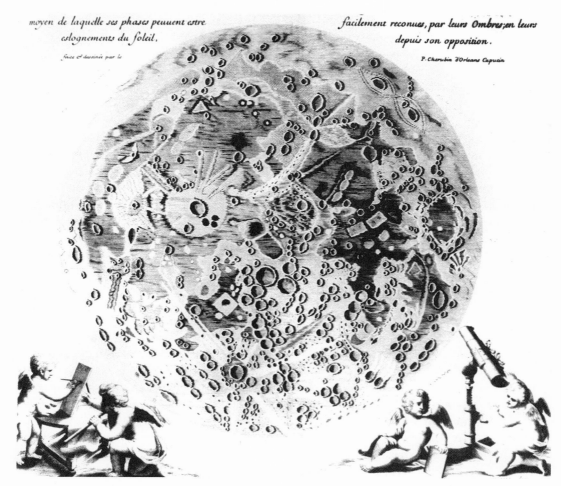

Telescopes and systematic observation greatly increased astronomers' knowledge of the moon's surface. (National Library of Medicine)

system, and revealed how much was still unknown about the skies. The universe as revealed by the telescope was incomparably more vast than that of Ptolemy, Tycho, or Copernicus. As Galileo pointed out, the telescope did not resolve the stars from points of light to small disks, as they did the planets. He concluded from this that the stars were much smaller than had been thought. As it happened, they were much farther away.

Kepler, the successor of Tycho as astronomer to the Holy Roman Emperor Rudolf II, had been taught Copernicanism at the University of Tübingen by Michael Maestlin (1550–1631), one of the few sixteenth-century Copernicans. Kepler's great innovations in astronomy were his laws of planetary

motion. The most radical of these was his description of the path of the planets as ellipses with the Sun at one focus. His abandonment of the idea that heavenly motion had to be uniform and circular was a dramatic break with the entire tradition of astronomy, as embodied in the Ptolemaic, Tychonic, and Copernican systems. Kepler was adamant that true astronomy was not about finding mathematical formulae for planetary movement but about discovering the true nature of the universe. He put forth not merely a description of planetary orbits but a magnetic mechanism that he believed explained their elliptical shape and rates of movement.

Kepler's elliptical orbits were slow to catch on, ignored by both Galileo and René

Descartes, who although not an astronomer had a great deal of influence on astronomical theory. Elliptical orbits did attract some interest in England, notably from the brilliant young astronomer Jeremiah Horrocks (1618–1641). Horrocks was one of two astronomers, the other being his friend William Crabtree (c. 1610–1644), to observe the transit of Venus across the face of the Sun in 1639. Horrock rejected Kepler's magnetic theory of planetary motion, and although his improvements to Kepler's system were not published until well after his death, they became quite influential in the 1660s, when Robert Hooke revived Horrock's theory that the bob of a pendulum in an oval orbit is a model for planetary motion. Ismael Boulliau (1605–1694) adopted Kepler's elliptical orbits with a different law of motion. His tables, although not as good as Kepler's *Rudolphine Tables* (1627), were influential in popularizing "elliptical astronomy."

In the Catholic world, the development of astronomical theory was dealt a heavy blow by the condemnation of Copernicanism in 1616 and the subsequent trial of Galileo in 1633. Outside France, where the church was more liberal, Catholic Europe went over to the Tychonic system. The Jesuits produced several distinguished astronomers during the seventeenth century, notably Galileo's foe and Europe's leading authority on sunspots, Christoph Scheiner (1573–1650), and Giambattista Riccioli (1598–1671), who invented the modern system of naming the geographical features of the Moon. (This is why one of the biggest craters is named after Riccioli's fellow-Jesuit astronomer Christoph Clavius.) Riccioli attacked Copernicanism while admitting its mathematical superiority. Despite their observational talents, Jesuit and other Catholic astronomers were reluctant to accept Copernicanism, and as a result, much of the Catholic world, especially Italy, was less important in astronomy in the mid-seventeenth century than earlier, although the Italians did continue to make the best telescopes.

The rest of the seventeenth century saw a refinement in telescopic astronomy. By the 1640s, the Galilean telescope was giving way to the "astronomical" or Keplerian telescope capable of higher magnifications. These led to the discovery of the phases of Mercury, the surface features of the planets, and the satellites of Saturn. (Christiaan Huygens's explanation of the mysterious appendages of Saturn as part of a ring system in 1658 attracted much interest.) The various micrometers in use from the 1630s made the telescope suitable for very precise observation, and telescopes were also added to traditional astronomical instruments such as quadrants. More accurate pendulum clocks also affected astronomy.

All of this was making serious astronomy ever more expensive. Astronomy had always attracted the interest of the state, in calendar making—the greatest Jesuit astronomer, Christoph Clavius, was a leader in the Gregorian reform of the calendar—in astrology—both Kepler and Tycho provided horoscopes for their royal patrons—and in navigation. Astrology was waning in importance, partly due to the reluctance of astrologers to convert to the Copernican system, and navigation was gaining. Galileo put forth a scheme for solving the classic navigational problem of determining the longitude based on the satellites of Jupiter, which he tried to sell successively to the king of Spain and Spain's archenemy, the Dutch Republic, but it proved impracticable given the difficulty of precise celestial observation from a moving ship. Much of the mapping of the Moon carried out in the seventeenth century was inspired by equally fruitless hopes that the Moon could provide an accurate clock for navigators.

Late seventeenth-century astronomy was dominated by state-sponsored enterprises put forth for navigational and cartographic ends. These, notably the Paris and Greenwich Observatories, differed from Tycho's observatory in being institutional rather than individual. The last great private astronomer was

Johannes Hevelius, and his refusal to use telescopic instruments made him an anachronism by the time of his death in 1689. The shift from private to state astronomy also further marginalized women astronomers who, like Hevelius's two wives, had worked as assistants to husbands, fathers, and brothers within astronomical households but were not employable at state institutions. The great astronomical achievements of the late seventeenth century include the discovery of four satellites of Saturn and other planetary work by Gian Domenico Cassini at the Paris Observatory, the star catalogs of Hevelius and the Greenwich astronomer John Flamsteed, and the cometary analyses of Isaac Newton and Edmond Halley. The greatest theoretical innovation, of course, was Newton's derivation of Kepler's laws of motion from his theory of universal gravitation, although this did not immediately vanquish Cartesian astronomical theories. In the eighteenth century, astronomy would be transformed by the reflecting telescope and the systematic application of Newtonian physics to celestial motions.

See also Astrology; Brahe, Tycho; Copernicanism; Copernicus, Nicolaus; Flamsteed, John; Galilei, Galileo; Halley, Edmond; Hevelius, Johannes; Navigation; Observatories; Planetary Spheres and Orbits.

References

North, John. *The Norton History of Astronomy and Cosmology.* New York: Norton, 1995.

Taton, Rene, and Curtis Wilson, eds. *Planetary Astronomy from the Renaissance to the Rise of Astrophysics. Part A: Tycho Brahe to Newton. The General History of Astronomy,* ed. M. Hoskin, vol. 2. Cambridge: Cambridge University Press, 1984.

Toulmin, Stephen, and June Goodfield. *The Fabric of the Heavens: The Development of Astronomy and Dynamics.* New York: Harper, 1961.

Atomism

The theory that the world is composed of tiny, indivisible, impenetrable bits of matter—atoms—surrounded by empty space had roots in the thought of the classical Greeks, notably Democritus (c. 460–c. 370 B.C.) and Epicurus (341–270 B.C.). The principal source for ancient atomism in the scientific revolution was the epic poem of the Roman Epicurean Lucretius (c. 96–c. 55 B.C.), *Of the Nature of Things,* a text widely printed in the humanistic revival of the sixteenth century. Classical atomism attempted to explain the physical properties of various substances by differences in their atoms: solid substances had atoms that held together with hooks whereas liquid substances had round atoms that easily slid past each other. Atomism played little role in medieval physics, both because it was rejected by Aristotle and because it was incompatible with the doctrine of transubstantiation, which relied on an Aristotelian distinction between the substance of a thing and its qualities as perceived by the senses, which atomism denied. To atomists, the sense properties of an object, such as the consecrated bread, were caused by the atoms of which it was composed; therefore, what appeared as bread had to be bread, and not the flesh of God as the doctrine of transubstantiation required. The bad reputation of the ancient Epicureans as atheists also helped discredit atomism. In the sixteenth century, atomism continued to attract little attention, and that mostly from fringe figures. Giordano Bruno upheld an atomism in which the atoms are divinely endowed with a tendency toward organization, and Paracelsians interested in the doctrine of the emergence of matter from seeds came close to an atomism that viewed the atoms as individually active.

Although it has been argued that the real reason for the trial of Galileo was atomism rather than Copernicanism, the evidence for this is slim. The man who really put atomism on the agenda for seventeenth-century scientists was the French priest Pierre Gassendi. His program of restoring the good name of Epicureanism led to his presentation of an atomistic theory in terms of the mechanical philosophy, unlike Bruno's vitalistic theory. He also argued that atoms aggregate to form other small bodies, called molecules. Gassendi's Christianity led him to reject specific features of Epicurean atomism, such as the belief in the eternity of atoms and their infinite number. However, Gassendi's atomism was still suspect in Catholic countries where the church continued to exert power over intellectual life, such as post-Galilean Italy, and in Catholic but more liberal France it was defeated by Cartesianism, which like Aristotelianism denied the void and believed matter infinitely divisible.

In the mid-seventeenth century, the development of the barometer and the air pump supported atomism by providing evidence of the existence of a void. English science became atomism's stronghold. Its principal English herald was the physician and virtuoso Walter Charleton (1620–1707), whose *Physiologia Epicuro-Gassendo-Charletoniana* (1654) explicated an atomistic physics. The philosopher Henry More (1614–1687) attempted to synthesize Platonism and atomism. Robert Boyle, although suspicious of the dogmatism of classical atomism, applied an atomistic, or "corpuscular," philosophy modified by alchemical ideas, including those of Johannes Baptista van Helmont about self-directed matter to chemistry, and Isaac Newton endorsed atomism in the "Queries" appended to his *Opticks* (1704). Newton's innovation was to speculate that the atoms might be held together, not by hooks or even by mutual contact, but by forces analogous to gravity, and that solid bodies might be mostly empty space with scattered atoms held together by these immaterial forces, rather than being tightly packed together.

See also Epicureanism; Gassendi, Pierre; Matter.

References
Hall, A. Rupert. *The Scientific Revolution, 1500–1800: The Formation of the Modern Scientific Attitude.* London: Longmans, 1954.
Pullman, Bernard. *The Atom in the History of Human Thought.* Translated by Axel Reisinger. New York and Oxford: Oxford University Press, 1998.

B

Bacon, Francis (Lord Verulam) (1561–1626)

One of the first people to conceive of the progress of natural philosophy as a distinguishing feature of the age in which he lived was the English lawyer, politician, and philosopher Francis Bacon. Although Bacon did not perform significant scientific work himself, he wrote an influential series of tracts in English and Latin, setting forth a new scientific program to reform intellectual life. He called this reform a "great instauration." A pioneer in the use of the English language for high-level natural philosophy, Bacon emphasized empiricism, the gathering of data from many sources without presuppositions as a prelude to theorizing, as well as collaboration between investigators. He also set forth a vision of scientific and technological improvement for utilitarian ends. "The true and lawful goal of the sciences is none other than this: that human life be endowed with new discoveries and powers," he wrote in *The New Organon* (1620). For Bacon, natural philosophy was about doing things rather than merely understanding nature. Although Bacon was interested in alchemy, he rejected the magician as a model for the active natural philosopher bent on improving the human condition, believing that most magic was based on an arrogant and impious attitude

rather than a humble respect for nature. (Bacon did sometimes speak of the natural-philosophical investigator as *coercing* nature—he called it "binding Proteus"—but his usual approach to nature was respectful.) This respect for nature included ridding the mind of unexamined assumptions—what Bacon called "idols."

From a powerful family of lawyers and civil servants, Bacon was educated at Trinity College, Cambridge—although he did not take a degree—and at the Inns of Court, the London institutions for legal training. At Cambridge, Bacon was dissatisfied with the Scholastic Aristotelianism that dominated the curriculum. Although sometimes speaking respectfully of Aristotle himself, Bacon set forth a new philosophy that he claimed would correct the emphasis of ancient Greek and modern Scholastic philosophy on the study of words rather than things. He had an immense store of humanist knowledge, which he put to use in *Of the Wisdom of the Ancients* (1609), a series of studies of the allegorical meanings of mythological deities that was actually a treatise on natural philosophy. For most of his career, which took him to the highest level of the English legal system as lord chancellor, Bacon would pursue natural philosophy as an avocation. He could devote full time to it only after his removal from office by

An undated engraving by E. Scriven (1775–1841) of Francis Bacon, influential British social philosopher and writer in the early sixteenth and late seventeenth century. (National Library of Medicine)

impeachment in 1624. He died from a cold caught while stuffing a goose with snow to see if it could be preserved.

Bacon's most influential published work of natural philosophy was *The Advancement of Learning* (1605), published in an expanded Latin edition as *De Dignitate et Augmentis Scientarum* (1623). In this and other works, such as *The New Organon* (1620), Bacon proposed a new form of inductive logic, reasoning from particular instances to general statements, to replace Aristotelian deductive logic. Bacon also set forth an ambitious program for the classification of all sciences, seeing no absolute division between the physical and human sciences. His utopia, *New Atlantis* (1627), sets forth a vision of a cooperative scientific institute called Solomon's House on the mythical island of Bensalem. Solomon's House is occupied by a hierarchy of natural philosophers and investigators—the investigators gathering information and the philoso-

phers analyzing it. Solomon's House is closely allied with the state, a central element in Bacon's vision. King James I of Great Britain (r. 1603–1625) Bacon's patron and employer for the most successful phase of his career, was the dedicatee of *The Advancement of Learning* and *The New Organon* and was known to relish being addressed as "Solomon." Bacon was suspicious of non-state-affiliated natural philosophy.

When Bacon himself did natural philosophy, it was not as innovative as he claimed. His natural philosophy was influenced by the Aristotelian tradition as well as by Bernardino Telesio, whom Bacon called the "first of the moderns." Bacon was not particularly progressive in his approach to the content of science; for instance, he doubted the Copernican claim of the Sun's centrality and rejected Galileo's claim to have discovered satellites of Jupiter. One weakness of Baconian method is the difficulty of getting from facts to theories. William Harvey, who was Bacon's physician, dismissed him as writing philosophy "like a Lord Chancellor," resulting in a system devoid of practical use. Baconian science in practice can degenerate to mere fact gathering, as in Bacon's own *Sylva Sylvarum; Or, a Naturall Historie* (1627). Bacon was also suspicious of mathematics, possibly due to its connection with magic. The Baconian ideal of fact gathering and collaboration, however, attracted interest in a number of places, eventually inspiring the foundation of permanent scientific associations such as the Royal Society.

See also Baconianism; Hartlib Circle; Personifications and Images of Nature; Renaudot, Théophraste; Telesio, Bernardino; Utopias.

References

Briggs, John C. *Francis Bacon and the Rhetoric of Nature.* Cambridge: Harvard University Press, 1989.

Martin, Julian. *Francis Bacon, the State, and the Reform of Natural Philosophy.* Cambridge and New York: Cambridge University Press, 1992.

Quinton, Anthony. *Francis Bacon.* Oxford: Oxford University Press, 1980.

Baconianism

The ideology of Baconianism is difficult to pin down. It is in some ways a vulgarization of the complex thought of Bacon and also draws from sources other than Bacon, such as the experimentalism of William Gilbert or the empiricism of Pierre Gassendi. Unlike other "isms" such as Aristotelianism or Cartesianism, Baconianism is not a scientific theory but a theory about science. Characteristics of Baconianism include an emphasis on cooperation between natural investigators, a foundation in empiricism and experimentalism, opposition to the ancients and particularly Aristotle, resistance to theorizing, and a belief that human powers can be improved through the progress of science and technology. None of these beliefs was unique to Baconianism during the scientific revolution, but Bacon's work did provide a powerful synthesis, one usable both by mechanical philosophers and by alchemists and Paracelsians. When actually used as a guide to scientific method, however, Baconianism tended to degenerate into the heaping up of isolated facts and experimental results. The greatest English scientist of the generation following Bacon's, William Harvey, regarded Bacon's philosophy as useless.

Many of Bacon's works were written in Latin for an audience of European intellectuals. Baconianism did have some influence on the Continent, as in the projects of Théophraste Renaudot and the educational philosophy of the millenarian Czech John Amos Comenius (1592–1670). However, the vast majority of avowed Baconians were to be found in England, beginning with the opening of the English Civil War in 1640. There were many reasons why Puritans found Baconian ideology attractive. Many, although not all, Puritan intellectuals shared Bacon's anti-Aristotelianism because of their hatred of medieval Catholic Scholastic theology, which was Aristotelian. More importantly, Bacon's emphasis on renewal and progress fit with the millenarian and reforming tendencies of Puritanism. This ideology is particularly characteristic of the natural philosophers of the Hartlib circle, who hoped to combine a Baconian organized inquiry into nature with technological advance and social reform. Natural philosophers and scientists more moderate in religion, such as John Wilkins, also adopted Baconianism during the Civil War and the period of Puritan rule in the 1650s.

Once it was shorn of its connection with millenarianism and social reform, Baconianism survived the change from Puritan rule to the Restoration of the British monarchy and Church of England in 1660, although some conservatives attacked it as subversive. For the Restoration natural philosophers of the Royal Society, Baconianism had several advantages as the society's quasi-official ideology. It was English, and praising Bacon offered an opportunity to express patriotism and assert the compatibility of natural philosophy with English culture. Harking back to "Bacon, the lord chancellor" helped purge organized natural philosophy of any politically rebellious taint it had acquired from the Puritan Revolution. The Baconian emphasis on cooperation was useful for any organization. Perhaps most important, Baconianism was neutral toward the various natural philosophical schools, and Cartesians, chemical philosophers, and even Aristotelians could work together under the Baconian aegis of natural philosophy and progress. The chief propagandists and defenders of the Royal Society and experimental science generally—Henry Power (1623–1668), Joseph Glanvill (1636–1680), and Thomas Sprat (1635–1713)—all identified the society's program as Baconian. Robert Boyle praised Bacon, and presented his own science as the gathering of facts rather than theorizing on them. Robert Hooke attempted to develop a scientific method, or "Philosophical Algebra," on a Baconian basis.

While Baconianism remained central to natural historical disciplines such as botany and zoology, Bacon's suspicion of mathematics made him an increasingly outmoded guide in areas such as astronomy and physics. By the early eighteenth century, Newtonianism had replaced Baconianism as England's dominant scientific ideology. However, the empirical, technological, and utilitarian bent Baconianism imparted to English science remained one of its distinguishing characteristics.

See also Hartlib Circle.
References
Jones, R. F. *Ancients and Moderns: A Study of the Rise of the Scientific Movement in Seventeenth-Century England.* 2d ed., revised with new preface. Gloucester, MA: P. Smith, 1975.
Webster, Charles. *The Great Instauration: Science, Medicine, and Reform, 1626–1670.* New York: Holmes and Meier, 1976.

Ballistics

See War.

Barometers

The barometer, an instrument for measuring atmospheric pressure, began its development with an experiment in 1643 by two disciples of Galileo Galilei, Evangelista Torricelli (1608–1647) and Vincenzo Viviani (1622–1703). Torricelli, knowing that water could not be pumped more than approximately 32 feet higher than its external level, was interested in how far a column of mercury could be raised. The "Torricellian experiment" involved filling with mercury a glass tube about two yards long, sealed at one end, then covering the open end with a finger and inserting the tube, open end downward, into a container of mercury. The mercury in the tube sank to a level about 30 inches higher than the mercury in the container. Torricelli hypothesized that the column was sustained by the pressure of the atmosphere on the mercury in the container.

Torricelli's death ended his involvement with the problem, but Blaise Pascal heard about the experiment from Marin Mersenne, and in 1648 tested Torricelli's theory by having barometers set up at different points on the side of a mountain. As the altitude grew higher and the atmosphere thinned, the height of the column of mercury diminished. Pascal believed that a systematic correlation of barometric measurements with the known heights of the places where the measurements were made could make barometers instruments for measuring height, but variations in atmospheric pressure made this impossible. Barometric experiments became popular in the late seventeenth century. The dependence of the height of the column on external pressure was proved experimentally by Robert Boyle, who placed a barometer in an air pump and showed that the column fell as air was withdrawn.

Influenced by reports of Torricelli's experiment, Otto von Guericke (1602–1686) developed a more cumbersome water-based barometer. Although this barometer proved to be a technological dead end, Guericke was the first to systematically correlate changes in atmospheric pressure with changes in the weather. Subsequent developments were aimed at making mercury barometers more precise and portable. For example, the open vessel of mercury could be dispensed with by using a tube in the shape of the letter *J*, with the short end open to the atmosphere. Lighter liquids could be mixed with the mercury to increase the sensitivity of the barometer. Robert Hooke attached a weight floating in the mercury to the arrow of a dial, thereby devising a barometer in which slight changes in the height of the column were exaggerated and made easier to measure.

See also Vacuum; Weather.
Reference
Middleton, W. E. Knowles. *Invention of the Meteorological Instruments.* Baltimore: The Johns Hopkins University Press, 1969.

Beeckman, Isaac (1588–1637)

Although he published only his dissertation, the Dutch physician and schoolmaster Isaac Beeckman contributed to the mechanical philosophy through his theories and his influence on the French natural philosophers René

Descartes and Pierre Gassendi. From a technical background—his father owned a candle factory—he studied for the Reformed Protestant ministry but abandoned theology to take an M.D. from the University of Caen. However, he never practiced medicine. He met Descartes in Breda in 1618, and they corresponded and became close friends. Beeckman greatly advanced Descartes's understanding of mechanics. A quarrel broke out in 1628 when Descartes thought Beeckman was boasting of being his teacher. Their relationship was restored, but without its former warmth. When Beeckman met Gassendi in 1629, the two men conversed on mechanics and Copernican astronomy; thereafter, they maintained a correspondence on these and other topics.

Beeckman was an expert technician, consulted by government bodies and individuals on technological questions such as the repair of water works. He founded a Collegium Mechanicum in Rotterdam, where he lived from 1620 to 1627, to discuss technical issues. His atomistic and mechanical natural philosophy emphasized material interactions and spurned those qualities, characteristic of Aristotelianism, which cannot be pictured or represented. Beeckman formulated the principle of inertia, denying that a moving object has an unrepresentable quality called impetus and claiming instead that it continues to move as long as nothing makes it stop. Divisions in the Dutch Reformed Church forced Beeckman to move to Dordrecht, where he was rector of the Latin school there in 1627. There he had a laboratory and a simple observatory and learned to grind lenses for telescopes. In 1635 he served on a Dutch committee to evaluate Galileo's plan to use the satellites of Jupiter to determine longitude.

See also Descartes, René; Mechanical Philosophy.
Reference

Van Berkel, Klaas, Albert Van Helden, and Lodewijk Palm, eds. *A History of Science in the Netherlands: Survey, Themes, and Reference.* Leiden, the Netherlands: Brill, 1999.

Bernoulli Family

The Bernoulli brothers, Jakob (1655–1705) and Johann (1667–1748), contributed to the advance of mathematics, particularly the Leibnizian calculus and probability. From a family of apothecaries in Basel, they founded a mathematical dynasty extending to the late eighteenth century. Jakob traveled extensively but settled down in his home town, attaining the chair of mathematics in Basel in 1687. He and his younger brother studied mathematics together as young men, and they were the first to follow Leibniz in calculus. Johann took an M.D., and in 1695 he gained the chair of mathematics and physics at the University of Groningen through the intervention of Christiaan Huygens, a personal friend. Then, from collaborators, the brothers became rivals, as Jakob claimed that Johann was only his student. The quarrel between the brothers grew increasingly bitter, culminating in a complete breakdown of relations after 1697. In 1705, Johann succeeded his deceased brother as professor of mathematics at Basel.

The Bernoulli brothers pioneered the application of the Leibnizian calculus to geometrical and mechanical problems, and their work substantially contributed to the victory of Continental Leibnizian over British Newtonian calculus. In probability theory, Jakob devised the law of large numbers for relating the distribution of outcomes in a large number of trials to the probability of a single outcome, which he referred to as the "golden theorem." Johann laid the foundation for the calculus of variations and was Leibniz's most distinguished supporter in his struggle against Newton over the calculus. (Unfortunately, Johann supported Cartesian against Newtonian physics as well as Leibnizian against Newtonian mathematics, and his prestige substantially delayed the acceptance of Newtonian physics on the Continent.) He became Europe's most eminent mathematician, turning down offers of chairs from the leading universities and being admitted to all of Europe's leading scientific societies, even the Royal Society, a Newtonian stronghold. Johann's three sons, Daniel (1700–1782), Johann (1710–1790), and Nicholas (1695–1726), and the brothers' nephew Nicholas Bernoulli (1687–1759) went on to distinguished mathematical careers.

See also Cycloid; Leibniz, Gottfried Wilhelm; Mathematics.
Reference
Grattan-Guiness, Ivor. *The Norton History of the Mathematical Sciences: The Rainbow of Mathematics.* New York: W. W. Norton: 1998.

Bible

The relationship of the Bible to the scientific revolution was complex and ambiguous. Early modern European culture was steeped in the Bible, and the disputes of the Reformation and Counter-Reformation gave it an even more central cultural role. Some natural philosophers and scientists, particularly in the sixteenth century, found that the Bible offered a liberating alternative to the established natural philosophy based on the work of pagans such as Aristotle. The Bible was also metaphorically paired with the "book of nature," also written by God. There were even a few abortive attempts, mostly in the Protestant world, to create a purely biblical natural philosophy. The Bible offered a way of presenting natural philosophy—as in commentaries on Genesis or in the "hexamaeral" literature, which organized information in a six-part structure based on the six days of creation. There were also works of natural history that described the plants and animals of the Bible. Athanasius Kircher believed that Noah's ark was the prototypical natural history collection. In some cases, however, the most notorious being the trial of Galileo Galilei, the Bible and the scientific revolution seemed to clash head-on.

Both Protestant and Catholic Europe saw the Bible as intellectually authoritative, but what this meant was open to interpretation. This period witnessed the growing popularity of literal readings of the Bible, as opposed to the allegorical and mystical interpretations that had been common in the Middle Ages. For Catholics, particularly following the decree of the Council of Trent on the Bible in 1546, the power to interpret the Bible lay ultimately in the hands of the church. Questioning the literal truth of any biblical statement, regardless of its relevance to theological doctrine, was forbidden. Although the Spanish Augustinian Diego de Zúñiga wrote a commentary on the book of Job, published in 1584, that endorsed the Copernican theory, the vast majority of biblical exegetes, Protestant and Catholic, initially found Copernicanism incompatible with the plain meaning of the words in such passages as that describing Joshua's halting of the Sun at the battle of Gibeon. Galileo, an enthusiastic user of the "two books" metaphor that distinguished between the Bible and nature as two of God's works, and the Carmelite monk Paolo Foscarini (1565–1616) attempted to render Copernicanism compatible with the Bible. They distinguished between the sphere of religious knowledge, in which the Bible held supreme sway, and that of natural knowledge, where certainty could be attained through other means. Galileo and Foscarini believed in "accommodation," the doctrine that God had adjusted the biblical text in accordance with the intellectual limitations of its hearers. Thus the halting of the Sun at Gibeon was really a halting in the rotations of the Earth, but it was presented in the Bible as a halting of the Sun for the ancient Jewish hearers and readers of the original text, ignorant of true natural philosophy. In 1616, the church denied this distinction with the condemnation of Copernicanism as incompatible with the Bible and the banning of Foscarini's book, entitled *Letter Concerning the Opinion of the Pythagoreans and Copernicus about the Mobility of the Earth and Stability of the Sun.* The greatest Catholic theologian of the time, Cardinal Robert Bellarmine (1542–1621), admonished Galileo not to teach or publicly endorse Copernicanism. Bellarmine believed every word of the Bible to be infallible and adopted his own anti-Aristotelian natural philosophy from biblical sources.

Protestantism also emphasized literal interpretation in this period, but Protestant churches lacked the institutional strength of the Catholic Church that enabled it to claim a monopoly on interpretation. Furthermore, Protestant societies were somewhat more open to individual interpretations. Early modern Protestant literalism was not the word-for-word "inerrancy" of modern fundamentalists but allowed for God's "accommodation" of truth to the limited capacities of human beings, a doctrine endorsed by major Protestant thinkers such as John Calvin (1509–1564). The growing awareness of the historical context of the biblical writings with the rise of critical biblical scholarship in the late seventeenth century, preeminently associated with Baruch Spinoza and the French Oratorian Richard Simon (1638–1712), also made naive literalism less tenable.

Whatever their stances on biblical literalism, few early modern scientists viewed the Bible as simply irrelevant to natural knowledge. Even Galileo sometimes used biblical arguments to make a scientific case. Johannes Baptista van Helmont used the Book of Genesis to argue that water was the original element. Some understood biblical events in scientific terms or treated the Bible as a source for accurate information on the history of the world. This was particularly true in seventeenth-century England, where many scientists and virtuosi were interested in the "Mosaic Cosmology," based on Genesis. The Cambridge professor Henry More (1614–1687) went so far as to find the Cartesian philosophy of nature in the five books of Moses. One biblically inspired work, Thomas Burnet's (c. 1635–1715) *Sacred Theory of the Earth* (1681 and 1689) provoked controversy. Burnet attempted to give a detailed natural-philosophical explanation for the great events in the Bible, past and future, from creation to the Last Judgment. For example, he explained Noah's flood as the release of sub-

terranean waters (rainfall alone seemed unable to account for the amount of water necessary to flood the Earth) and the apocalypse as the release of subterranean fire. Conservative leaders in the Church of England believed that Burnet's natural explanations diminished the role of divine power. Here intellectually conservative religious authorities argued for a separation of the Bible from science, while natural philosophers such as Burnet and John Ray, author of *Miscellaneous Discourses Concerning the Dissolution and Changes of the World* (1692), argued that they were joined. The greatest scientific biblical interpreter in the late seventeenth century was Isaac Newton, who devoted many years to unraveling the mysteries of biblical chronology and understanding the figurative language of the prophetic books.

See also Book of Nature; Chronology; God; Religion and Science; Trial of Galileo.

References
Blackwell, Richard J. *Galileo, Bellarmine, and the Bible: Including a Translation of Foscarini's Letter on the Motion of the Earth.* Notre Dame, IN: University of Notre Dame Press, 1991.
Harrison, Peter. *The Bible, Protestantism, and the Rise of Natural Science.* Cambridge and New York: Cambridge University Press, 1998.

Blood Transfusions

The first blood transfusions were carried out as experiments in the 1660s. By 1664, the Royal Society experimenters were injecting poisons directly into the bloodstream of dogs to observe their effects. Dr. Richard Lower (1631–1691), the best-known English physiologist, performed the first recorded blood transfusions, between dogs, at Oxford in 1665. On November 14, 1666, two physicians whom Lower had coached performed a similar experiment at a meeting of the Royal Society. This set off a craze for transfusion in England and on the Continent in the following years. Transfusions took place between individuals of the same species and between different species. There was speculation that

This 1667 engraving shows the procedures and instruments used for blood transfusion. (National Library of Medicine)

1642–1692) to speculate on the probability of patients developing coats of wool. The series of human transfusions came to an abrupt end when Denis transfused a man named Anthony du Mauroy Saint Amant, who suffered from periodic insanity. The first two transfusions of calf's blood, calves being considered calm and placid animals, seemed to alleviate Saint Amant's condition, but a third transfusion at the behest of his wife was followed by his death. This led to a messy legal affair. Saint Amant's wife charged the physicians with having killed him, while Denis charged Madam Saint Amant with having killed her husband. Denis was cleared in court, but by royal decree, the approval of the notoriously conservative medical faculty of the University of Paris was to be required for all future transfusions. Human transfusions ceased in both France and England immediately thereafter, although they remained a target for satirists.

Reference

Nicolson, Marjorie Hope. *Pepys' Diary and the New Science.* Charlottesville: University Press of Virginia, 1965.

transfusion could cause one individual to adopt the traits of another—for example, that transfusing the blood of a young individual could rejuvenate an old one.

The first human blood transfusion was carried out by the French physician and medical professor Jean-Baptiste Denis (1643–1704) in 1667. A young man suffering from a violent fever was transfused with the blood of a lamb, and he seemed to be feeling better afterward. A healthy porter, also transfused with lamb's blood, likewise suffered no reported ill effects. The English, a little resentful of French priority, followed with two successful transfusions of lamb's blood into a poor and slightly mad divinity student named Arthur Coga in late 1667, in the hopes that it would be therapeutic. This operation prompted the satirist Thomas Shadwell (c.

Book of Nature

One metaphor that legitimized the study of the natural world in Christian Europe was that of nature as a vast book, to be read by the devout natural philosopher as a divine revelation. Nature was God's second book, after the Bible. The book of nature did not teach the way of salvation, which was a monopoly of the Bible, but neither would the Bible be regarded as an authority in natural philosophy. The "two books" metaphor was very widely employed in natural theology, as it justified the study of nature in religious terms and provided a clear way to apply the knowledge of nature to religious questions. Robert Boyle used it to argue against the direct application of biblical texts to questions of natural philosophy. But the metaphor had uses outside natural theology as well.

Galileo Galilei used the idea of the two books to argue for the independence of natural philosophy from biblical authority, asserting that the two books were separate but equal in their authority. He also claimed that the language of the book of nature was mathematical. The book of nature metaphor was also employed by astrologers to explain their "readings" of the stars. A metaphor similar to the book of nature was the theater of nature.

See also Bible.
Reference
Brooke, John Hedley. *Science and Religion: Some Historical Perspectives.* Cambridge: Cambridge University Press, 1991.

Books of Secrets

Hundreds of printed and manuscript books of the wonders and secrets of nature circulated in early modern Europe, promoting an active view of nature as something to be used and manipulated through techniques and formulae rather than passively observed. Books of secrets, most numerous in Italy, drew from the lore of various crafts, from natural magic, and from popular beliefs, explaining for example how to make dyes or soaps or how to prepare medicines. Some of their authors, or "professors of secrets," became celebrated figures, whose names were attached to books they had nothing to do with. The books found an audience from many segments of the literate population. Although it is difficult to trace a direct link between the books of secrets and the scientific revolution, the books shared many of the ideological features of early modern science. They assumed that the sphere of human knowledge was expanding and that nature's secrets were only secret until they were found out through experimentation. They also assumed that the resultant knowledge should not be restricted to a social or intellectual elite, but widely publicized.

See also Popularization of Science.
Reference
Eamon, William. *Science and the Secrets of Nature: Books of Secrets in Medieval and Early Modern Culture.* Princeton: Princeton University Press, 1994.

Botanical Gardens

The botanical garden, which maintained and presented thousands of different plants for their intellectual interest rather than for their beauty, originated in the early modern period from a variety of sources. Europeans were interested in collecting and growing the plants of the New World and the farther reaches of Asia, of which they had just recently become aware. Princes and nobles viewed gardens as one way of displaying their wealth and power over nature. Physicians and apothecaries needed gardens for the cultivation of herbal medicines. Some believed that a well-ordered garden containing a huge variety of plants in some way recreated the Garden of Eden and could even lead to the recovery of the power over nature that Adam possessed when he named the plants.

The first large botanical garden incorporating plants from the recently encountered areas of the world was founded at the University of Pisa in 1543 by the medical professor Luca Ghini (c. 1490–1556). It was quickly followed by the great garden at the University of Padua, founded in 1545. (The founding of a garden was only an early stage in its creation, and it could take decades for the original plan to be fulfilled in actuality.) Padua's garden was connected to Padua's medical preeminence, and the supervisor of the garden was the professor of pharmacology. Similarly, the other great university botanical gardens of early modern Europe were usually connected to outstanding medical faculties. Besides Padua, they include the gardens of the Universities of Montpellier (founded 1598) and Leiden (founded 1577), the leading medical schools of France and the Netherlands respectively. A relatively late

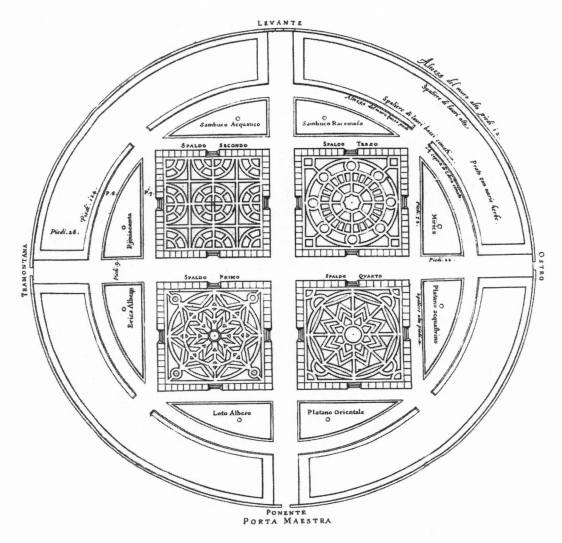

The plan of the botanical garden at the University of Padua. (Art Resource)

foundation was the botanical garden at the University of Uppsala in Sweden, founded in 1657 by the Swedish botanist Olof Rudbeck. One exception to the connection between strong medical faculties and university botanical gardens was the University of Oxford, undistinguished as a medical school yet possessed of a large botanical garden founded in 1621 on land donated by Henry, Lord Danvers (1573–1644) that had formerly been a Jewish cemetery. Oxford's garden was originally only loosely connected to the university; Danvers gave a 99-year lease of the site to the gardener Jacob Bobart (1599–1680),

including the right to sell the fruits and vegetables it produced. In 1669, the university took tighter control of the garden, and for the first time appointed a professor of botany, Robert Morison (1620–1683). As medical resources, all these university gardens were expansions of the "physic gardens," which existed to provide medicinal plants. The English Society of Apothecaries, for example, founded a physic garden at Chelsea in 1673. Such gardens, however large, were more functional and lacked the encyclopedic ambitions of the botanical gardens.

The botanical garden was always laid out

in a geometrical pattern that varied from garden to garden. In one popular pattern, plants were arranged according to continent of origin, with quarters devoted to Europe, Asia, Africa, and the Americas. Straight lines and right angles were considered preferable to following the contours of the land.

The most significant garden not associated with a university was the Royal Botanical Garden in France, founded in 1626 but not established until 1640. Covering 18 acres and a variety of different habitats, the Royal Botanical Garden was one of France's leading scientific institutions, operated under the direction of the king's physician. The leading spirit of the garden in the reign of Louis XIV (r. 1643–1715) was Guy Crescent Fagon (1638–1718), appointed superintendent in 1664. Fagon supervised the collection of plants from many parts of the world, including the introduction of tea and jasmine sent by Jesuit missionaries in China. The greatest botanist associated with the garden was Joseph Pitton de Tournefort (1656–1708), also a member of the Royal Academy of Sciences and professor of botany at the Royal Botanical Garden from 1683. Tournefort collected over 1,000 new plants in a visit to the Middle East from 1700 to 1702, and he devised one of the earliest systems for classifying plants based on their characteristics.

See also Botany; Rudbeck, Olof.
References
Duval, Marguerite. The King's Garden. Translated by Annette Tomarken and Claudine Cowen. Charlottesville: University Press of Virginia, 1982.
Prest, John. The Garden of Eden: The Botanic Garden and the Re-Creation of Paradise. New Haven, CT: Yale University Press, 1981.

Botany

During the scientific revolution, the science of botany emerged from medicine and the humanistic revival of ancient texts. It was transformed by the flood of new information coming from the world outside Europe and by the rise of new scientific approaches in the seventeenth century. Botany had ancient roots in the work of Aristotle (384–322 B.C.), Theophrastus (c. 372–c. 287 B.C.), Dioscorides (A.D. c. 40–c. 90), and Pliny the Elder (A.D. 23–79). Humanists of the fifteenth and sixteenth centuries recovered, edited, and translated these texts, as well as producing commentaries on them. It was soon found that this work could not be accomplished purely by textual editing; an understanding of the classic knowledge of plants was required. The problem of identifying the plants the ancients discussed led to independent botanical research, and it also revealed that the ancients seemed to have been unaware of a vast number of plants, particularly in northern Europe. This number was radically expanded by increased European knowledge of the outside world.

The principal practical application for botanical investigation in the sixteenth century remained medical. The first university chairs in the study of plants were established in the Italian medical schools, beginning at Padua in 1533. University botanical gardens were set up in the same period. The leaders in early-sixteenth-century botany were Italy and Germany, where a number of Lutheran botanists created impressive and innovative herbals. Otto Brunfels (1489–1534) and Leonhard Fuchs (1501–1566) transformed botanical illustration with highly accurate woodcut illustrations. Valerius Cordus (1515–1544), a professor at the University of Wittenberg, created a standard verbal formula for plant description. In Italy, the dominant figure of early-sixteenth-century botany was Luca Ghini (c. 1490–1556), the botanical professor first at the University of Bologna and then at the University of Pisa. Ghini promoted botany and botanical gardens and seems to have introduced the "dry garden," a technique for preserving plants by drying them. He created a massive herbarium, or collection of these dried plants. Gifts and exchanges of dried plants would help knit together the European botanical community during the scientific revolution.

Ghini's pupil, Andrea Cesalpino (1519–1603), was a professor at Pisa for 40 years and wrote *Sixteen Books of Plants* (1583), which contained a theoretical discussion of plants and descriptions of about 1,500, arranged in a classification system Cesalpino invented. This scheme, inspired by Aristotle and Theophrastus, was based on the characteristics of the plants rather than their medical uses, as was common in other herbals. Cesalpino, a skilled and diligent observer, viewed the reproductive systems of the plants as appropriate for classification. For example, he divided those plants that reproduce without seeds, such as ferns, from those that have seeds. Further divisions were based on such things as the positions and numbers of the seeds. Ghini also taught the first significant English botanist, William Turner (1508–1568), author of *A New Herball* (1551).

The number of foreign plants known to Europeans, compiled in medically oriented works such as Garcia d'Orta's *Colloquies on the Herbs and Drugs of India* (1563) and Francisco Hernández's *Four Books of the Nature and Virtues of the Plants of Mexico and New Spain* (1615), continued to grow through the early modern period. The French physician Gaspard Bauhin's (1560–1624) *Pinax* (1623) included over 6,000 plants. One serious problem for botanists was that different names could refer to the same plant or the same name to different plants, and Bauhin attempted to introduce a standard nomenclature for plants. He introduced the custom of using two words to describe a plant, one to describe the group it belongs to, the other to describe the plant itself. This would develop into the now-familiar genus-species way of naming living things. Bauhin's contemporary, the German Joachim Jungius (1587–1657), built on Cesalpino's approach by discussing plant forms in a theoretical manner rather than simply describing existing plants.

Plant anatomy and physiology was transformed in the second half of the seventeenth century. William Harvey's discovery of the circulation of the blood in animals led Johann Daniel Major (1634–1693) in 1660 to propose that sap circulates in plants, an idea that led to a great deal of research. The rise of the mechanical philosophy led to the desire to explain the processes within plants in mechanical terms. The microscope enabled much closer observation and the discovery of previously unknown parts of plants. Robert Hooke included observations of vegetable matter in his *Micrographia* (1665). Hooke's observations of the pores in cork led him to make the first description of cells in plants. As so often happened with Hooke's observations, this description was fully developed by others, notably Nehemiah Grew and Marcello Malpighi, who established the universality of cellular structure. Members of the French Royal Academy of Sciences largely avoided microscopy; their favored method of plant analysis was chemical distillation. The results proved confusing and hard to interpret. But the academy was reluctant to abandon it, as members believed that distillation would provide clues to the nutritional values and medicinal uses of plants.

The late seventeenth century saw several projects to systematize descriptions of plants into vast and theoretically exhaustive catalogs. John Ray, Olof Rudbeck, and the Royal Academy of Sciences all began universal plant compendia. All these efforts failed, dogged by bad luck—as in the fire that destroyed much of Rudbeck's work or the shifts in government policy to which the academy was vulnerable—but also stymied by the sheer vastness of the task. Ray came closest to success, with the three installments of his *History of Plants* in 1682, 1688, and 1704 dealing with 19,000 plants. Ray also had the most sophisticated classification scheme, set forth in *Botanical Method* (1682), which introduced the concept of species. Building on the work of his ancient and modern predecessors, Ray attempted to distinguish between the essential and accidental characteristics of plants, warning against classification based on only one or two factors. Ray's contemporary Joseph Pitton de Tournefort (1656–1708), professor of botany at the Royal Botanical Garden of France and the leading French botanist of his time, best known for his study of the plants of the eastern Mediterranean, introduced a classification scheme based on genus rather than species but had less success than Ray.

The German Rudolf Jakob Camerer (1665–1721), known by the latinization Camerarius, first demonstrated plants' sexual nature. Camerarius was a medical professor and director of the Botanical Garden at the University of Tübingen. The question of whether plants reproduce sexually had been asked since the time of Aristotle and Theophrastus. Grew and Ray had supported the idea of sexual reproduction in plants, but they were unable to demonstrate it. Camerarius settled the question by establishing the fertilizing role of pollen through experiment. It took longer for his discoveries to reach other botanists because of the 1694 publication of his *Letter on the Sex of Plants* in an obscure journal, *Transactions of the Tübingen Academy*. However, his discovery paved the way for Carl Linnaeus's (1707–1778) modern system of botanical classification.

See also Botanical Gardens; Grew, Nehemiah; Herbals; Ray, John; Rudbeck, Olof.

References

Arber, Agnes. *Herbals, Their Origin and Evolution: A Chapter in the History of Botany, 1470–1670.* 2d ed. Cambridge: Cambridge University Press, 1938.

Morton, A. G. *History of Botanical Science: An Account of the Development of Botany from Ancient Times to the Present Day.* London: Academic Press, 1981.

Stroup, Alice. *A Company of Scientists: Botany, Patronage, and Community at the Seventeenth-Century Parisian Royal Academy of Sciences.* Berkeley and Los Angeles: University of California Press, 1990.

Boyle, Robert (1627–1691)

Robert Boyle made major contributions to physics and chemistry, and he also played the role of the classic "Christian Virtuoso" in Restoration England. The fourteenth child and seventh son of Richard Boyle (1566–1643), an English land speculator in Ireland who became earl of Cork, Boyle was educated mostly by tutors and never attended a university. As a young man, Boyle was chiefly interested in humanistic moral philosophy, but after his return to England from a grand tour of the Continent in 1644, he became progressively more fascinated by natural philosophy. Dwelling at a manor his father had bought at Stalbridge in Dorset, Boyle was involved with the Hartlib circle, to which he had possibly been introduced by his sister, Lady Katherine Ranelagh (1615–1691), who shared many of Boyle's intellectual interests. After moving to Oxford in the mid-1650s, Boyle became a member of the Oxford group surrounding John Wilkins. He also became acquainted with Robert Hooke, whom he hired as an assistant in the performance of experiments, the beginning of a long-lasting association.

After the Restoration of Charles II (r. 1660–1685) to the British throne in 1660, Boyle was among the founders of the Royal Society. He served several times on its council, but he turned down an offer of the presidency in 1680 because he did not wish to take the oath of office. In 1668 Boyle, a lifelong bachelor, moved to London permanently, living in Lady Katherine's Pall Mall house. Among the best-known scientists in Restoration England, Boyle possessed a number of significant advantages for the study of natural philosophy. His wealth enabled him to maintain assistants and a laboratory with expensive equipment, such as the air pump. His aristocratic social standing and respected personal character put his word beyond doubt. Although not interested in day-to-day politics, Boyle was in favor at Charles's court and received several prestigious and lucrative appointments. As a leader of English science, Boyle received dedications from many significant scientific writers, including Thomas Sydenham and John Wallis. He also received many distinguished foreign visitors interested in English science, a practice that sometimes interfered with his own work.

Whatever his distractions, Boyle was among the most prolific of seventeenth-century scientists. In addition to his many contributions to *Philosophical Transactions,* hardly a year went by without the publication of at least one Boyle book. His ability to finance the translations of his works into Latin gave him a higher European profile than other contemporary English scientists. Boyle's own authorized editions were frequently plagiarized and pirated on the Continent, which made him somewhat paranoid and sometimes led him to rush into print. His publications cover chemistry, physics, medicine, natural theology, and religion.

Boyle's most influential text in chemistry, *The Skeptical Chymist* (1661), is a prolix and confusing work patched together from manuscripts. Its principal purposes were to attack the positions held by some contemporary chemists and alchemists and to raise the intellectual and social status of chemistry, until then a discipline many natural philosophers despised. Boyle attacked both those chemists who had too pragmatic and technological an approach and those who erected vast cosmological systems on a chemical basis. Descriptions of *The Skeptical Chymist* as an attack on alchemy as a whole are not accurate. Boyle distinguished between alchemical traditions; Paracelsians bore the brunt of his attack, whereas he admired both the followers of Johannes Baptista van Helmont and many of those traditional alchemists who concentrated on the making of gold.

Like many of Boyle's controversial writings, *The Skeptical Chymist* attacked the representatives of various positions under pseudonyms —Boyle shied away from direct confrontation. Alchemy interested him throughout his career, and he was involved with a number of English and Continental alchemists and alchemical groups. He was also influential in getting the English Parliament to repeal an old law against the alchemical manufacture of gold and silver. Another chemical work, *Experiments and Considerations Touching Colours* (1664), introduced an early version of the lit-

Long after his death, Robert Boyle remained a symbol for the alliance of science and Christian piety. In this 1740 engraving, his portrait is bathed in heavenly light and an air pump and other lab equipment are also depicted. (National Library of Medicine)

mus test for identifying acids and alkalis.

Boyle's "corpuscularian" matter theory had roots in both atomism and alchemy. It held that tiny particles of matter coalesce to form corpuscles, the basic building blocks of material substances, and it explained the properties of matter by the nature and structure of these corpuscles. Interaction between corpuscles was deemed largely mechanical, but Boyle did not believe that all material interactions could be explained strictly in mechanical terms in the manner of Descartes and the Cartesians. Although identified as a "mechanical philosopher," Boyle was not a dogmatist. He was very suspicious of all-encompassing systems such as Aristotelianism, which he attacked early in his career, Paracelsianism, and Cartesianism. Rather than system building, Boyle preferred investigating particular phenomena through experiment and the gathering of observations in the Baconian tradition. Experiment in particular obsessed Boyle. He believed that the experimenter, as opposed to the arrogant system-builder, approached nature with the proper humility.

The piece of experimental equipment with which Boyle was most identified was the air pump. Hooke built one for him that represented the state of the art in air-pump manufacture. Boyle's *New Experiments Physico-Mechanicall Touching the Spring of the Air, and Its Effects* (1660) recounted a number of air-pump experiments. This led to a controversy with Thomas Hobbes, one of the very few controversies involving Boyle. In the course of this dispute, Boyle published for the first time what came to be known as "Boyle's law": that the pressure exerted by air varies inversely with its volume. Boyle's law was not originally discovered by Boyle but by the English Catholic Cartesian Richard Towneley (1629–1707) and his associate Henry Power (1623–1668), but Boyle verified it with the aid of Hooke. Boyle also became a champion of the existence of vacuum, which in England was often referred to as the *vacuum Boylianum*.

Boyle was a devout Christian who devoted much of his fortune to spreading the gospel, sponsoring translations of the New Testament into several languages, including Irish, Turkish, Malayan, and Lithuanian. He hoped that knowledge of the philosopher's stone would lead to communication with angelic spirits, thus demonstrating the reality of the spiritual world. Boyle's will established the Boyle Lectures to defend Christianity from the alleged onslaughts of Deists and atheists. He rebutted suggestions that natural philosophy leads to atheism in works such as *The Christian Virtuoso* (1690), written to show that there was no conflict between religion and experimental philosophy.

See also Air Pumps; Alchemy; Boyle Lectures; Chemistry; Hooke, Robert; Mechanical Philosophy; Religion and Science; Royal Society.

References
Hunter, Michael, ed. *Robert Boyle Reconsidered.* Cambridge: Cambridge University Press, 1994.
Maddison, R. E. W. *The Life of the Honourable Robert Boyle, F.R.S.* London: Taylor and Francis, 1969.
Principe, Lawrence. *The Aspiring Adept: Robert Boyle and His Alchemical Quest, Including Boyle's "Lost" Dialogue on the Transmutation of Metals.* Princeton: Princeton University Press, 1998.

Boyle Lectures

One of the principal venues for natural theology and popular science in late-seventeenth-century England was the Boyle Lectures, founded in Robert Boyle's will. With his death in 1691, Boyle left £50 annually to pay for a series of London lectures for the defense of Christianity against atheists, Deists, pagans, Muslims, and Jews, although in practice the lecturers ignored the latter three groups. Boyle's will specifically barred the lecturers from discussing those issues that divided Christians themselves. The Boyle Lectures were dominated by Anglican proponents of "reasonable religion," many of whom used natural-theological arguments to demonstrate God's existence and providential care.

The first Boyle Lectures were delivered in 1692 by Isaac Newton's fellow Cambridge man Richard Bentley (1662–1742). After consulting with Newton himself on the possible theological uses of Newton's Principia (1687), Bentley used the Newtonian concept of gravity to demonstrate that purely mechanical explanations for attraction are insufficient and that explaining gravity requires assuming the existence of an active God. Other eminent Newtonians such as Samuel Clarke (1675–1729) and William Whiston (1667–1752) also gave Boyle Lectures. Some Boyle lecture series, including Bentley's, were extremely popular, being published as books that went through a number of editions and were translated into other languages. In the early eighteenth century, Boyle Lecturers tended to put less emphasis on Newtonian physics and to make more use of biological arguments based on the design of living things. The ingenious design of living things, both in themselves and as they ministered to the uses of human beings, was used to demonstrate the existence of God by Boyle Lecturers such as the Reverend William Derham (1657–1735), whose lectures in the years 1711–1712 were published as Physico-Theology (1713), a work reprinted many times in the eighteenth century.

See also Popularization of Science; Religion and Science.

Reference
Jacob, Margaret. *The Newtonians and the English Revolution, 1689–1720.* Ithaca: Cornell University Press, 1976.

Brahe, Tycho (1546–1601)

Among those who made first-rank contributions to the scientific revolution, the person with the highest social status was Tyge Brahe, a Danish nobleman usually known by the Latin form of his first name, Tycho. The Brahes were a leading family, distantly related to the Swedish royal house. Tycho became the greatest astronomical observer of the pre-telescope era, as well as an influential astronomical theorist. He was educated at the University of Copenhagen and a number of German Lutheran universities, where he lost most of his nose in a duel but gained an interest in astronomy, unusual and somewhat shocking in the anti-intellectual Danish nobility. After publishing an influential work on the new star of 1572, which he insisted was located above the sphere of the Moon, Tycho was granted the island of Hven by the Danish King Frederick II (r. 1559–1588) in 1576. There he established an observatory, called Uraniborg, meaning "the fortress of Urania," named for the muse of astronomy.

At Uraniborg, Tycho made the best and most systematic observations ever made with the naked eye, using improved instruments, many of which he designed himself and had built in the instrument shop that was part of the Uraniborg complex. Tycho's observations provided astronomers with the best data to date on the position of the celestial bodies, and Uraniborg itself became famous, an attraction for visitors to Denmark, including royalty. Tycho received extensive financing from the crown, and Denmark set the standard in terms of state support for scientific research at the time.

Astronomy did not exhaust Tycho's scientific interests, and like many Danish scientists and physicians, he was a Paracelsian. Uraniborg included several alchemical furnaces used by Tycho and his sister Sophie (1559–1643), the only member of his family to share his scientific interests and a frequent visitor to Uraniborg. The complex also included a printing press that printed Tycho's works, and he eventually acquired his own paper mill.

Tycho's greatest contribution to astronomical theory was the Tychonic model of the solar system, based on a stationary Earth; in this system, the Moon and Sun orbit the Earth and the other planets orbit the Sun. Tycho rejected the Copernican model, as he saw no reason to accept the radical idea of a moving Earth. Although the Tychonic system was mathematically equivalent to the Copernican and incorporated some of the advantages of Copernicanism in terms of simplicity, it was not produced all at once by simply inverting Copernicanism but was worked out by Tycho over a period of years before being published in his *Of More Recent Phenomena of the Ethereal World* (1588). Tycho had a nasty feud with the German astronomer Nicholas Reimers, known as Ursus (1551–1600), over Tycho's claim that Ursus had stolen his theory while visiting Uraniborg and published it in distorted form in his *Fundamental Astronomy* (1588). However, Tycho's colleagues confirmed his authorship of the Tychonic model, which was quite influential in the seventeenth century, displacing Ptolemaicism and becoming the chief rival to Copernicanism.

Tycho ruled Hven as a feudal lord, and the

Tycho Brahe, one of the most influential pretelescope astronomers, is pictured here with his mural quadrant, which was mounted on the west wall of his observatory of Uraniborg. This device enabled the elevation of a heavenly body to be more accurately measured than before. (Ronan Picture Library and Royal Astronomical Society)

labor that went into Uraniborg included that of peasants who owed him two days of work per week in household labor dues. He was not a popular lord, taking a grasping and high-handed tone with the peasantry. What led to the end of Uraniborg, though, was not peasant resentment but a change in Danish royal policy. The death of Frederick II in 1588 was followed by a regency dominated by friends and relatives of Tycho, under which he continued to do quite well. However, when King Christian IV (r. 1588–1648) took power in 1596 at the age of 19, he was determined to retrench Danish finances and challenge the power of the great noble families, of whom Tycho was so conspicuous a representative. After a brief struggle, Tycho left Uraniborg and Denmark forever in the summer of 1597.

After traveling through Germany, Tycho went to Prague, the capital of the Holy Roman Empire, causing another flare-up of his feud with Ursus, then imperial mathematician, who greeted Tycho's impending arrival with scurrilous attacks on his personal and professional character. Ursus fled shortly after Tycho's arrival in June 1599, and Tycho succeeded him as imperial mathematician to Rudolf II, with the young Johannes Kepler as his assistant and eventual successor. After Tycho died, Kepler obtained his log of observations, which became the data on which he built his astronomical theories.

See also Astronomy; Kepler, Johannes; Observatories; Planetary Spheres and Orbits; Priority; Rudolf II.

References

Dreyer, J. L. E. *Tycho Brahe: A Picture of Scientific Life and Work in the Sixteenth Century.* Edinburgh: A. & C. Black, 1890.

Thoren, Victor, with contributions by John R. Christianson. *The Lord of Uraniborg: A Biography of Tycho Brahe.* Cambridge: Cambridge University Press, 1990.

Bruno, Giordano (1548–1600)

The most notorious magician, heretic, and natural philosopher of the sixteenth century, Giordano Bruno lived a life full of wanderings and wrote publications in both Latin and Italian. The son of a poor Spanish soldier in Naples, Bruno was born in the town of Nola and was frequently referred to by contemporaries as "the Nolan." He did not receive a university education (although he taught at the Universities of Wittenberg, Oxford, and Toulouse), and the dominant intellectual passion of his life was a hatred of the Scholastic Aristotelian philosophy taught at the universities. He investigated many alternatives, from the ancient traditions of Platonism, Epicureanism, Hermeticism, and Pythagoreanism, to the Jewish Kabbalah, to the new science of Copernicus, and he combined elements from all of them into a syncretistic natural philosophy of his own. Although he did not have the observational and mathematical skills of an astronomer, Bruno was one of the very few sixteenth-century natural philosophers who took Copernicanism as an accurate description of physical reality, rather than a mathematical scheme for making astronomical calculations. He went beyond Copernicus, though, in setting forth an idea that had ancient precedents: that the universe is infinite and full of stars that, like the Sun, have planets orbiting around them. Bruno denied the notion that it is celestial spheres that carry the planets in their orbits, and he criticized Copernicus for retaining them in his system. However, he was unable to come up with an alternative explanation for planetary motion. Bruno also endorsed an atomist matter theory, although unlike later mechanist atomists, he regarded atoms as endowed with divine power. Bruno viewed mathematics as a search for truth in numbers, and he attacked trigonometry as based on approximations.

After an admittance into the Dominican order in 1563, Bruno was the subject of a proceeding for heresy in 1576. He left Naples for Rome, but the proceeding followed him there so he escaped from Rome, leaving behind his monastic vows and wandering first through Italy. In 1579, he left Italy behind for a series of journeys through France, England, and central Europe. Bruno's ingenuity and literary style made him well-suited to the court environment, and he served in the courts of Henry III of France (r. 1572–1589) and the Holy Roman Emperor Rudolf II, as well as spending two years in the household of the French ambassador to Elizabeth of England.

On his return to Italy in late 1591, Bruno was handed over by the Venetian authorities to the Inquisition, which after a lengthy investigation burned him in Rome in 1600 for heresy. Bruno was the only natural philosopher of the period whose intellectual radicalism had the effect of repudiating Christianity itself in favor of a universal reform based on improved natural knowledge. He became a martyr to the idea of freedom of thought, much admired by the freethinkers of the early Enlightenment, and his vitalistic natural philosophy influenced thinkers such as William Gilbert, Thomas Harriot, and Tommaso Campanella.

See also Copernicanism; Infinity; Kabbalah; Magic; Religion and Science.
References
Gatti, Hilary. *Giordano Bruno and Renaissance Science.* Ithaca: Cornell University Press, 1999.
Yates, Frances Amelia. *Giordano Bruno and the Hermetic Tradition.* London: Routledge and Kegan Paul, 1964.

Bureau of Address
See Renaudot, Theóphraste.

C

Cabala

See Kabbalah.

Calculus

See Leibniz, Gottfried Wilhelm; Mathematics; Newton, Isaac.

Calendar

See Chronology; Gregorian Reform of the Calendar.

Cambridge University

Although through most of the scientific revolution the curriculum of Cambridge University remained traditionally Aristotelian, it played a significant role in the rise of English science, culminating in the tenure of Isaac Newton as Lucasian Professor of Mathematics. The English universities had a hard time during the English Reformation and were generally behind their Continental rivals in the sixteenth century. There were, however, some stirrings of intellectual life at the time.

Cambridge had an active mathematical and scientific culture that, unfortunately, left little record because it existed outside the individual institutions or "colleges" that com-

posed the university. In addition, some of the colleges themselves promoted science. St. John's College, founded in 1511, emphasized mathematics, and the Padua-educated physician John Caius (1510–1573), who refounded Gonville and Caius College in 1558, served as its master and introduced new anatomical techniques. Later Cambridge physicians, notably Francis Glisson (1597–1677), Regius Professor of Physick from 1636 to 1677, adopted William Harvey's explanation of the circulation of the blood. Medical study led to an interest in natural history among students such as John Ray. After 1650, the Cambridge Platonists, notably Henry More (1614–1687) and Ralph Cudworth (1617–1688), encouraged the spread of Cartesian natural philosophy in Cambridge, although they later turned against it as tending to materialism and atheism.

The Lucasian Professorship of Mathematics, in imitation of Oxford's Savilian professorships, was founded in 1663 by the will of Henry Lucas (d. 1663), an English politician. Its first incumbent was Isaac Barrow (1630–1677), the noted mathematician and teacher of Newton, who held the post until 1669 when he relinquished it to Newton. Newtonian natural philosophy and Lockean epistemology took over the Cambridge curriculum by the first decades of the eighteenth century, and Cambridge would become known for its mathematical emphasis, as opposed to Oxford's humanism.

See also Newton, Isaac; Universities.
References
Feingold, Mordechai. *The Mathematicians' Apprenticeship: Science, Universities, and Society in England, 1560–1640.* Cambridge: Cambridge University Press, 1984.
Gascoigne, John. *Cambridge in the Age of the Enlightenment: Science, Religion, and Politics from the Restoration to the French Revolution.* Cambridge: Cambridge University Press, 1989.

Campanella, Tommaso (1568–1639)

Tommaso Campanella was an Italian Dominican friar from Calabria whose life was dedicated to, among other goals, overthrowing Aristotelianism and replacing it with an eclectic natural philosophy drawing on magical, biblical, Platonic, and empirical sources. He spent most of his working life in prison.

Campanella was one of the last important natural philosophers in Europe to operate within the framework of Renaissance magic and Platonism. As a Dominican, he received a thorough training in Aristotle but rejected him in favor of eclectic reading in ancient and contemporary sources, notably Bernardino Telesio, who remained a major influence. Campanella's anti-Aristotelianism required him to deny that the medieval Dominican philosopher Thomas Aquinas (c.1225–1274), who he revered, was an Aristotelian.

In 1589, Campanella arrived in Naples, the leading city of southern Italy, where he entered the natural-philosophical circle of the della Portas. After a brief imprisonment in 1592, when he was called upon to renounce Telesianism, he journeyed north to Padua, where he met Galileo Galilei. Arrested and tortured by the Inquisition on several charges, Campanella was sent to Rome, released, and then imprisoned again on charges of heresy, after which he was sent back to Calabria. There, in 1598, he was involved, in ways that are unclear, with a massive revolt, partially inspired by millenarian hopes. Upon the defeat of this undertaking, he was arrested and imprisoned, saving himself from execution as a relapsed heretic by feigning madness, a pretense maintained under extreme torture.

Campanella's decades-long imprisonment eventually grew less severe, and he was able to read the works of contemporary natural philosophers such as Tycho Brahe, William Gilbert, and Galileo and to circulate his own manuscripts. He corresponded with Galileo, and in 1616 he composed a Latin *Apologia pro Galileo* on Galileo's troubles with the church. The Dominican did not fully share Galileo's Copernicanism, remaining uncommitted to any astronomical theory, but he did defend free inquiry in natural philosophy. Campanella's fame spread throughout Europe even as his body was confined, and during his imprisonment many of his works, including the *Apologia,* were published in Germany by Lutheran Rosicrucians.

Campanella believed in an authoritarian, universal Catholic monarchy guided by magic and natural philosophy. In his utopia, *The City of the Sun,* composed in 1602, he put forth a vision of an ideal society where the very layout of the city and the breeding of its residents were governed by the precepts of magic, astrology, and natural philosophy. Briefly freed in 1626, Campanella was soon rearrested and assigned to the relatively pleasant confinement of the Roman Inquisition, from which he was released in 1629. He went to France in 1634, appearing to the mathematical and mechanical natural philosophers of the time like a relic from another age. He feuded with Pierre Gassendi, whose atomism he disliked, and also with the hater of magicians Marin Mersenne. Although a disappointment to the French government, he did receive an irregularly paid French pension and cast the horoscope of the newborn future Louis XIV.

See also Magic; Politics and Science; Religion and Science; Trial of Galileo; Utopias.

Reference

Headley, John. *Tommaso Campanella and the Transformation of the World.* Princeton: Princeton University Press, 1997.

Capitalism

Many phenomena of early modern Europe, including the Renaissance, the Protestant Reformation, and the rise of absolute monarchy, have been interpreted by historians as related to the rise of capitalism, sometimes identified as the "rise of the middle class." The scientific revolution has been no exception. Interpreting the rise of science as an aspect of the rise of capitalism is particularly appealing because the transfer of scientific leadership from northern Italy to the countries of the North Atlantic—England, France, and the Dutch Republic—parallels the transfer of leadership in capitalist advance. Unsurprisingly, the most persistent exponents of the view that the scientific revolution was caused by the rise of capitalism have been Marxist historians. The first to assert this was the Russian Boris Hessen. In a paper of 1931 titled "The Social and Economic Roots of Newton's 'Principia,'" Hessen argued that Isaac Newton's approach to physical problems had been conditioned by the needs of the growing capitalist economy, such as the need to increase the carrying capacity of ships. Crude and dogmatically Marxist, Hessen's argument was mostly dismissed by serious historians of science. But it did pose the question, however inadequate its answers.

In the decades following Hessen, the connection between science and capitalism was examined by both Marxist and non-Marxist scholars. The Austrian Marxist Edward Zilsel (1891–1944) asserted that the root of scientific advance in early modern Europe could be found in the interaction between academic natural philosophers and artisans. The economic rise of artisans with the expansion of the economy and the dissolution of the restrictions of the old guild system forced academic natural philosophers to take notice of artisan empiricism and craft knowledge, abandoning the ancient and medieval prejudice against manual operations. Zilsel argued that capitalism based on free labor was necessary to this process, as manual labor would also be despised in any society built on unfree, slave, or serf labor. The growth of a capitalist money economy also contributed to the growing tendency among early modern scientists to see nature in mathematical terms. The group of scientists who founded the experimental tradition at the turn of the seventeenth century—William Gilbert, Galileo Galilei, and Francis Bacon—did so by combining the observational skills of the artisan with the theoretical orientation of the natural philosopher.

For the British crystallographer, Stalinist, and historian of science John D. Bernal (1901–1971), the drive of early modern science—its experimentalism and observationalism but not its specific content—rested on the revolution of the rising bourgeoisie who overthrew the feudal economy and put the capitalist economy in its place. The later seventeenth century saw the creation of the science-based technology on which the capitalism of Bernal's own day rested—a capitalism he was confident would soon be overthrown by scientific communism. A non-Marxist, the American sociologist Robert Merton (b.1910), returned to Hessen's paper, much of which he adopted in *Science, Technology and Society in Seventeenth-Century England* (1938), where he argued that a scientist's choice of subject was strongly influenced by the technological and economic needs of early modern society. On the other hand, other historians, notably A. Rupert Hall (b. 1920), argue that the correlation in time between the growth of capitalism and the scientific revolution in itself proves nothing, and that the real scientific revolution was an intellectual movement having little to do with the "external" circumstances of the society in which scientists lived. These historians have generally placed less emphasis on observation and experiment, and more on the theoretical approaches available to natural philosophers.

Determining the actual relations between the scientific and capitalist revolutions, besides the fact that they took place in the same space and time, is a challenging task. Some connections seem to be clear. Much of the active mathematical life of Renaissance Italy was fostered by the spread of mathematical schools catering to the needs of businesses for accurate reckoning and accounting. The use of Arabic numbers in the fifteenth and sixteenth centuries was spurred by the needs of Italian and European business and would eventually greatly ease the calculations of scientists. Two important sixteenth-century mathematicians and mechanicians, Niccolò Tartaglia (c. 1499–1557) and Simon Stevin, were bookkeepers. Works of science were sometimes addressed to a commercial audience. Stevin's *The Tenth* (1585), which set forth his system of decimals, was addressed to bankers and merchants, and the works of Galileo are full of metaphors drawn from the countinghouse.

European expansion and the necessities of navigation, such as the longitude problem, also brought together scientists and merchants. The English mathematicians Robert Recorde (c. 1510–1558) and John Dee were advisers to merchant companies, and some of Gilbert's magnetic work followed on magnetic experiments by the sea captain Robert Norman. London's Gresham College, established to provide technical training for London merchants, stands as an expression of this integration of commercial and scientific culture in Elizabethan England. The commercial expansion of Europe also contributed to the increase in natural knowledge of remote places. The natural historian of Indonesia, Georg Eberhard Rumph, was an employee of the great Dutch East India trading company. At home, the dissemination of scientific information was greatly advanced by the capitalist printing industry.

On a more modest level, capitalism contributed to the scientific revolution by keeping individual scientists employed. Some important scientists, such as the brewer and mayor of Magdeburg Otto von Guericke (1602–1686) and the burgher of Delft Antoni van Leeuwenhoek, spent their lives in a decidedly middle-class and capitalist milieu. The chemist Johann Rudolf Glauber (1604–1670), a more marginal figure, supported himself by manufacturing tartaric acid and mineral salts for the commercial market. Science itself produced capitalist businesses of modest scale, ranging from Italian telescope makers to Dutch importers of exotic natural history items for collectors.

Other connections between early modern science and capitalism are more tenuous. Science was not a particularly "middle-class" phenomenon. Scientists came from a wide social spectrum, with many supplied by Europe's traditional intellectual classes— the clergy, the nobility and gentry, and the legal profession. Many scientists avoided the capitalist "market" economy, preferring to function in the aristocratic "patronage" system, although by the late seventeenth century patronage declined and the market rose as a means of support for scientists. Science itself was not thought to express particularly capitalist or middle-class values. In fact, it was often argued that only gentlemen with an independent income from the possession of land had the necessary disinterest to successfully practice science. Tradespeople interested in material gain could not be trusted, as it was assumed that they would attempt to keep for their own benefit whatever secrets of nature they discovered. Many of Robert Hooke's difficulties at the Royal Society were caused by the fact that the society's aristocratic leaders looked down on him as a tradesman. Even Glauber had to disassociate himself from his for-profit activities to assert his right to speak as a natural philosopher.

Scientific programs designed for material improvement, of which the most famous if not the most practical was Francis Bacon's, were usually addressed to rulers of states rather than individual capitalist entrepreneurs. Many of the projects that brought together scientists and businessmen for specific technical purposes failed. Gresham College never fulfilled its purpose of providing technical training for London merchants, making its mark more as an institution for scientific research. The Royal Society's efforts to draw up "histories of trades," reducing the practices of various types of businesses to writing so they could be improved, also failed to significantly advance either science or capitalism.

See also Exploration, Discovery, and Colonization; Gresham College; Technology and Engineering.

References
Cohen, H. Floris. *The Scientific Revolution: A Historiographical Enquiry.* Chicago: University of Chicago Press, 1994.
Hadden, Richard W. *On the Shoulders of Merchants: Exchange and the Mathematical Conception of Nature in Early Modern Europe.* Albany: State University of New York Press, 1994.
Jacob, Margaret. *Scientific Culture and the Making of the Industrial West.* New York: Oxford University Press, 1997.

Cardano, Girolamo (1501–1576)

Girolamo Cardano was a highly prolific Italian physician and mathematician, author of over 200 works on a variety of scientific, religious, and occult subjects. The illegitimate son of Fazio Cardano, a lawyer, geometer, and friend of Leonardo da Vinci, Girolamo received a B.A. from the University of Padua in 1518 and an M.D. from the University of Pavia in 1526. His treatise on algebra, *The Great Art* (1545), contains new applications of geometrical methods to the solution of equations of the third degree, where the unknown number is cubed. He also went beyond geometry to discuss algebraic methods for the solution of fourth-degree equations that could not be geometrically represented, methods actually invented by his student,

Lodovico Ferrari (1522–1565). Cardano also described solutions using negative and imaginary numbers, which cannot be represented by lines. This work led to a dramatic quarrel with Niccolò Tartaglia (c. 1499–1557), who accused Cardano of stealing his technique for solving equations of the form $x^3 + ax = b$. Cardano was a habitual gambler, and wrote the first work on probability in relation to games, a short pamphlet not published until long after his death.

After a lengthy and difficult beginning in the medical profession, Cardano established a successful practice in Milan. Despite his voluminous writings, Cardano made few original contributions to medicine beyond giving the first clinical description of typhus. His medical philosophy was an eclectic mix of Galenism, Aristotelianism, Hippocratism, and the work of the Arabic and medieval commentators. He was also influenced by the new anatomy of Andreas Vesalius. Cardano had a naturally eclectic temperament and was unwilling to exclude any possible source of knowledge. He shared the medical humanist regard for Hippocrates (c. 460–c. 377 B.C.), regarding him as a man of almost divine knowledge and producing massive commentaries on the Hippocratic texts. However, Cardano also attacked the anti-Arabism of the medical humanists, defending medieval Muslim physicians such as Avicenna (980–1037). Cardano's natural philosophy, expressed in his popular compendia *Of Subtlety* (1550) and *On the Variety of Things* (1557), was marked by fascination with wonder and occultism, expressed in a basically Aristotelian framework. An astrologer, he firmly believed in natural magic and the significance of dreams and portents. He was also an ingenious mechanician, inventing a commonly used suspension device.

Cardano's life was eventful. A celebrated physician, he was often invited to the courts of kings and rulers. In 1552, he journeyed to Scotland to cure the archbishop of Edinburgh, acquiring in the process a taste for Scottish beer. His eldest son was executed in

1560 for poisoning his wife. This devastated Cardano, who himself was arrested and imprisoned by the Inquisition in 1570 for publishing the horoscope of Jesus Christ. Released in 1571 but barred from teaching, he lost the position as professor of medicine at the University of Bologna he had held since 1562. Cardano spent his final years in Rome practicing medicine and writing an autobiography, the first scientist to do so.

See also Mathematics.

References
Ore, Oystein. *Cardano: The Gambling Scholar.* Princeton: Princeton University Press, 1953.
Siraisi, Nancy G. *The Clock and the Mirror: Girolamo Cardano and Renaissance Medicine.* Princeton: Princeton University Press, 1997.

Cartesianism

In the 50 years after the death of René Descartes in 1650, Cartesianism became the predominant school of natural philosophy in France and most of Europe. Although there were many differences between individual Cartesians, all were mechanical philosophers, reducing natural phenomena to matter and motion. Despite Descartes's mathematical genius, Cartesian physics remained qualitative rather than mathematical. Cartesians held that the world is composed of vortices of matter. Following Descartes in accepting Copernican astronomy, they believed the planets are carried in their orbits around the Sun in these vast whirlpools of matter. Smaller vortices were invoked to explain other natural phenomena, such as magnetism. Cartesians, unlike atomists, claimed that matter fills all space, with no vacuum. They also claimed that the non-material human soul interacts with the body through the pineal gland. Animals, not possessing souls, have no consciousness or feeling, being essentially living machines. Methodologically, Cartesians also viewed experiment as secondary to logical deduction as a mode of scientific inquiry. Many Cartesians deviated from Descartes by abandoning his belief in certain knowledge and making ingenious explanations for natural phenomena with the goal of plausibility rather than certainty.

Cartesianism had several advantages over its principal competitors in France, the Aristotelianism of the universities and the rival atomistic mechanical philosophy of Pierre Gassendi. In the 1650s and 1660s, a small band of dedicated Cartesians edited and published Descartes's letters and unpublished works, organized conferences, and finally, in 1667, orchestrated the dramatic return of Descartes's body from Sweden and its burial in Paris, keeping Descartes in the public eye. Descartes's tendency to separate the soul from the body was popular with French women intellectuals, as it meant that their mental and spiritual capacities were not hindered by their "inferior" female bodies. A French Cartesian man named Francois Poullain de La Barre (1647–1723) wrote a treatise arguing for women's equality on Cartesian grounds in 1673; quite influential, it was translated into English as *The Woman as Good as the Man* (1677). Cartesianism also had an advantage among women in that it was expressed in elegant French rather than the Latin of the Aristotelians and Gassendi, and it swept the woman-run Parisian salons, a key advantage in its victory in French culture as a whole.

Cartesians were also prominent in the academies and informal scientific societies of France. Although the Royal Academy of Sciences originally banned dogmatic Cartesians along with Jesuits, many academicians, notably Christiaan Huygens, were basically Cartesian in their natural philosophy, and the academy steadily drifted in the direction of Cartesianism. This process was accelerated after the reorganization of the academy in 1699, after which dogmatic Cartesians were admitted, and by the early eighteenth century the academy was a Cartesian stronghold. The Oratorian order, a rival to the Jesuits in teaching, became strongly Cartesian, producing the most important Cartesian philosopher after Descartes, the academician Nicholas Malebranche. Cartesianism was also spread by public lectures and demonstrations, notably the famous Wednesday evening lectures of Jacques Rohault (1618–1672), in which a physical phenomenon, such as the barometer, would first be demonstrated and then explained in Cartesian terms of "subtle matters" before an audience of upper-class Parisian men and women. In the closing decades of the seventeenth century, Cartesianism as opposed to Aristotelianism had the intangible quality of being "modern."

However difficult the struggle for the support of the Parisian upper class, Cartesians faced a much longer and tougher battle to displace Jesuit-backed Aristotelianism from its leading position in the French universities. (Cartesianism was already very strong in the French Protestant schools before their suppression by Louis XIV [r. 1643–1715].) Into the eighteenth century, all theology professors at the conservative University of Paris were required to sign a formal rejection of Cartesianism. One problem the Cartesians faced when dealing with church and university authorities was the association of Cartesianism with Jansenism, a reform movement in the French Catholic Church that violently opposed the Jesuits and was persecuted by the French government in the late seventeenth century.

Cartesianism was always theologically suspect as materialist, but Cartesian natural philosophers made it religiously acceptable by eschewing Cartesian metaphysics. Furthermore, unlike other challengers to Aristotelianism, Cartesianism was ideal for pedagogical purposes because of its emphasis on deduction from first principles. Despite the best efforts of the Jesuits, who got Descartes's works put on the Catholic Church's Index of Forbidden Books in 1663, Cartesianism was being taught in French universities by the 1690s. Outside France, Cartesianism spread from the Dutch universities, where it was already influential during Descartes's lifetime, to those of Geneva,

Scotland, Scandinavia, and Germany. Although Cartesianism was influential in England as well, the conservative English universities remained Aristotelian. Some English thinkers, notably the Cambridge Platonist Henry More (1614–1687), initially an enthusiastic Cartesian, feared that Cartesian mechanism was conducive to atheism.

By the early eighteenth century, Cartesianism faced a new rival in Newtonianism (although Newton himself had originally been a Cartesian). The struggle between the two would last well into the eighteenth century, and Cartesianism's emphasis on logic and deduction from first principles remains influential in French science.

See also Descartes, René; Malebranche, Nicholas; Mechanical Philosophy; Physics; Vacuum.
References
Lennon, Thomas M. *The Battle of the Gods and Giants: The Legacies of Descartes and Gassendi, 1655–1715.* Princeton: Princeton University Press, 1993.
Sutton, Geoffrey V. *Science for a Polite Society: Gender, Culture, and the Demonstration of Enlightenment.* Boulder, CO: Westview Press, 1995.

Cartography

Cartography, the science of mapping, was transformed in the sixteenth century by the humanist revival of the ancient cartographical tradition, by the vast increase in geographic information available to Europeans as a by-product of European expansion, and by the introduction of the printed as opposed to the hand-drawn map. The most common form of world map in the Middle Ages depicted Asia, Europe, and Africa huddled around Jerusalem, at the center of the world. (This was known to be a flat representation of a spherical world—no educated person in the Middle Ages believed in a flat Earth.) The continents were shown as crude outlines, with no attempt to represent geographic detail. More detailed maps were the *portolans,* charts of particular areas of coast for use by sailors. By the later Middle Ages,

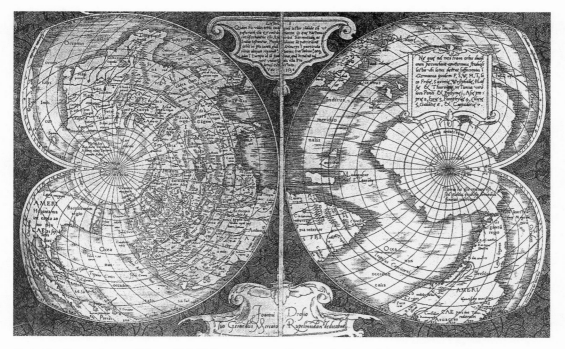

Gerhardus Mercator's map of the world (1538) was constructed by using the famous "Mercator projection," which eventually allowed cartographers to display the whole world on a rectangular map. (Library of Congress)

portolans were being compiled into more detailed and extensive maps of particular areas, but these maps were not widely distributed and were vulnerable to corruption in copying and recopying.

The revival of the system developed by ancient Greek geographer and astronomer Claudius Ptolemy (A.D. 90–168) provided a more sophisticated, geometrical way of mapping the geography of an area, although Ptolemy's focus on the Mediterranean and Indian Oceans rather than the European continent made his usefulness for specific geographic data limited. The creation of new maps was stimulated by new knowledge, beginning with the charts of the coast of West Africa and the Atlantic Islands produced by Portuguese seafarers in the fifteenth century. The development of printing made it possible, for the first time, to produce large numbers of identical maps. The most important cartographers of the sixteenth century were two Flemish friends, Gerhardus Mercator (1512–1594) and Abra-

ham Oertel, known by the latinized name Ortelius (1527–1598). Both issued printed maps. Mercator projected the Earth's spherical surface onto the flat surface of a map, a projection that still bears his name. The term "atlas" is derived from his reference to the ancient Greek Titan who held the world on his shoulders, found in his later three-part collection of world maps, *Atlas* (1585, 1590, 1595).

Ortelius published *The Theater of the World* (1570), a printed and bound collection of mostly European maps. He did not create the maps himself, but collected and improved them, reducing them to a common format and providing a bibliographic commentary that amounts to a history of cartography. With the prestige gained from *The Theater of the World,* Ortelius became official geographer to Philip II (1527–1598), the king of Spain and sovereign of Flanders, whose globe-spanning empire was a preeminent consumer of cartographic services.

Maps were instruments not only of knowledge, but also of power. More accurate domestic maps were produced to satisfy the taxation and defense needs of government. Europeans invading other European countries needed the most accurate maps possible, and in the colonial realm, mapping both facilitated invasion and functioned as a symbolic expression of power—to map a country and to name its geographical features was to control it. Although the earliest printed maps did not include political boundaries, national maps contributed to a sense of national unity—in 1579, one of the European countries with the most highly developed sense of nationhood, Elizabethan England, also produced Europe's first national compilation of regional maps, by Christopher Saxton. On a smaller scale, landlords wanted maps of their holdings, and the early modern period saw the growth of the surveying profession.

In addition to steadily integrating more new geographic knowledge, seventeenth-century cartography became a more mathematically based and cooperative discipline, with the gradual acceptance of the Mercator projection as the standard. One of the most famous state-sponsored mapping projects, incorporating the most modern and accurate techniques, was the never-completed map of France, which occupied much of the energy of the Royal Academy of Sciences in its early years.

See also Exploration, Discovery, and Colonization; Geography.
Reference
Boorstin, Daniel. *The Discoverers: A History of Man's Search to Know His World and Himself.* New York: Random House, 1983.

Cassini, Gian Domenico (1625–1712)

Gian Domenico Cassini was the most important astronomical observer of the planets during the late seventeenth century and the dominant personality of the Paris Observatory. Of obscure background, Cassini was educated at a Jesuit college in Genoa. He acquired as a patron a Bolognese senator who was an amateur astronomer and astrologer and who allowed Cassini to use his observatory at Panzano. At his patron's instigation, Cassini became professor of astronomy at the University of Bologna in 1650, and in addition to astronomy he worked on hydraulic engineering projects in Italy. Cassini was part of the talent from many parts of Europe drawn to the Paris of Louis XIV (r. 1643–1715). He was invited there in 1668 to help set up the Paris Observatory. Arriving in 1669, he never returned to Italy, obtaining French citizenship in 1673 and marrying into an aristocratic French family. His descendants would dominate the Paris Observatory well into the nineteenth century.

Cassini is best known for his discovery of the empty space between the rings of Saturn, called "Cassini's division." While still in Italy, he measured the rotations of Mars and Jupiter and constructed tables of the motions of Jupiter's satellites. In Paris he discovered four moons of Saturn. Following a successful project to measure the parallax of Mars at the opposition of Mars in 1672, Cassini estimated the distance between the Earth and Sun as 87 million miles, a much more accurate figure than any previously. He was involved in a controversy over the shape of the Earth, maintaining that it was flattened at the equator when Newton and others maintained, correctly, that it was flattened at the poles. During his Italian period, Cassini endorsed the church-approved Tychonic system of astronomy, and even in the freer environment of Paris, his embrace of Copernicanism was lukewarm.

See also Paris Observatory.
Reference
Wolf, A., with the cooperation of F. Dannemann and A. Armitage. *A History of Science, Technology, and Philosophy in the Sixteenth and Seventeenth Centuries.* 2d ed., prepared by Douglas McKie. London: Allen and Unwin, 1950.

Gian Domenico Cassini is shown here looking at the Paris Observatory, of which he was the first head. (Courtesy of the Galileo Project)

Causation

During the scientific revolution, the definition of causation was much more narrow and reductionist than it had been in medieval

Aristotelian natural philosophy. Medieval natural philosophers inherited from Aristotle the idea of the "four causes," which they called formal, material, efficient, and final. Causation was central to their idea of explanation. Eliminating the other three causes to concentrate on the efficient cause, early modern scientists ended by moving away from the search for causes as the basis of the scientific endeavor.

Of the four causes, the formal cause is the form that a thing takes. For example, the formal cause of a statue is its shape, without which it would not be a statue. The material cause is the substance out of which a thing is made. The material cause of a bronze statue is bronze. The efficient cause is the actor that causes the thing to take its particular shape or form. Efficient causes are individuals rather than universals. The efficient cause of the statue is the sculptor. The final cause is the purpose for which the thing exists. The final cause of the statue is to be erected in a temple.

This is a trivial example, and Aristotle was aware that natural phenomena could not all be slotted into a rigid formula of causes. Scholars in the Middle Ages accepted this division of causes, adding to it the idea of God as the First Cause, the original and ultimate cause of what happens in the universe. The question of how God relates to the causes of particular events was debated throughout the medieval and early modern periods. Medieval philosophers also developed a logic of inquiry based on reasoning either from effects to causes or from causes to effects, a logic that was preserved to the early modern period. This method, developed by the Italian Aristotelians of the University of Padua and the Collegio Romano, was the basis of the scientific method of Galileo Galilei and other early modern scientists.

Although the Aristotelian system of causation continued to influence early modern scientists—William Harvey's explanation of the circulation of the blood is structured according to the four causes—it was abandoned during the course of the seventeenth century. Early modern mechanical philosophies reduced causation in the physical world to mechanical interaction. This process was initiated by René Descartes, the most influential participant in the debate over causation. Descartes eliminated the final cause from consideration by identifying it with the essentially unknowable purposes of God. (The attribution of final causes to the phenomena of nature was particularly reviled by early modern natural philosophers—Francis Bacon compared the final cause to a barren virgin and claimed that it does not advance the sciences, but corrupts them.) The idea of a final cause persisted in the argument that God had designed the universe with providential purposes in mind, but this was not considered a scientific explanation. Descartes also denied that formal and material causes are causes at all, and he set forth the "causal principle"— that there must be a similarity between a cause and an effect—although he did not always adhere to it.

Once the principle of inertia was established, what mechanical philosophers needed to explain was change in a body's motion, explained for strict mechanical philosophers by the impact of other bodies. This resulted in elaborate causal explanations, based on the size, shape, and motions of different kinds of particles, for phenomena such as magnetism and gravity. This presented the problem, though, of the causal aspect of mind-body relations. Given Descartes's causal principle and the dualism he maintained between the spiritual mind and the material body, there seemed no possibility of mind-body interaction. Only the most radical materialists—such as Thomas Hobbes, who denied that mind is a substance different from matter—could reduce the causation of physical by mental events to mechanical interactions.

The most radical solution to the mind-body causation problem within Cartesianism was the occasionalism of Nicholas Malebranche, which built on certain aspects of Descartes's work to claim that God is the only cause and that all things caused are caused by the immediate power of God. In this view, whatever appears to us as physical causal relations are in reality occasions rather than causes—water freezes on the occasion when it is cold, but God and not the temperature is the cause. However useful in philosophy and theology, occasionalism clearly could not be practically applied to scientific questions, as Malebranche knew.

Unlike the metaphysicians, the natural philosophers of the late seventeenth century largely abandoned the quest for a theory of causation in favor of an investigation of particular causes. Some English natural philosophers made a virtue of *not* seeking causes. The Royal Society propagandist Joseph Glanvill (1636–1680) claimed that we can only know that *a* causes *b* if *a* continually accompanies *b*, but that even so, we can never be certain that the relationship we have identified is a causal relationship. The major causal question of late-seventeenth-century science was that of the cause of gravity. Isaac Newton's failure to assign a mechanical cause to gravity aroused great opposition among leading scientists such as Christiaan Huygens and Robert Hooke. Newton's gravity was ridiculed as an "occult," or hidden, cause. Newton's friend John Locke, who sought a philosophical approach to causation that was reconcilable with contemporary scientific practice, was skeptical as to the ability of the human mind to grasp true causes.

See also Aristotelianism; Cartesianism.
References
Clatterbaugh, Kenneth. *The Causation Debate in Modern Philosophy, 1637–1739.* New York: Routledge, 1999.
Wallace, William A. *Causality and Scientific Explanation.* 2 vols. Ann Arbor: University of Michigan Press, 1972 and 1974.

Cavendish, Margaret, Duchess of Newcastle (1623–1673)

Although culturally constrained by her gender and the lack of education it implied, Margaret Cavendish produced, among other writings, several volumes of an original and eclectic natural philosophy. Born into a wealthy family as Margaret Lucas, she became a maid of honor to Queen Henrietta Maria (1609–1669). In 1664, the English Civil War forced the queen into exile in Paris, and Margaret accompanied her there. In Paris, she married another Royalist in exile, William Cavendish, Marquess of Newcastle (1592–1676), 30 years her senior.

In Paris, Margaret Cavendish met leading natural philosophers, including René Descartes, Thomas Hobbes, and Pierre Gassendi. She published the first work by an Englishwoman on natural philosophy, *Philosophical Fancies,* in 1653, on a visit to England on matters connected with her husband's estate. Her other significant natural philosophical works include *Philosophical and Physical Opinions* (1655; reissued in 1668 as *Grounds of Natural Philosophy*) and a utopia, *The Description of a New World, Called the Blazing World* published as an appendix to her *Observations Upon Experimental Philosophy* (1666). These were published at her husband's expense because as a married woman, she could have no money of her own under English law.

As a natural philosopher, Margaret Cavendish accepted atomism but rejected the mechanical philosophy in favor of one that saw the universe and its material component parts as alive, intelligent, and self-acting. She expressed doubts about the reliability of the microscope. Cavendish, who with her husband returned to England after the Restoration of Charles II in 1660, corresponded with distinguished natural philosophers including Hobbes and Christiaan Huygens, to whom she gave a set of her books. Personally flamboyant and eccentric, with a gift for self-promotion that was viewed as inappropriate for a woman, she was ridiculed as "Mad Madge." Cavendish arranged to visit the Royal Society in 1667, causing a furor among the Fellows. Given Cavendish's social rank, they could not refuse, and she was admitted to a viewing of an experiment performed by Robert Boyle, the only time before 1945 that a woman attended a Royal Society meeting.

See also Women.
Reference
Jones, Kathleen. *A Glorious Fame: The Life of Margaret Cavendish, Duchess of Newcastle, 1623–1673.* London: Bloomsbury, 1988.

Celestial Spheres

See Planetary Spheres and Orbits.

Chemistry

Although chemistry in the scientific revolution did not change dramatically, as did astronomy and physics, it did undergo significant transformations. One of these was the formation of chemistry or "chemical philosophy" from a number of traditions dealing with substances and their properties. These included alchemy, the practical traditions associated with miners and craftspeople, and medicine. (The word "chemistry" itself was formed by removing from "alchemy" the prefix *al-*, seen by humanists as an Arabic and therefore illegitimate definite article, leaving the original Greek "chemia.") All of these traditions received classic treatment during the early modern period: for the alchemical tradition, the high point came in the early sixteenth century with Paracelsus; for the practical and technological tradition, it came with Georgius Agricola and his *Of Metallic Things* (1556) and from the Italian engineer and papal official Vannoccio Biringuccio (1480–1537) and his *Pyrotechnics* (1540); for the pharmaceutical tradition, it came with the Swiss naturalist and humanist Konrad Gesner (1516–1565) and his widely translated *Of Secret Remedies* (1552). Yet these traditions, which employed distinct vocabularies, remained largely separate and even hostile, with Agricola and Biringuccio, for instance, denouncing the symbolic language of esoteric alchemy and the fraudulent practices of alchemists. What they did have in common is that they were all low-status disciplines from the point of view of established, Aristotelian natural philosophy. The use of furnaces and equipment in chemical processes marked them as "base" sciences.

The Paracelsian "chemical philosophy" of the late sixteenth and early seventeenth century was allied to magic and Hermeticism by thinkers such as Oswald Croll and Robert Fludd. Another side of the Paracelsian legacy was developed by the "iatrochemists," medical chemists who studied chemical remedies for diseases. Some completely separated a core of "chemical" knowledge from magical or mystical Paracelsianism. This was accompanied by a desire to make chemical knowledge public, whereas in the alchemical tradition it was often passed down from master to student and veiled in an opaque language of symbols and metaphors, with talk of "green lions" and "chemical weddings." The Ramist anti-Paracelsian schoolmaster Andreas Libavius published the first textbook of chemistry, *Alchemia,* in 1597. This would be frequently abridged, adapted, or translated throughout the century. The textbook tradition would be carried on principally in France, by a succession of professors at the Royal Botanical Gardens who were mostly interested in training apothecaries. The culminating work of this tradition was Royal Apothecary Nicholas Lémery's (1645–1715) *Course of Chemistry* (1675), which combined Paracelsian and Cartesian influence and went through over 30 editions.

However, the most influential chemist of the early seventeenth century was not a professor or textbook writer, but a nobleman, Johannes Baptista van Helmont. Van Helmont developed the Paracelsian tradition further, putting forth a new system based on a single element, water. He and his disciples, such as the Leiden professor Franciscus Sylvius (1614–1672), described nature as governed by chemical processes. Van Helmont, for example, described digestion as caused by the action of acids, as opposed to the Galenist theory that it is caused by heat. Sylvius reduced all disease to an imbalance either of acid or alkali, although there was still no

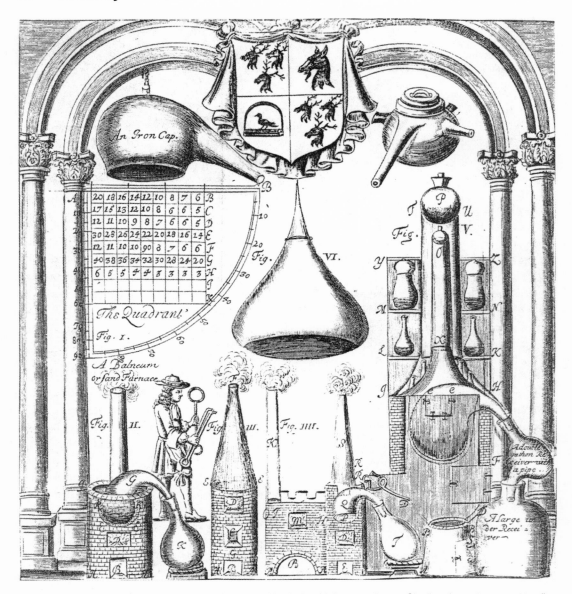

This unsigned engraving from 1692 depicts William Yworth's idealized laboratory design of his "Academia Spagirica Nova"; his coat of arms appears at the top. (British Library)

definition of either, other than that an acid reacts with an alkali and vice versa.

The new mechanical philosophies of René Descartes and Pierre Gassendi originally had little to say about chemical problems, which interested neither philosopher, both being inheritors of the Aristotelian disdain for chemistry. Robert Boyle made it his mission to bring together chemistry and the mechanical philosophy, legitimizing the study of chemistry as appropriate for natural philosophers. This task was made easier by Boyle's high social standing and reputation for piety—if Robert Boyle thought chemistry worthwhile, few mainstream English scientists would deny it. Boyle's mechanical philosophy did not mean that he explained all chemical processes in mechanical terms, as he was also deeply influenced by the Helmontian "chemical philosophy." Indeed, he studied alchemy and believed that the transmutation of metals was at least possible, and his *The Skeptical Chymist* (1661) attacked the arrogance of both Aristotelians and Paracelsians in believing that they knew the fundamental nature of matter.

Boyle called for a Baconian-inspired chemical program putting experiment and the compilation of experimental knowledge ahead of chemical theorizing. He denied all elemental theories—from the Aristotelian four-element system (earth, air, fire, water) to the Paracelsian three-element (salt, sulfur, mercury) to the Helmontian one-element system (water)—in favor of a "corpuscularian" theory, which combined atomism with a belief in self-directed matter. While pre-Boyle elemental theories saw all substances as combinations of all elements, Boyle asserted that not all elements are present in all substances. Later chemists, although they never abandoned the notion of an element, were much less concerned with reducing all elements to a small number. Boyle's example led to much chemical theorizing in Britain, notably that of the physician John Mayow (1641–1679). His *Five Medico-Physical Treatises* (1674) developed a theory, originally put forward by Robert Hooke, that explained a variety of natural phenomena, from breathing to weather, by the interaction of sulfurous and nitrous particles. In addition to the activities of Boyle and his disciples, chemistry also attracted interest because of the separation of phosphorus from urine, first accomplished in 1669 by the German alchemist and physician Hennig Brand (d. c. 1692) and later duplicated by another German, Johann Kunckel von Löwenstjern (c. 1630–1703), and by Boyle. Phosphorus's glow made it popular for exhibitions and demonstrations.

Even though the only part of Isaac Newton's extensive work on chemistry and alchemy published during his lifetime was a paper on the nature of acids and some passages in the *Opticks* (1704), his attempt to explain chemical interactions in terms of attractions and repulsions would have major influence on eighteenth-century chemistry. The dominant chemist at the close of the scientific revolution, however, was the German physician and professor of medicine at the University of Halle, Georg Stahl (1660–1734). Stahl worked in a German tradition in which his immediate predecessor was the alchemist and government official Johann Becher (1635–1682). Becher had described something he called "fatty earth," an intellectual descendant of Paracelsus's sulfur, which Becher believed responsible for combustion. Stahl renamed this substance "phlogiston" (chemistry still lacked a standardized nomenclature, and chemists commonly renamed substances) and claimed that it was emitted from burning matter. Substances like sulfur that burn leaving little solid residue were supposedly rich in phlogiston. Various forms of phlogiston theory would dominate eighteenth-century chemistry (chemistry began to be taught as a discipline in medical schools around 1700) until the Frenchman Antoine-Laurent Lavoisier (1743–1794), the founder of modern chemistry, demolished it.

See also Alchemy; Boyle, Robert; Libavius, Andreas; Matter; Paracelsianism.

References

Beretta, Marco. *The Enlightenment of Matter: The Definition of Chemistry from Agricola to Lavoisier.* Canton, MA: Science History Publications, 1993.

Brock, William H. *The Norton History of Chemistry.* New York: W. W. Norton, 1992.

Hannaway, Owen. *The Chemists and the Word: The Didactic Origins of Chemistry.* Baltimore: Johns Hopkins University Press, 1975.

Chronology

Chronology, the science of reconciling different calendars and assigning precise dates to historical events and processes, attracted the attention of many early modern scholars and drew on a huge range of knowledge. The passion for chronological precision was linked to the drive to create more and more accurate clocks and watches, the desire for an accurate accounting of time. It was also linked to religious issues, specifically the finding of the correct date for Easter, which had been an obsession of the church for over a millennium. Even the correct interpretation of biblical prophecies often required an accurate accounting of past as well as future time. The simpler aspects of chronology were well-known; many more early modern people possessed the ability to calculate the date of Easter than do now. There were many manuals for the "computus"—the art of reconciling calendars and knowing dates. But many more complex problems, like the nature of the ancient Greek and Roman calendars, awaited solution.

Biblical, rabbinical, and humanistic learning as well as mathematics and astronomy were all necessary to the chronologist. Early modern chronologers faced a bewildering array of calendars and temporal problems. The Julian calendar widely used in the West had become increasingly out of synchronization with the solar year. Dates in the Julian calendar had to correlate with dates in a bewildering variety of calendars, from those relatively well-known to Europeans, such as the Islamic lunar calendar, to ancient calendars both solar and lunar, and even the calendars of the New World (the chronological achievements of the Aztecs and Mayans were among the few aspects of their societies to win unreserved praise from European scholars).

Chronologers were limited by the need to arrange their calendars of human history within the biblical time frame of a universe about 6,000 years old. It was in the early seventeenth century that Archbishop James Ussher (1581–1656) arrived at the popular date of 4004 B.C. for the Creation. Although there was a variety of alternative dates, fitting the ancient evidence into the biblical narrative of early human history was always a challenge.

Chronology and science were linked in many ways. The understanding of Aristotle's ancient texts of natural history could turn on the interpretation of a date or month in the

year Aristotle claimed marked a particular event in the life cycle, such as the blooming of a plant or the spawning of a fish. But the main links between science and chronology were astronomy and astrology. Eclipses offered one hope for immovable events on which to hang uncertain time, and the astrologically based chronological scheme based on "great conjunctions" of Saturn and Jupiter continued to wield great influence in the Renaissance.

One of the greatest early modern chronologers, Joseph Justus Scaliger (1540–1609), author of *On the Correction of Times* (1583), approached chronology through humanism rather than through astronomy. Scaliger, a professor at the University of Leiden, was the most learned and influential humanist scholar of his day, but had little background in astronomy. This did not prevent him from claiming authority in the interpretation of ancient astronomical texts such as *Astronomica* by the Roman poet Marcus Manilius (1st c. A.D.). Scaliger's claims to authority, combined with his lack of technical astronomical knowledge, set off quarrels with the astronomers of his day. (Scaliger, a Protestant, also loathed the Gregorian calendar and referred to its chief defender, the Jesuit Christoph Clavius, as a fat German.)

Scaliger's work set the humanistic understanding of ancient chronology on a much firmer foundation and influenced chronological thinking for centuries. He also invented a complex system called the Julian period after his father, Julius Caesar Scaliger (1484–1558). The Julian period was a cycle of 7,980 years produced by multiplying the 28-year solar cycle, after which days of the year recur on the same day of the week, by the 19-year lunar cycle, after which phases of the Moon recur on the same day of the solar year, and by the 15-year indiction cycle used by ancient Roman tax collectors. Each day of the resulting cycle was numbered. The Julian cycle provided a means of dating that was independent of calendars, and it eventually became standard for astronomers—somewhat ironically, given Scaliger's dislike for the science.

The standard B.C./A.D. method of dating was the product of the Jesuit chronologist Domenicus Petavius (1583–1652), author of *On the Doctrine of Time* (1627). Earlier systems had assigned the years before Christ as "years of the world," leaving the problem of how to assign the beginning date of the calendar. Petavius solved the problem by ignoring it. A skilled astronomer, Petavius did not believe in the traditional date for Christ's birth, but he accepted it as the fixed point his dating system needed.

Other early modern astronomers also worked on chronological problems. Tycho Brahe attempted to collaborate with the prickly Scaliger, providing him with precise astronomical data that he was unable to use. Johannes Kepler, a skilled chronologer, denounced Scaliger's astronomical errors and was responsible for the modern dating of the birth of Jesus in 4 B.C. Chronology also fascinated Isaac Newton, who was more focused on biblical studies and less focused on humanism than Scaliger or most other chronologers. Newton's posthumously published *The Chronology of Ancient Kingdoms Amended* (1728) placed ancient chronology in a prophetic framework.

See also Astrology; Astronomy; Bible.
References

Crosby, Alfred W. *The Measure of Reality: Quantification and Western Society, 1250–1600.* Cambridge: Cambridge University Press, 1997.

Grafton, Anthony. *Joseph Scaliger: A Study in the History of Classical Scholarship.* Vol. 2, *Historical Chronology.* Oxford: Oxford University Press, 1993.

Wilcox, Donald J. *The Measure of Times Past: Pre-Newtonian Chronologies and Relative Time.* Chicago and London: University of Chicago Press, 1987.

Circulation of the Blood

The path of the blood through the body was one of the oldest and most complex problems in medicine, and its eventual solution by William Harvey was considered the most important medical and biological discovery of the scientific revolution. Galenism, the dominant school of medical thinking during the Renaissance, divided blood into two types. In the Galenic system, the venous, originating in the liver and carried in the veins, delivers nourishment. Some venous blood is drawn from the liver to the heart via the vena cava, a movement powered by the heartbeat. In the heart, venous blood is combined with air to form arterial blood, which carries vitality through the body not by the pumping of the heart but by the pumping of the arteries themselves in synchronization with the heart. The problems for Galenic physiology were how the venous blood crosses from the right ventricle of the heart to the left and how air gets from the lungs to the heart. Galen theorized that the wall separating the ventricles has tiny pores through which the venous blood seeps, and that the pulmonary vein carries air from the lungs to the left ventricle and then carries vapors back to the lungs to be exhaled.

In the sixteenth century, anatomists, notably those associated with the university of Padua, began to chip away at this picture. Andreas Vesalius questioned the idea of pores of the heart and the role of the vena cava. Believing that the Holy Spirit affects the soul as the air affects the blood, the Spanish theologian and physician Michael Servetus (1509–1553) asserted that the blood circulates from the right ventricle to the lungs via the pulmonary artery, where it is mixed with air and sent back down to the left ventricle through the pulmonary vein. However, because Servetus was a notorious heretic, hated both by Catholics and by the Calvinist Protestants of Geneva who burned him, his book was nearly completely destroyed. More influential was the Padua anatomist Realdo Colombo (c. 1510–1559), who also described the "lesser" or pulmonary circulation. Another Paduan, Fabricius of Acquapendente (1533–1619), presented the first full discussion of the valves of the veins, arguing that the valves prevent the lower extremities of the body from being flooded with blood.

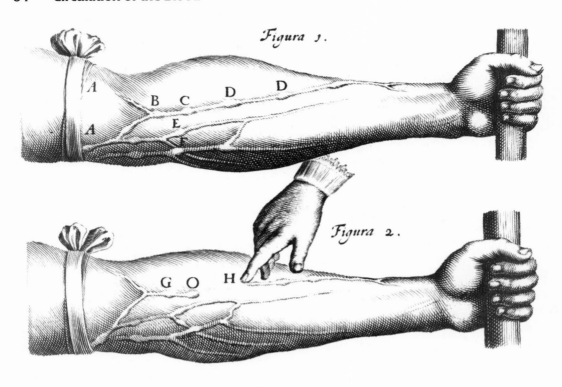

This illustration of Harvey's work shows how the circulation can be experimentally demonstrated. (National Library of Medicine)

Finally, Fabricius's pupil William Harvey set forth in his Latin *On the Motion of the Heart* (1628) a complete theory whereby all the blood, rather than a portion of it, is transformed from venous to arterial via the pulmonary circulation, and whereby the heart, a muscle that acts in the body as a pump, more active in contraction than dilation, drives the blood through the system of arteries and veins—a single system centered on the heart rather than Galen's dual heart-liver system. In Harvey's system, the function of the valves of the veins is to allow only one-way transit of the blood to the heart from the veins. As an Aristotelian, Harvey viewed the heart as more important than the liver, and as an anatomist, he noted the vast quantities of blood that circulate with each heartbeat, far more than could be replenished by the liver.

Harvey's theory of the circulation was rejected vehemently by conservative Galenists at university medical faculties such as that of Paris, most notably by Jean Riolan the Younger (1580–1657), who set forth a rival circulation scheme, bypassing the lungs and restricting circulation to the major veins and arteries. Other Galenic physicians and surgeons ignored circulation as irrelevant to medical practice. However, despite Harvey's own Aristotelianism, his theory was taken up in modified form by mechanical philosophers such as René Descartes. Descartes denied that the blood is expelled by the contraction of the heart, claiming instead that it is expelled by being vaporized by the heart's heat. From a vastly different position, the English magician and physician Robert Fludd integrated the circulation into a pantheistic view that, like Servetus's system, emphasized the connection between God and the individual through the lungs.

Harvey's system was refined by subsequent anatomists. Richard Lower (1631–1691), in his *Study of the Heart* (1669), demonstrated that the transformation of venous to arterial blood takes place in the lungs rather than the heart, as Harvey had believed. The principal problem for the Harveian system, how the arterial blood circulates back into the veins as venous blood, was solved by Marcello Malpighi's microscopic observation of the capillaries in the lungs of a frog, described in his *On the Lungs* (1661). The discovery of the circulation, universally accepted by the late seventeenth century, became a stock example for those who maintained the superiority of modern to ancient science, and Harvey became a scientific hero, particularly in England.

> *See also* Fludd, Robert; Harvey, William; Malpighi, Marcello; Physiology.
> *References*
> Bylebyl, Jerome J., ed. *William Harvey and His Age: The Professional and Social Context of the Discovery of the Circulation.* Baltimore: Johns Hopkins University Press, 1979.
> French, Roger. *William Harvey's Natural Philosophy.* Cambridge and New York: Cambridge University Press, 1994.
> Hall, A. Rupert. *The Scientific Revolution, 1500–1800: The Formation of the Modern Scientific Attitude.* London: Longmans, 1954.

Clavius, Christoph (1537–1612)

Christoph Clavius was the leading astronomer and mathematician of the Jesuit order in the late sixteenth and early seventeenth century, a founder of the Jesuit mathematical and scientific tradition. Although a friend of Galileo Galilei, he was also the last major astronomer to support Ptolemaicism. Born in Germany, he was received into the Jesuit order by its founder, Ignatius of Loyola (1495–1556), in 1555. Clavius was educated at the University of Coimbra in Portugal, and at the Collegio Romano he later took the theological degree required for all full members of the order. He taught mathematics at the Collegio for many years, working to advance mathematics and mathematical disciplines, such as astronomy, to an intellectual footing equal with that of natural philosophy.

In 1570, Clavius published a *Commentary on the Sphere of Sacrobosco* (John of Sacrobosco's *Sphere* was a standard medieval astronomical work) and in 1574 he published an edition of Euclid's *Elements*. These and other works by Clavius would become standard textbooks, frequently revised and reprinted and even translated into Chinese by Jesuit missionaries. Clavius served on the papal commission that eventually produced the Gregorian reform of the calendar, explicating the reform and defending it from its Protestant detractors. He also defended the Ptolemaic system from its opponents, whether those who denied the reality of the celestial spheres or Copernicans. Clavius was not averse to adopting innovations within an overall Ptolemaic framework, and he did accept some of Copernicus's mathematical innovations, if not his cosmology. Clavius's loyalty to Ptolemaic astronomy exceeded his loyalty to Aristotelian physics, and he was willing to consider the possibility that the matter of the heavens is changeable. Although initially skeptical of the reliability of telescopic data, Clavius along with other astronomers at the Collegio Romano confirmed Galileo's telescopic discoveries in 1610 and 1611, and there are indications that at the end of his life, Clavius believed these discoveries required extensive revisions in the Ptolemaic system. He did not believe, however, that they required the abandonment of celestial spheres or the adoption of the Copernican or Tychonic system.

> *See also* Collegio Romano; Gregorian Reform of the Calendar; Jesuits.
> *Reference*
> Lattis, James M. *Between Copernicus and Galileo: Christoph Clavius and the Collapse of Ptolemaic Cosmology.* Chicago: University of Chicago Press, 1994.

Clitoris

The clitoris reentered male scientific and medical knowledge during the scientific revolution. It had been described by ancient anatomists, but the knowledge had been lost during the Middle Ages, partly due to the difficulty of translation and confusion with the labia. The French professor, physician, and publisher Charles Estienne (c. 1504–1564) was the first modern anatomist to describe it, in his *Dissection of the Parts of the Human Body* (1545). Estienne associated the clitoris with urination. His description had no discernible influence, and priority in the "discovery" of the clitoris would be keenly contested by two Italians. Gabriele Fallopio (1523–1562), best known for his discovery of the fallopian tubes, described it in *Anatomical Observations,* written around 1550 but not published until 1561. The ambitious and unscrupulous Realdo Colombo (c. 1510–1559) took advantage of the time between Fallopio's writing and his publication to claim credit for the discovery in his *On Anatomy* (1559). Contemporaries generally judged Fallopio to have been the discoverer, but Colombo did innovate in his emphasis on the role of the clitoris in female sexual pleasure.

It took some time for the difference between the clitoris and the labia to become fully established in the medical literature. The establishment of the clitoris as a distinct organ of sexual pleasure tremendously affected the way male medical specialists thought about female sexuality, particularly in France. The clitoris's resemblance to the penis cast doubt on the standard approach to the vagina as an inside-out penis. Some, although not all, writers on the subject found the idea that women had two penis-equivalents questionable. The clitoris offered a way to explain hermaphrodites without having to admit the possibility of indeterminate gender; many hermaphrodites could now simply be explained as women with enlarged clitorises. But the clitoris had the greatest intellectual impact on the medical understanding of sex between women. "Tribades" had been previously understood as anatomically normal women who perform sexually either by rubbing their genital regions together, or, much more shocking to medical men, using artificial instruments for penetration. Now women who penetrated other women were defined as monstrous, possessing freakishly enlarged clitorises. Clitoridectomy was recommended, and sometimes practiced, as a way of ending this behavior.

See also Sexual Difference.
Reference
Park, Katherine. "The Rediscovery of the Clitoris: French Medicine and the Tribade, 1570–1620." In David Hillman and Carla Mazzio, eds., *The Body in Parts: Fantasies of Corporeality in Early Modern Europe.* New York: Routledge, 1997: 170–193.

Clocks and Watches

Accurate timekeeping was a European obsession in the early modern period, and the quest for accuracy involved many European scientists. Increasing the ability to measure smaller and smaller intervals of time also changed the nature of scientific endeavor. In the sixteenth century, clocks were at best approximate. Clocks were reset daily at noon, and they often had to be adjusted when running fast or slow. Relatively accurate clocks were expensive luxury items, each one individually crafted. They were not very portable, nor were they useful for measuring small, precise intervals of time. Despite their limitations, late medieval and Renaissance spring-driven European clocks were the best in the world at the time. German clocks were included in the tributes some German states paid to the Ottoman sultan, and Italian clocks were among the gifts Jesuit missionaries brought to the emperor of China in the seventeenth century. (The Chinese court remained an avid collector of European clocks and clockwork mechanical toys through the eighteenth century.) The small, portable clock, or watch, was introduced in Europe in the late fifteenth century.

One of the factors driving the demand for more accurate clocks was the need of scientists, particularly astronomers, for accurate time measurement. Tycho Brahe, for exam-

This spring-driven clock (c. 1695) was made by Thomas Tompion, an English clockmaker who worked with Robert Hooke and provided the clocks for the Greenwich observatory. (Bridgeman Art Library)

ple, was never satisfied with the accuracy of his clocks. When Galileo Galilei needed a precise timekeeping device for his experiments on acceleration, he had to use a waterclock rather than a mechanical clock. Physicians also wanted precise timekeepers to measure pulses.

The first major clockmaking innovation of the scientific revolution was the pendulum clock, which cut variation to about 15 seconds a day. Although the idea was first put forth by Galileo, Christiaan Huygens devised the first pendulum clock, working closely with a professional clockmaker, Amsterdam's Salomon Closter. Clockmaking was not purely altruistic—serious money was involved, and Huygens was frustrated by his inability to obtain a French patent on the device.

One problem with pendulum clocks was that they required stable platforms to maintain accuracy. They could not keep time on a ship or be adapted to a watch mechanism. Accurate timekeeping at sea was particularly important, because of its relevance to a ship's determination of its longitude. The most promising alternative to the pendulum clock and the second major innovation in clock-

making during the scientific revolution was the spring balance. The spring balance was the subject of a nasty priority dispute between Robert Hooke and Huygens. Huygens recorded in his notebook a design for a spring balance on January 23, 1675, shortly afterwards communicating his discovery to the Royal Society in London and exhibiting a watch to the Royal Academy of Sciences in Paris. Hooke indignantly claimed that he had had the idea in 1658, but that he had abandoned it due to the lack of financial backing. Hooke also produced a spring balance watch in 1675, working with the great clockmaker Thomas Tompion (1639–1713), one of those responsible for the eighteenth-century British superiority in watchmaking. Hooke's complaint was justified in that his basic idea had been communicated to Huygens by Robert Moray (1608–1673) and may have inspired Huygens's clock design.

Spring balance clocks and watches led to the introduction of minute and second hands, finally providing a timekeeper accurate enough for scientific and medical purposes. However, they were still not accurate enough for navigation. The technical demands of clock design were exceeding the ability of scientists such as Hooke and Huygens to contribute further to the problem. The final solution to the problem of the "marine chronometer" awaited the English clockmaker John Harrison (1693–1776) in the eighteenth century.

See also Navigation; Technology and Engineering.
Reference
Landes, David. *Revolution in Time: Clocks and the Making of the Modern World.* Cambridge: Harvard University Press, 1983.

Clockwork Universe
See Personifications and Images of Nature.

Collegio Romano
Founded in 1551 by Ignatius of Loyola (1495–1556), the Collegio Romano in Rome stood at the apex of Jesuit educational

institutions. In 1584, the Collegio moved from its originally cramped quarters to a spacious new building through the patronage of Pope Gregory XIII (pope 1572–1585). (The successor institution to the Collegio today is the Gregorian University.) Logic, natural philosophy, and mathematics played a central role in the Collegio's scholastic but innovative curriculum. The Collegio was a preeminent center for astronomy in the Catholic world. Christoph Clavius, an alumnus of the Collegio as well as a professor, ran an informal group for advanced mathematical and astronomical study at the Collegio in the early seventeenth century, and he attempted to mediate the disputes between mathematicians and natural philosophers. Galileo Galilei was influenced early in his career by the logic and mechanics taught at the Collegio, and he was honored for his telescopic discoveries with a solemn convocation there in 1611. Astronomers there built astronomical telescopes shortly after Galileo did and independently of him.

The Collegio was frequently visited by distinguished visitors to Rome, and it was the first institutional home of the museum of Athanasius Kircher, its most eminent professor in the sciences in the mid-seventeenth century. Kircher was appointed professor of mathematics, physics, and oriental languages on his arrival in Rome in 1633, and he remained associated with the Collegio after he resigned his position eight years later. Graduates of the Collegio who distinguished themselves in the sciences include the missionary Matteo Ricci (1552–1610), responsible for much of the introduction of Western science into China, the mathematician Gregorius St. Vincent (1584–1667), and the natural historian Filippo Buonanni (1638–1725), a pupil of Kircher's.

See also Clavius, Christoph; Jesuits; Kircher, Athanasius; Papacy.

Reference
O'Malley, John W., S.J., and Garvin Alexander Bailey, Steven J. Harris, and J. Frank Kennedy, S.J., eds. The Jesuits: Cultures, Sciences, and the Arts, 1540–1773. Toronto: University of Toronto Press, 1999.

Comets

In early modern Europe, comets both presented dramatic apparitions and posed important scientific questions. As prodigies, comets could be seen as divine warnings or as phenomena to be analyzed astrologically according to the sign and house of their appearance. The most scientific question comets posed was whether they were atmospheric phenomena, as Aristotle had claimed, or celestial phenomena farther from Earth than the Moon's orbit, as believed by alternative ancient authorities, notably the Roman Stoic Seneca (c. 4 B.C.–A.D. 65) in his Natural Questions.

The Aristotelian theory that comets are hot, dry, gaseous exhalations of the Earth had the advantage of offering a "rational" explanation for the association of comets with disasters, ascribing plagues and wars to the irritating qualities of comet gas. However, the great comet of 1577, as bright as Venus, seemed to demonstrate the falsity of the Aristotelian theory of comets, as studies of its parallax, most notably by Tycho Brahe, demonstrated its position above the Moon.

The next question was whether comets are ephemeral or regularly recurring phenomena. Johannes Kepler believed that comets are ephemeral objects, on straight-line trajectories. In a German treatise on the comet of 1607, he expressed the idea that space is full of temporary comets, only a few of which can be seen from Earth, formed out of fatty globules in the ether. He repeated this theory in a more elaborate Latin work published in connection with the three dramatic comets of 1618, the first to be observed with the telescope.

These comets touched off a debate between Orazio Grassi (1583–1654), Jesuit professor of mathematics at the Collegio Romano, and Galileo Galilei. In a Latin work largely based on Tycho's theories, Grassi claimed that comets are celestial phenomena that travel in great circles and shine by reflected sunlight. Galileo, through his friend and student Mario Guiducci, responded by attacking Grassi's arguments without seriously putting forth a comet theory of his own. Galileo's incidental attack on Tycho drew Kepler into the battle in defense of Tycho's reputation. The 1618 comets were also looked back on as presaging the Thirty Years War, which began that year.

Advances in cometary studies tended to depend on the presence of actual comets, and only a few appeared in the following decades. René Descartes did include in *The World* a discussion of the physical nature of comets, asserting that they are dead suns covered over with sunspots and forced in their particular paths by vortices. The next impressive comet to appear to the European scientific community was in 1664, quickly followed by a less spectacular one in 1665. The development of telescopic astronomy meant that these comets were much more closely observed and their courses carefully plotted. Comet positions were also observed and recorded from North America by Massachusetts clergyman Samuel Danforth (1626–1674).

The comet of 1680 was the first to be discovered through the telescope, by the German astronomer Gottfried Kirch (1639–1710). John Flamsteed theorized that after disappearing, the comet would reappear, having circled the Sun, and he was proved correct. Isaac Newton, who was fascinated by comets, responded by suggesting that the two appearances of the comet were actually different comets, and he then developed a technique for plotting the orbits of comets from three observations. Newton eventually changed his mind about the 1680 comet, and a discussion of the comet's orbit around the Sun figures prominently in his *Principia* (1687). The 1680 comet also prompted the most exhaustive attack on the belief that comets are divine signs, as well as on astrology, namely, *Thoughts on the Comet* (1682) by the French Protestant Pierre Bayle (1647–1706). This work, which covered many subjects besides comets, was reprinted several times in French and was translated into English in 1708. The belief in comets as divine signs had been declining among Europe's educated population for many years previously.

The comet of 1682 is best known as Halley's comet, but Edmond Halley was not particularly interested in it at first. Not until 1695 did he begin to work out comet orbits in collaboration with Newton. In 1696, he declared to a meeting of the Royal Society that the 1607 and 1682 comets were the same comet, and in *Synopsis of Cometary Astronomy* (1705) he made a dramatic prediction of its return in 1758.

The discovery of the periodicity of comets did not mean they lost a human or providential meaning. Newton claimed that they are divinely ordained for the purpose of replenishing the stars and planets with fluids and vital spirits. He suspected that novae are old stars replenished when comets fall into them. Such a nova in the Sun could destroy all life on Earth, heralding the Last Judgment. The water from a comet's tail could have caused Noah's flood, or, as Halley claimed, the flood could have been caused by a comet crashing into Earth.

See also Astronomy; Halley, Edmond; Kirch née Winkelmann, Maria; Planetary Spheres and Orbits; Prodigies.

References
Genuth, Sara Schechner. *Comets, Popular Culture, and the Birth of Modern Cosmology.* Princeton: Princeton University Press, 1997.
Van Nouhuys, Tabitta. *The Age of Two-Faced Janus: The Comets of 1577 and 1618 and the Decline of the Aristotelian World View in the Netherlands.* Leiden, the Netherlands: Brill, 1998.
Yeomans, Donald K. *Comets: A Chronological History of Observation, Science, Myth, and Folklore.* New York: Wiley, 1991.

Compasses

The magnetic compass, originally invented by the Chinese, was introduced to Europe or independently invented there in the Middle Ages. Used by surveyors, astronomers, and sailors, it was one of the three great technological inventions that Francis Bacon saw as marking off the modern age from antiquity (the others are gunpowder and printing, also originally Chinese). The compass enabled improved navigation not only in local waters but also on the long-distance voyages of exploration of the fifteenth and sixteenth centuries. Compasses were mysterious in their operations, a classical example of "natural magic," and were sometimes feared for this reason. Sailors noticed variations in the behavior of compasses at different points in the globe—now understood as the "magnetic variation" or declination caused by the difference between the magnetic and the true north. Compass needles did not point to the true north. It was also discovered in the sixteenth century that compass needles dip when suspended at their center of gravity. This was described by the English sailor and compass-maker Robert Norman in his *The Newe Attractive* (1581). The experiments Norman described were the starting point of William Gilbert's work on magnetism, which succeeded in explaining the variation by demonstrating that the Earth is a magnet. Gilbert and later successors such as Edmond Halley hoped to extend the navigational uses of compasses by using the dip and variation to establish longitude, although this never came to fruition.

See also Gilbert, William; Navigation.
Reference
Wolf, A., with the cooperation of F. Dannemann and A. Armitage. *A History of Science, Technology, and Philosophy in the Sixteenth and Seventeenth Centuries.* 2d ed., prepared by Douglas McKie. London: Allen and Unwin, 1950.

Conway, Anne (c. 1630–1679)

The foremost woman philosopher of the seventeenth century, Anne Conway was born into the influential Finch family, receiving an education of unusual quality for a woman of the time. In 1651 she married Edward, third viscount and first earl of Conway (1623–1683), a fine scholar in his own right. Her married life was spent at Edward Conway's estate in Warwickshire, which she made an intellectual center.

Throughout her life, Anne Conway suffered from intense headaches, for which she sought medical aid from such noted physicians and healers as William Harvey, Theodore de Mayerne (1573–1655), Valentine Greatrakes (1629–1683), and Francis Mercury van Helmont. All failed to alleviate her suffering—indeed, some worsened it—but van Helmont remained in the Conway household for several years, setting up an alchemical laboratory, carrying on philosophic conversation, and introducing Anne Conway to the Kabbalah. Shortly before her death, she and van Helmont converted to Quakerism, much to the dismay of her husband and her old friend, the Cambridge Platonist Henry More (1614–1687).

Van Helmont published Conway's posthumous and anonymous work, *The Principles of the Most Ancient and Modern Philosophy* (1690), which for a long time was erroneously attributed to him. Conway's philosophy, influenced by Cambridge Platonism and the Kabbalah, opposed the mechanical philosophy associated with Thomas Hobbes, Baruch Spinoza, and especially René Descartes. Conway asserted that spirit and matter are not distinct substances, but that matter is simply a more condensed and grosser form of spirit. Conway's vitalism, which asserted that all matter is somehow alive, none of it acting purely mechanically, was similar to that of Gottfried Wilhelm Leibniz, who acknowledged his debt to her book.

See also Helmont, Francis Mercury van; Kabbalah; Women.
References
Merchant, Carolyn. *The Death of Nature: Women, Ecology, and the Scientific Revolution.* San Francisco: Harper & Row, 1980.
Nicolson, Marjorie Hope, ed. *The Conway Letters: The Correspondence of Anne, Viscountess Conway, Henry More, and Their Friends.* Rev. ed., with an introduction and new material, edited by Sarah Hutton. Oxford: Clarendon Press, 1992.

Copernicanism

Not until a century after it was first put forth by Nicolaus Copernicus in his *On the Revolutions of the Celestial Spheres* (1543) did the doctrine that the planets circle the Sun and the Earth rotates win wide acceptance throughout Europe. The Copernican system itself was substantially modified in this period.

The principal obstacles Copernicanism faced in the sixteenth century were not the religious criticisms that would later be more prominent, but objections based on common sense and Aristotelian physics. A moving Earth, not just revolving around the Sun but also rotating, seemed to defy the evidence of the senses. Displacing the Earth from the center of the universe also threatened a basic Aristotelian tenet, the idea that heavy objects fall to Earth because they seek their natural place at the center of the universe. Although the Copernican system required fewer mathematical devices than did the previous Earth-centered Ptolemaic system, it also seemed less elegant. While everything revolves around the Earth in Ptolemaicism, in Copernicanism the Moon continues to revolve around the Earth while the planets revolve around the Sun, and some denied that the universe could have two centers of motion. The fact that the apparent positions of the stars do not move when the Earth moves—called the lack of "stellar parallax"—also forced Copernicus and subsequent Copernicans to posit a much larger universe with the stars at a much greater distance from Earth than was acceptable.

For these reasons, only about a dozen sixteenth-century thinkers can be identified as accepting Copernicanism as a true picture of the universe. However, many astronomers accepted Copernicanism as a valid system for making astronomical calculations. Making calculations, rather than speculating on the structure of the universe, was considered by many to be the true task of astronomy anyway. This computational use of Copernicanism was particularly influential at the Lutheran University of Wittenberg. One widely used set of astronomical tables based on Copernican assumptions, the *Prutenic Tables* (named after the duke of Prussia, to whom they were dedicated), was issued by the astronomer and Wittenberg professor Erasmus Reinhold (1511–1553) in 1551.

Copernicanism did present religious problems, as the Bible's view of the universe was Earth-centered. For example, in the biblical narrative of the battle of Gibeon, Joshua is described as making the Sun stand still, not the Earth. As early as 1540, one of Copernicus's followers argued that the Bible was adapted to the scientific understanding of its time, but little opposition in the sixteenth century, whether by Catholics or Protestants, was based on the alleged incompatibility of Copernicanism with the Bible.

Interpretations varied among those few who did accept Copernicanism as a true picture of physical reality. Some, closely following Copernicus himself, kept the crystalline spheres of traditional astronomy, simply shifting the center of the finite universe from the Earth to the Sun. The more intellectually radical Italian magician Giordano Bruno put a Sun-centered planetary system into an infinite universe with an infinite number of other stars and planets following the same arrangement. The most influential astronomical system originating in the second half of the sixteenth century, however, was not Copernican at all: Tycho Brahe posited that the Sun and Moon orbit the Earth and the other planets orbit the Sun. This system, known as the Tychonic system, is mathematically equivalent to Copernicus's, combining its advantages for calculation with a more believable stable Earth.

The most important Copernicans in the early seventeenth century were the German Johannes Kepler and the Italian Galileo Galilei. Kepler, who imbibed Copernicanism from his teacher Michael Maestlin (1550–1631), made the most important modification in the Copernican system by abandoning circular orbits in favor of elliptical ones, creating a far more elegant and accurate system.

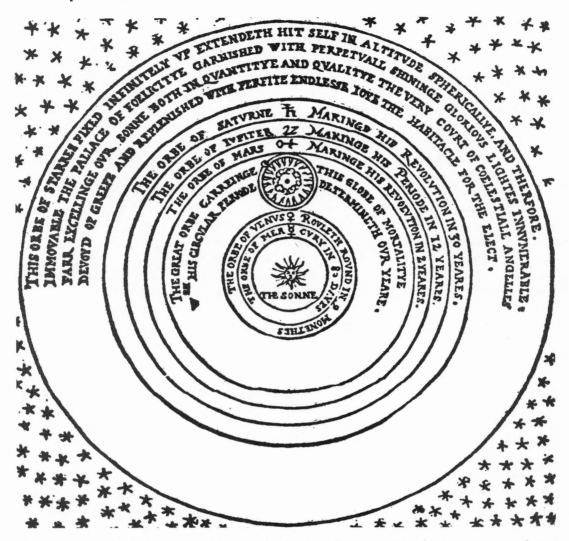

Thomas Digges extended the Copernican model to claim that the sphere of the stars, and therefore the universe, was infinite. This diagram of the universe is found in his work A Perfit Description of the Caelestiall Orbes *(1579).*

His *Epitome of Copernican Astronomy,* published in three sections between 1618 and 1621, was an influential textbook. Galileo, the first astronomer to make extensive use of the telescope, discovered the satellites of Jupiter, which made it clear that the cosmos has more than one center of motion. More importantly, he discovered the phases of Venus, incompatible with the Ptolemaic system, although compatible with the Tychonic as well as the Copernican systems. Galileo also published one of the most influential works of Copernican propaganda, the *Dialogues on the Two Chief World Systems* (1632), somewhat mislead-

ingly defining the debate as one between Copernicanism and Ptolemaicism rather than between Copernicanism and Tychonicism.

The great blow to Copernicanism was the adoption of an anti-Copernican stand by the Catholic Church. First, in 1616, Copernicus's *On the Revolutions* was placed on the Index of Forbidden Books, prohibited among Catholics "until corrected." Then in 1663 came the trial and condemnation of Galileo. These events were the culmination of an internal Catholic struggle over Copernicanism and its increasingly apparent religious implications. As a result of this struggle, Copernicanism was forbidden in Catholic countries and communities where the Inquisition and the strongly Tychonic Jesuit order were strong. Copernican beliefs could be presented only surreptitiously, or as mere hypotheses. The decree had little to no effect in Catholic France, which did not have an Inquisition. The condemnation of Copernicanism may have even speeded its acceptance the Protestant world, which could now see it as anti-Catholic.

For Copernicanism to be widely accepted outside astronomical circles, it needed to overcome the objections based on Aristotelian physics. In many areas of Europe, particularly in university circles, Copernicanism was accepted as part of the move from Aristotelian to Cartesian or Newtonian physics. After Galileo's condemnation, René Descartes himself had refused to publish his Copernican *The World,* but he and his followers were still Copernicans. By the late seventeenth century, Copernicanism was accepted throughout the scientific community in the Protestant world and France. It was also presented for the general educated public in the popularizing works of John Wilkins, Bernard Le Bouvier de Fontenelle, and Christiaan Huygens. In the eighteenth century, Copernicanism conquered the last Catholic holdouts, even the notoriously reactionary university culture of Spain. It also became a worldwide phenomenon, studied in a culture as remote from Europe as Japan.

See also Galilei, Galileo; Kepler, Johannes; Planetary Spheres and Orbits; Trial of Galileo.

References
Dobryzyicki, Jerzy, ed. *The Reception of Copernicus' Heliocentric Theory: Proceedings of a Symposium Organised by the Nicholas Copernicus Committee of the International Union of the History and Philosophy of Science.* Dordrecht, the Netherlands, and Boston: Reidel, 1972.
Koyre, Alexander. *The Astronomical Revolution: Copernicus-Kepler-Borelli.* Translated from the French by Dr. R. E. W. Maddison. Ithaca: Cornell University Press, 1973.
Kuhn, Thomas S. *The Copernican Revolution: Planetary Astronomy in the Development of Western Thought.* Cambridge: Harvard University Press, 1957.

Copernicus, Nicolaus (1473–1543)

Nicolaus Copernicus (the latinized form of "Mikolaj Kopergnick") initiated the most dramatic innovation of the scientific revolution, the shift from a Ptolemaic Earth-centered universe to a Sun-centered universe. He was born in the city of Toruń in Poland to a middle-class family with significant connections in the Catholic Church. His father died when he was ten, and the young Copernicus was adopted by his maternal uncle Lucas Waczenrode, a powerful cleric. Copernicus, destined for the church, was educated at the Universities of Cracow, Bologna, Padua, and Ferrara. It was at Cracow, where he attended from 1491 to 1494, that Copernicus acquired his interest in astronomy. He studied medicine at Padua, and sometime during his stay in Italy he learned Greek well enough to later publish a translation of some ancient verses from Greek to Latin.

On his return to Poland around the beginning of 1506, Copernicus practiced medicine and held various jobs in the church, and in 1512 he settled in as a canon of the cathedral at Frauenberg (Frombork). He used one of the towers of the wall surrounding the cathedral for astronomical observations, even though he was not a particularly skilled observer. Although he lacked publications and had no experience teaching at a university, Copernicus acquired a reputation as an astronomer, and in 1514 he was invited to Rome to help reform the calendar. Although he refused the invitation, his name remained known in Rome, and by 1533, Pope Clement VII (1478–1534) had heard of Copernicus's astronomical theories. Copernicus was a busy churchman and politician at a time when the area where he lived was disputed, first by the military order of the Teutonic Knights and the king of Poland and then by the Catholic Church and the leaders of the Protestant Reformation. He wrote a small treatise on economics, attacking the adulteration of coins with base metals.

All this time, Copernicus was working out his astronomical theory, which various people in Rome were urging him to publish. However, it was the young German Lutheran astronomer Georg Iserin, known as Rheticus (1514–1574), a mathematics professor at the University of Wittenberg, who made the work of Copernicus known. Visiting Copernicus in 1539, Rheticus urged him to publish the manuscript. Copernicus entrusted his Latin treatise *On the Revolutions of the Celestial Spheres* to Rheticus for publication, but the young German was called to a university position in Sweden and left the manuscript with the Lutheran clergyman Andreas Osiander (1498–1552) to see it through to publication. Osiander blunted the radicalism of Copernicus's tract by adding a preface suggesting that Copernicus's theories were only aids to mathematical calculation rather than descriptions of physical reality.

On the Revolutions of the Celestial Spheres, dedicated to the pope, was published with Osiander's preface in 1543, the year of Copernicus's death. It set forth the Copernican hypothesis, arguing for the daily revolution of the Earth and its annual motion around the Sun. A Sun-centered universe has ancient Greek precedents, both in the mystical tradition of the Pythagoreans, who put the Sun at the center to honor it, and in the astronomy of Aristarchus of Samos (c. 217–145 B.C.). Some medieval scientists had also broached the idea, if only to refute it. However, Copernicus was the first to work out heliocentrism mathematically. (Technically, Copernicus's system is not heliocentric; he conceived the planets as rotating around a hypothetical point at the center of the Earth's orbit rather than the physical Sun.) Copernicus's system did explain certain celestial phenomena better than did Ptolemaicism; for example, the fact that Mercury and Venus are never seen far from the Sun is easy to explain on the principle that they orbit the Sun more closely than does the Earth. The retrograde motion of the planets, when they appear to be moving backward, is also easier to explain in the Copernican system, where this phenomenon is explained as resulting from the

Earth's passing them in orbit around the Sun.

Despite this break with the past, Copernicus's system was shaped in many ways by the Ptolemaic astronomy with which he was familiar. The structure of *On the Revolutions* was modeled closely on that of Ptolemy's *Almagest.* The Copernican system also used, at least as an explanatory device, solid crystalline spheres in which the planets are embedded, and like the Ptolemaic system, it required a complicated system of corrections to reconcile observations with a theory based on circular orbits. Copernicus viewed the principle of uniform circular motion as central to his astronomy—the circle being the most perfect of all shapes. By eliminating the equant—a mathematical device of Ptolemaic astronomy whereby the sphere that contains the planet is considered to rotate around a point other than its center—Copernicus proclaimed that he was more faithful to the principle of circular motion than was Ptolemy. The system was complicated further because Copernicus uncritically accepted previous observations, mostly Greek and Islamic, many of which were inaccurate or had been corrupted while being copied from manuscript to manuscript. Accounting for these imperfect observations required further adjustments. Copernicus's theory was not really much less cumbersome than the Ptolemaic system, as it still retained epicycles and deferents.

See also Astronomy; Copernicanism; Planetary Spheres and Orbits.

References

Armitage, Angus. *Copernicus: The Founder of Modern Astronomy.* London: Allen and Unwin, 1938.
Kuhn, Thomas S. *The Copernican Revolution: Planetary Astronomy in the Development of Western Thought.* Cambridge: Harvard University Press, 1957.
Westman, Robert S. "Proof, Poetics, and Patronage: Copernicus's Preface to *De Revolutionibus.*" In Robert S. Westman and David Lindberg, eds., *Reappraisals of the Scientific Revolution.* New York: Cambridge University Press, 1990.

Correspondence

During the scientific revolution, scientific information was communicated and connections between scientists were maintained by correspondence. Letters served a number of purposes. They announced new discoveries or theories, they reported observations and posed challenges, and they generally kept natural philosophers, particularly those unfortunate enough to dwell in small communities distant from great cities, courts, and universities, abreast of what others in the "republic of letters" were doing. Letters that flattered the recipient or that enclosed gifts could be pleas for patronage, or could set up a reciprocal obligation between equals. Some prestigious intellectuals honored their recipients by the mere act of sending a letter. Some manuscript letters, such as Galileo Galilei's *Letter to the Grand Duchess Christina*, written in 1615 on the relationship of scientific inquiry to the Bible, were designed to make a public statement. These practices were not limited to natural philosophers but were characteristic of many areas of European intellectual and cultural life. Letter writing was a valued skill in early modern Europe, and there were a number of manuals designed to teach it.

A letter or a correspondence was made most public, of course, by publication in a printed edition. Publication of a correspondence could be a bid for prestige, a technique pioneered by sixteenth-century humanists. The publication of René Descartes's correspondence in the decade and a half after his death was a move by French Cartesians to establish Descartes at the center of French philosophical and scientific discourse. Other high-profile correspondences, such as the Leibniz-Clarke correspondence, were clearly intended from the beginning to be published. The controversy over sunspots between Galileo and the Jesuit Christoph Scheiner (1573–1670) was carried out in the form of published letters that the two astronomers wrote to the German merchant and scientific amateur Marcus Welser.

One of the earliest scientists to leave a large body of correspondence was Tycho Brahe. Astronomers and astrologers were particularly active letter-writers, as it was often necessary to gather observations made from a number of different places. The major surviving collections of scientific correspondence, however, are from the seventeenth century, when postal conditions and transport had begun to improve. Particularly notable are the collections made by "intelligencers," men who made it their mission to facilitate scientific communication and often amassed vast collections of correspondence, with letters numbering in the thousands from all parts of Europe and beyond. Notable among them are the Frenchmen Nicolas-Claude Fabri de Peiresc, who more or less originated the role, Marin Mersenne, whose smaller and more select group of correspondents included the leading mechanical philosophers of France and Italy, the mathematician Claude Mylon (c. 1618–c. 1660), and the astronomer Ismael Boulliau (1605–1694). In mid-seventeenth-century Rome, Athanasius Kircher sat at the center of the Jesuit network of scientific correspondents. In England, intelligencers included Samuel Hartlib (c. 1600–1662), the millenarian enthusiast who believed his work was a preparation for the Second Coming, and the astrologer and alchemist Elias Ashmole (1617–1692), whose letters and papers are now in the Bodeleian Library at the University of Oxford. The most important English intelligencer was the Secretary to the Royal Society Henry Oldenburg, like Hartlib a German residing in England, who offered English natural philosophy a connection to the European continent and vice versa.

Intelligencers were often afflicted by a tension between the ideal of the scientific communication of ideas and the reality of a proprietary approach to those same ideas. Intelligencers were involved in priority disputes and were sometimes accused of having been accomplices to intellectual theft. Oldenburg in particular was often the subject of such accusations, partly because of his relatively obscure background, which left him more vulnerable than nobles such as Peiresc. Archives of letters and other papers could be manipulated in disputes between scientists, the most notorious example being the use that Isaac Newton and other Newtonians made of the Royal Society archive in the priority struggle with Gottfried Wilhelm Leibniz over the calculus.

Oldenburg redefined the role of intelligencer in that he presented himself as a representative of the Royal Society rather than an independent operator. He also embodied the transition from correspondence to the scientific periodical as the principal means of scientific communication, as many of the letters written to Oldenburg were published, and were intended to be published, in his journal *Philosophical Transactions*. The letter thus provided a rhetorical model for the published scientific paper. Although the scientific importance of correspondence diminished in the eighteenth century, it has never been extinguished.

See also Hartlib Circle; Leibniz-Clarke
Correspondence; Mersenne, Marin;
Oldenburg, Henry; Peiresc, Nicolas-Claude
Fabri de.
Reference
Hunter, Michael, ed. *Archives of the Scientific
Revolution: The Formation and Exchange of Ideas in
Seventeenth-Century Europe*. Woodbridge,
England: Boydell, 1998.

Courts

The courts of monarchs, princes, and prelates were important centers of culture in the early modern period, and this was true in the area of science as much as in art or literature. Although early modern courts varied in the approaches they took to science, there were some common features of courtly science.

Courtly science was patronage-based, with the ideal patron being the ruler him or herself. The higher the status of the patron, however, the more the demands placed on the client; the best patrons were interested in patronizing only the best clients, as the quality of the client reflected on the patron. Courts were also more open to new ideas than were universities, and they valued originality and innovation. For example, Paracelsian physicians were prominent in many European courts, while they were largely shut out of the universities. Court culture was international, and many scientists found patronage outside their native land, as Tycho Brahe did at the court of the Holy Roman Emperor after leaving Denmark.

The traditional hierarchy of disciplines meant less at courts; one reason Galileo left the University of Padua for the court of the duke of Tuscany was that as a mathematician, he was considered incompetent to pronounce on natural philosophy in a university, but he could do so in a court. Astronomers also found their status higher in courts than in universities. Courtly science valued discourse and conversation, as opposed to the authoritarianism of the university lecture format. Conversation was not meant to be conclusive, and the ingenious explanations and theories of natural philosophers were likely to be evaluated in terms of their entertainment value rather than their accuracy.

Court science emphasized the individuality of "curious" and "marvelous" phenomena rather than looking for universally valid generalizations. It was uncourtly to be overly loyal to a philosophical system, and university Aristotelians—"dogmatists"—were frequent targets of court mockery. Courtier-scientists liked to contrast the "freedom" of their intellects at court to the "slavery" of the Aristotelians to their master.

Courtly science often presented its goal as not simply understanding nature but exerting power over nature. Technical expertise, such as that required to carry out the prince's military and engineering projects, was highly valued. For example, many mathematicians were recipients of court patronage because of their expertise in such fields as ballistics and navigation. Alchemists, with their claims to enhance a prince's wealth through the making of gold or by some other means, were also very prevalent in courts. Astrologers, too, were considered useful to princes, and many early modern astronomers, including Tycho and Johannes Kepler, performed astrological services for their patrons.

Gifts were fundamental to the court patronage economy, given by both client and patron. The most famous courtly scientific gift during the scientific revolution was intangible: Galileo's gift of the moons of Jupiter, the "Medicean stars," to the Medici duke of Tuscany to serve as emblems of the family. The dedication of a book to a prince or great noble was a form of gift and was usually done in the hope of patronage. Princes gave, received, and collected exotic and rare objects as tangible manifestations of their power. Organized into "cabinets of curiosities," these courtly collections were among the ancestors of the museum, and they often included exotic animals, plants, and gems, along with antiquities.

Of course, the scientific interest of courts varied according to the interest and circumstances of each ruler, and courtly science changed over time. Some rulers were keen scientists themselves; Prince Wilhelm IV of Hesse-Kassel (1532–1592) was a skilled astronomer who made his court a leading scientific center. The Medici brothers, Ferdinand II (r. 1621–1670), Grand Duke of Tuscany, and Prince Leopold gathered around them the Accademia del Cimento and were seriously interested in science. The Holy Roman Emperor Rudolf II's occultism made him a major scientific patron. Others, such as Elizabeth I of England (r. 1558–1603) simply gathered scientists as part of an overall program of patronage, including writers, artists, musicians, and many others. Wealthy princes such as the king of France or the pope were able to create massive patronage empires, whereas the small German and Italian courts had to be much more selective about their clients. These smaller courts often were treated by ambitious clients as stepping stones, as in the case of Galileo, who graduated from the patronage of the duke of Tuscany to that of the pope.

By the late seventeenth century, courts were declining as scientific centers, giving way to the new academies and societies such as the Royal Society and the Royal Academy of Sciences. Earlier academies, such as the Accademia del Cimento, had functioned as extensions of the court, lacking a separate legal existence and meeting at the pleasure of the prince. These new academies, although founded under royal patronage, had an institutional existence independent from the court.

See also National Differences in Science; Papacy; Politics and Science; Rudolf II.

References

Biagioli, Mario. *Galileo, Courtier: The Practice of Science in the Culture of Absolutism.* Chicago: University of Chicago Press, 1993.

Moran, Bruce, ed. *Patronage and Institutions: Science, Technology, and Medicine at the European Court, 1500–1750.* Woodbridge, England: Boydell, 1991.

Croll, Oswald (c. 1560–1609)

The German Oswald Croll, sometimes known by the Latin version of his name, Crollius, was one of the most important systematizers and interpreters of Paracelsianism, prominent enough to be a target of the great anti-Paracelsian Andreas Libavius. Croll received an M.D. from the University of Marburg in 1582, after which he worked in France, Germany, and throughout central Europe as a tutor in noble houses and a physician. He was also involved in a shadowy way with the Protestant forces within the Holy Roman Empire.

Croll's Latin work, *Basilica Chymica* (1609), paid lip service to the ideal of cooperation between medical sects, but in essence it was an attempt to synthesize Paracelsianism with the Renaissance tradition of Hermetic and Kabbalistic high magic. Influenced by Calvinist theology, Croll also altered Paracelsianism to emphasize the distance between humanity and God, and he dropped Paracelsus's emphasis on the physician's cooperation with God in the work of redemption. *Basilica*

Chymica also included a number of medical-chemical recipes showing an unusually sophisticated knowledge of chemical reactions, in contrast to the emphasis on distillation found in the work of Paracelsus and in the classical alchemical tradition.

Basilica Chymica went through a number of editions in the seventeenth century, appearing in French, English, and German translations. Croll died in Prague, where he had been one of the natural and occult philosophers attracted to the court of Rudolf II.

See also Libavius, Andreas; Paracelsianism.
References
Debus, Allen G. *The Chemical Philosophy: Paracelsian Science and Medicine in the Sixteenth and Seventeenth Centuries.* 2 vols. New York: Science History Publications, 1977.
Hannaway, Owen. *The Chemists and the Word: The Didactic Origins of Chemistry.* Baltimore: Johns Hopkins University Press, 1975.

Cunitz, Maria (1610–1664)

The most prominent woman astronomer of the mid-seventeenth century was the German Maria Cunitz. Cunitz began the study of astronomy with her father, a physician and landowner in Silesia who chose to have his daughter educated in science, mathematics, medicine, and Latin, an exceptional decision at the time. Like most other women astronomers of the early modern period, Cunitz married another astronomer, but unlike most of them, Cunitz was the intellectually dominant partner in her marriage. Her main work was a simplification of Kepler's complex *Rudolphine Tables* (1627), which she designed for use by working astronomers. The book was *Urania Propitia* (1650), written in Latin and named after the classical muse of astronomy. In addition to providing simplified versions of Kepler's tables, the work also dealt with astronomical theory and practice. So impressive was *Urania Propitia* that Cunitz was referred to as the "second Hypatia," after the ancient Greek woman mathematician (c. 370–415). However, some denied that a woman could have

been capable of writing the book, and to counter these skeptics, subsequent editions appeared with a foreword by Cunitz's husband, denying that he had any part in it. A crater on Venus has been named after Cunitz.

Reference
Wertheim, Margaret. *Pythagoras' Trousers: God, Physics, and the Gender Wars.* New York: Random House, 1995.

Cycloid

The cycloid is a curve formed by a point on the circumference of a circle as it rolls along a straight line. During the seventeenth century, the cycloid attracted the attention of the most eminent mathematicians of the age, including Galileo Galilei, Blaise Pascal, Pierre de Fermat, Christopher Wren, the Bernoulli brothers, Christiaan Huygens, Gottfried Wilhelm Leibniz, and Isaac Newton, as well as a host of minor figures. So fascinating was the cycloid that it was referred to as the mathematical equivalent of Helen of Troy, and the contests and rivalries it provoked were as fierce, if not as bloody, as the Trojan War.

Galileo, who named the cycloid, began studying it at the end of the sixteenth century. In an effort to determine the ratio of the area of a single arc of the cycloid (the arc is the portion of the curve between two points where it touches the baseline) to the area of the generating circle, Galileo resorted to crude empiricism, making cycloids out of metal and then weighing them. He concluded that the ratio was approximately, but not exactly, three to one. As it turned out, the ratio is in fact exactly three to one, as was demonstrated in 1634 by Gilles Personne de Roberval (1602–1675), employing the method of indivisibles. Roberval had been urged to take up the problem in 1628 by Marin Mersenne, who had failed to solve it himself. Galileo's disciple Evangelista Torricelli (1608–1647), with whom he corresponded about the cycloid, also independently solved the problem. Roberval was challenged by René Descartes to draw a tangent to a cycloid, but he failed to do so, and this problem was solved by Pierre de Fermat. (The ever-gracious Descartes asserted that Roberval's solution to the area problem could have been found by "any moderately skilled geometer.") Girard Desargues (1591–1661) proposed cycloidal gear teeth in the 1630s, although making these was not possible, given the metalworking skills at the time.

The cycloid revived Pascal's interest in mathematics after a long period of religious obsession. Unable to sleep at night due to pain, Pascal worked on a general solution to the problem of determining the area of any segment of the cycloid, and he also examined the properties of the solid that is formed by rotating a cycloid. Delaying publication of his results, he published a challenge to other mathematicians to tackle the problems he had solved. Many eminent mathematicians, including John Wallis, failed, but those who succeeded included Huygens, Fermat, and Wren. Wren also demonstrated that the length of the arch was eight times the radius of the generating circle.

In 1673, Christiaan Huygens published his discovery of the tautochronous property of the cycloid. Under ideal conditions, the time it takes for a ball to drop from the curve of an inverted cycloid to the bottom will remain constant, no matter where along the curve it begins its descent. Huygens hoped to use this principle to make an accurate clock, one whose pendulum would trace a cycloidal path in its swing. This required cycloidal buffers on each side of the pendulum. However, this was never practical because of difficulties with the effects of wear and humidity on the thread from which the pendulum bob hangs.

Another fascinating property of the cycloid is that it is brachistochronous. If two points are not in a vertical or horizontal line, and one wishes to find the way of joining them that will enable a ball, under ideal, fric-

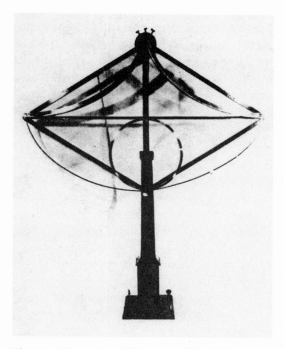

Christiaan Huygens's pendulum, pictured here, had cycloidal buffers to force the pendulum to follow a cycloidal path. (Conservatoire National des Arts et Metiers)

tionless conditions, to roll from the higher to the lower point in the shortest amount of time, the resulting curve is a segment of an inverted cycloid. This property was discovered by Johann Bernoulli (1667–1748), who made it the subject of a contest announced in the July 1696 issue of the German journal *Acta Eruditorum*. He made the contest especially tantalizing by claiming that the curve that had the brachistochronous property, which readers were challenged to find, was already well known. Many attempted to find the curve with the brachistochronous property, but only a very few of Europe's greatest mathematicians succeeded. These were Johann's brother Jakob Bernoulli (1655–1705), Gottfried Wilhelm Leibniz, the French nobleman and mathematician Michel de la Hopital (1661–1704), and Isaac Newton. Newton solved the problem in a single sleepless night after a full day's work at the mint, and he sent in his solution anonymously. Legend has it that upon seeing the solution and noting its English postmark,

Johann Bernoulli immediately knew it was Newton's, remarking that he knew the lion by the mark of its claw. The investigation of this problem led to the development of the calculus of variations.

See also Bernoulli Family; Fermat, Pierre de; Huygens, Christiaan; Mathematics; Pascal, Blaise; Wren, Christopher.

Reference
Cooke, Roger. *The History of Mathematics: A Brief Course.* New York: Wiley, 1997.

D

Decimals

The system of expressing fractions through a succession of tenths, which had obscure Arabic precedents, was put forward by Simon Stevin in his book *The Tenth* (published in 1585, translated from Dutch to English in 1608). He claimed that decimals would make calculations easier and enable people to identify the larger of two quantities at a glance, which is sometimes difficult with fractions that include large numbers. Stevin was an enthusiast for decimalization in general, urging the Dutch government to adopt a decimal currency and suggesting that the survivors of the old Babylonian base 60 system—the hours of the day and the degrees of a circle—also be decimalized. Stevin did not devise the modern decimal point but proposed that to indicate the order of magnitude, each decimal be accompanied by another numeral, an awkward and potentially confusing system. Following a suggestion of Francois Viète, however, Continental mathematicians in the seventeenth century adopted a comma to separate the integer from the decimal. The decimal point first appeared in the writings of the Scottish mathematician John Napier (1550–1617), and although decimals continued to be represented in many ways in the seventeenth century, the decimal point was the dominant convention in Britain.

See also Stevin, Simon; Viète, Francois.
Reference
Wolf, A., with the cooperation of F. Dannemann and A. Armitage. *A History of Science, Technology, and Philosophy in the Sixteenth and Seventeenth Centuries.* 2d ed., prepared by Douglas McKie. London: Allen and Unwin, 1950.

Dee, John (1527–1608)

John Dee was Elizabethan England's leading natural philosopher, with interests in mathematics, magic, and English colonial expansion. He received a B.A. from St. John's College, Cambridge, in 1545 and an M.A. in 1548. Dee studied mathematics and navigation informally at the University of Louvain from 1548 to 1551, and in 1550 he lectured publicly on Euclid in Paris. His major work, *Monas Hieroglyphica,* printed at Antwerp in 1564, expressed in obscure language a program for reforming the disciplines through a new form of writing based on the Kabbalah and hieroglyphics. Reprinted in Frankfurt in 1591, it gave Dee a European reputation and influenced the later Rosicrucian writings.

Dee turned down academic positions at Oxford and the University of Paris. As an intellectual operating outside the academy, his career was largely a quest for patronage, which he received from important English nobles such as Robert Dudley, first earl of

By showing him with a compass resting on a globe, this portrait of John Dee alludes to his contributions to navigation and cartography. (National Library of Medicine)

Leicester (c. 1532–1588), prominent in the court and government of Queen Elizabeth I (r. 1558–1603). He also succeeded in obtaining patronage, although never as much as he would have liked, from the queen herself, who appointed him royal astrologer, consulted him on mathematical, calendrical, navigational, and alchemical issues, and protected him from his enemies (given Dee's reputation as a magus, he was a target for religious persecution).

From 1566 to 1583, Dee and his family lived in a large house at Mortlake, equipped with a laboratory and a fine scientific and magical library, including many Paracelsian works. He worked closely with English sea captains and navigators, notably from around 1551 to the early 1580s as adviser to the Muscovy Company, which conducted trade with Russia. His *Mathematicall Preface* (1570) to the first English translation of Euclid expressed a Neoplatonic emphasis on mathematics and included a discussion on the usefulness of experiment. Dee's *Perfect Arte of Navigation* (1577) combined a treatise on geography with propaganda for English imperialism, written in hopes of patronage from Elizabeth.

Dee's disappointment with English patronage led him to accept an invitation from a Polish nobleman, Albert Lasky, to come to Poland in 1583. (Dee's neighbors marked his departure by storming his house and destroying much of his library and equipment.) From Poland, Dee journeyed to Prague in hopes of obtaining the patronage of the Holy Roman Emperor, Rudolf II. But that attempt failed, and Dee and his associate, Edward Kelly, spent the latter part of the 1580s wandering Europe as itinerant occultists. By this time, Dee was primarily interested in the conversations he and Kelly were having with angels through a crystal ball. On his return to England, he received a royal pension (which was never paid) and a position as warden of Christ's College, Manchester. He died penniless.

See also Courts; Magic.

References

Clulee, Nicholas H. *John Dee's Natural Philosophy: Between Science and Religion.* London: Routledge, 1988.

French, Peter. *John Dee: The World of an Elizabethan Magus.* London: Routledge, 1972.

Demonstrations and Public Lectures

As educated European men and women became more interested in science and its progress, public demonstrations of scientific principles, usually Cartesian or Newtonian, became popular entertainment. The production of dramatic effects had always had a prominent place in court science and in the meetings of scientific societies, and by the late seventeenth century, demonstrations expanded to the lecture halls, salons, and coffeehouses of Paris, London, and Amsterdam. The first great scientific demonstrator to perform for a general public was the French Cartesian physicist Jacques Rohault (1618–1672). He was a shrewd self-publicizer whose famous Wednesday afternoons, dating from the 1650s, combined lectures with dramatic demonstrations for the Parisian elite.

Rather than demonstrating the weight of the air with a regular mercury barometer, for example, Rouhault used a huge water barometer with a 30-foot iron pipe. English demonstrators were Newtonian rather than Cartesian and were also more frankly commercial, charging admission to their lectures. This provided a way for scientists to support themselves on the commercial market rather than depending on the universities or patronage. For example, the religious heretic William Whiston (1667–1752), after losing the position of Lucasian Professor of Mathematics at Cambridge due to his unconcealed opposition to the doctrine of the Trinity (which also kept him out of the Royal Society), was able to support himself in London in part as a lecturer and demonstrator. The first Newtonian lecturer was Francis Hauksbee (1666–1713), also the demonstrator of the Royal Society, who used mechanical devices to illustrate Newton's principles and later developed electrical demonstration apparatus.

By the second decade of the eighteenth century, scientific lectures and demonstrations had spread from London to the English provinces. The French Protestant and naturalized Englishman John T. Desaguliers (1683–1744), Hauksbee's successor as Royal Society demonstrator, helped spread Newtonianism (and Freemasonry) to the English provinces and the European Continent through his lectures and demonstrations. These formed the basis of his influential textbook, A Course of Experimental Philosophy (1734–1744), which was translated into Dutch and French. Desaguliers also developed an elaborate orrery, or mechanism for demonstrating the orbits of the planets. Although English lecturers appealed to a general public interested in science, they were more likely than French or Dutch lecturers to address themselves to the technological concerns of business. Scientific lecturing continued to grow throughout the eighteenth century as a form of education and popular entertainment.

See also Popularization of Science.
Reference
Sutton, Geoffrey V. Science for a Polite Society: Gender, Culture, and the Demonstration of Enlightenment. Boulder, CO: Westview Press, 1995.

Descartes, René (1596–1650)

René Descartes is remembered today principally as a metaphysician, but he thought of himself primarily as a mathematician and natural philosopher. As such, he was responsible for the classic formulation of the mechanical philosophy, analytic geometry, and a highly influential physical theory. His style of thinking has continued to influence science, particularly in France, to the present day.

Descartes's early life is obscure. He was born into a family of physicians and civil servants and received a solid humanistic and scholastic education at the Jesuit College of La Flèche from 1606 to 1614, taking a law degree from the University of Poitiers in 1616. He then went to the Dutch Republic to serve in the Dutch Army. In the Netherlands, he encountered Isaac Beeckman, who inspired his interest in mechanics and natural philosophy generally. However, he left shortly thereafter to spend the next decade wandering Europe (because of his family's wealth, finances were never a problem for Descartes). In 1619, in Germany, he had a vision of a new philosophy. Descartes, never a humble man, seems to have envisioned himself as a new Aristotle, with a philosophy universal in its application and suitable for university teaching. For this reason, throughout his career he sought the approval of the Jesuits, who were extraordinarily influential in European education.

In 1628, Descartes returned to the Dutch Republic, where he would remain for the next 20 years. He traveled frequently and maintained contact with French intellectual life through his correspondence with Marin Mersenne. Even though he lived in a Protestant society and held Copernican views

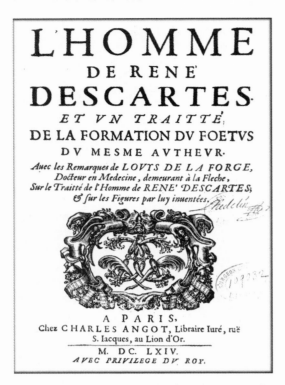

L'HOMME
DE RENE
DESCARTES·
ET VN TRAITTÉ,
DE LA FORMATION DV FOETVS
DV MESME AVTHEVR·
Auec les Remarques de LOVYS DE LA FORGE,
Docteur en Medecine, demeurant à la Fleche,
Sur le Traitté de l'Homme de RENE' DESCARTES;
& sur les Figures par luy inuentées.

A PARIS,
Chez CHARLES ANGOT, Libraire Iuré, ruë
S. Iacques, au Lion d'Or.
M. DC. LXIV.
AVEC PRIVILEGE DV ROY.

René Descartes's Treatise on Man *was published only after his death. (National Library of Medicine)*

that were condemned by the Catholic Church, Descartes remained a loyal Catholic. As a resident of the Protestant Dutch Republic, he was protected from the Inquisition, and he was shocked by the condemnation of his fellow Copernican, Galileo Galilei. Yet he never showed any interest in conversion. (Descartes had visited Florence during Galileo's residence there, but the two never met.)

Abandoning a treatise on the verge of publication which would have systematically expounded his natural philosophy, Descartes turned to metaphysics to find a religiously unimpeachable basis for natural knowledge. In 1637, he published *Discourse on Method,* setting forth his program for natural philosophy, and he also published three associated treatises that he claimed exemplified his method on geometry, optics, and meteorology, including matter theory. These works were in French rather than Latin, aimed at a public of educated men and women rather than university scholars. Descartes was the first European

male intellectual to think of women as an important part of his audience.

The *Discourse* sets forth the famous *cogito*—Descartes's argument that the process of thinking proves that the thinker exists. Descartes attempted to use this as a foundation both for metaphysical claims—that there is a logical proof of the existence of God—and for physical claims—that whatever can be logically deduced from known truths can be deemed certain. Despite his early interest in the work of Francis Bacon, Descartes was a rationalist who viewed logical consistency as prior to empirical observation. This metaphysics was further elaborated in *Meditations on First Philosophy,* published in 1641 with a number of objections, solicited by Mersenne, and replies by Descartes.

As a natural philosopher, Descartes set forth a vision of nature as mechanical, a "mechanical philosophy." The most systematic expression of this philosophy is his 1644 Latin textbook, *Principles of Philosophy.* Here he described the universe as full of matter, defined as that which occupies space (Descartes, like Aristotle, denied the possibility of a vacuum), and he claimed that everything that occurs in the material universe can be explained by the interaction of matter and motion. Descartes's picture of matter in motion is dominated by vortices, or whirlpools of matter. These large circular vortices carry the planets around the Sun. (Descartes knew Johannes Kepler's work in optics but ignored his theory of elliptical orbits.)

Descartes was a great mathematician, and along with his contemporary and detested rival, Pierre de Fermat, he founded analytic geometry, the branch of mathematics that represents geometrical forms as algebraic equations. Descartes used these powerful methods to solve long-standing problems in mathematics. He also introduced the still-existing convention of representing powers by numerical superscripts. This was an important contribution in making mathematics more abstract, as the previous convention of referring to second powers as squares and third powers as cubes made it hard to deal with fourth and higher powers.

In optics, by 1628 Descartes had independently rediscovered the sine law of refraction previously known to Thomas Harriot and Willebrord Snel (1580–1626). He described light as transmitted instantly through a material fluid, as a stick when pulled moves all along its length.

Descartes, who conducted dissections, claimed that he had never seen anything in a body that could not be explained mechanically. He hoped his philosophy would culminate in a vast reform of European medicine, and he devoted much work, particularly in the last years of his life, to physiology and psychology, publishing *The Passions of the Soul* in 1649. In Descartes's mechanical philosophy, animals lack souls and therefore can be treated solely as machines. His treatment of the human body as an automaton raised a question about the link between the body and the mind, which Descartes claimed was not reducible to matter. Generations of Cartesians would wrestle with this problem presented by the so-called Cartesian dualism. Descartes speculated that the link was located in the pineal gland. He accepted the circulation of the blood but rejected William Harvey's system, viewing the blood as driven by heat instead of by the pumping action of the heart.

Descartes's principal problem with Catholic authority was not his discreet Copernicanism but the incompatibility of his philosophy with transubstantiation, the doctrine that the priest by consecration changes the bread and wine into the actual body and blood of Jesus Christ. Since the Middle Ages, this process had been explained in Aristotelian terms, based on the distinction between substance and accident. Descartes abolished this distinction and was forced to come up with an alternative explanation for the process. His solution—that the properties of bread and wine apparent to the senses persist after consecration only on the surface—was unsatisfactory, and the problem remained to vex

Catholic Cartesians for decades. By the 1640s, Descartes also ran into trouble in the Protestant Netherlands. There he faced attack from intellectually conservative, university-based Aristotelian Calvinists who identified Cartesianism, which had gained an extensive and vociferous Dutch following, with their liberal Protestant enemies.

Although Descartes was not a courtier by nature and was quite concerned to avoid patronage, he eventually succumbed to the lure of the court and went to Stockholm in 1649 to tutor the brilliant young Queen Christina of Sweden (1626–1689) in philosophy. Unfortunately, she wanted to be tutored at 5:00 A.M. in one of the coldest winters in Swedish history, and Descartes died shortly thereafter. After Descartes's death, Cartesianism became the dominant school of natural philosophy in France, where his body was triumphantly returned in 1667, and his thought was widely influential elsewhere as well.

See also Cartesianism; Mathematics; Mechanical Philosophy; National Differences in Science.
References
Cottingham, John, ed. *The Cambridge Companion to Descartes.* Cambridge: Cambridge University Press, 1992.
Gaukroger, Stephen. *Descartes: An Intellectual Biography.* Oxford: Oxford University Press, 1995.

Digby, Sir Kenelm (1603–1665)

Sir Kenelm Digby was a distinguished natural philosopher, one of the first Englishmen to be influenced by René Descartes. From an aristocratic Catholic family (his father had been executed for his involvement with the Gunpowder Plot to blow up the Protestant king and Parliament in 1605), Digby was fascinated with chemical, alchemical, culinary, and medical recipes and experiments.

Following the death of his beloved wife, Venetia, in 1633, Digby retreated to Gresham College in London, where he spent the next two years mostly devoting himself to alchemical experiments. After leaving Gresham, he relocated to Paris, where he came into contact with the Mersenne circle. For the next 25 years, most of which he spent in France, Digby was an intermediary between French and British scientists. For example, he presided over the mathematical controversy between John Wallis and Pierre de Fermat. Digby's own major work of natural philosophy was *Two Treatises, in the One of Which, the Nature of Bodies; in the Other, the Nature of Man's Soule, Is Looked Into: In the Way of Discovery, of the Immortality of Reasonable Soules* (1644), which incorporated Cartesian ideas in a basically Aristotelian framework.

In 1657, Digby gave a presentation to a society of learned men and physicians in the southern French town of Montpellier. He spoke on the "powder of sympathy," on which he had been working for decades. This was a revival of the weapon salve, which had attracted much interest earlier in the century. Digby had used this powder to cure a friend of a sword-wound: he had taken a garter soaked with the wounded man's blood and immersed it in a container of water in which he had poured some of the powder. The patient's wounds were completely healed in under a week. Digby explained this cure in atomistic terms, claiming that the atoms of blood in the garter had attached themselves to the atoms of the powder, composed of vitriol and salt, and were drawn out of the water, carrying the healing vitriol to the blood in his friend's wound. Digby denied that there was anything magical to the cure. The powder of sympathy became very well-known in late-seventeenth-century England.

Digby returned to England following the Restoration of Charles II (r. 1660–1685) in 1660. He was a founding member of the Royal Society and served on its governing committee. In 1661, he read a paper to the society, later published as *A Discourse Concerning the Vegetation of Plants* (1661), that combined close observations of growing plants with the speculation that plants and animals might both be nourished by a substance in the air. However, he made few other contributions to the society, and his eclectic Aristotelian-Cartesian-alchemical approach to natural philosophy was becoming unfashionable.

Digby retained a reputation as a wonder-worker even after his death. Two collections of his recipes and cures were published posthumously in England. They included a famous cure for warts, which Digby believed to be caused by the heat and dryness of the Sun. The cure, then, was to bathe them in the cold and moist light of the Moon as reflected in a silver basin.

See also Magic; Weapon Salve.
Reference
Petersson, R. T. *Sir Kenelm Digby: The Ornament of England, 1603–1665.* London: Jonathan Cape, 1956.

Dissection and Vivisection

Although dissection already had a long tradition in the Middle Ages, it gained in prominence as a practice and source of knowledge during the scientific revolution. Medieval and Renaissance human dissection already differed from that practiced by the ancient physician Galen in that Galen had worked in a culture where human dissection was forbidden and so extrapolated from dissections of animals, mostly monkeys, to humans, whereas medieval and Renaissance dissectors worked directly on human bodies as well as animals.

There were two fundamental changes in the way dissection was done in the late fifteenth and sixteenth centuries. First, dissections moved from the lecture hall to the anatomical "theater," an originally temporary wooden building, constructed on the model of the ancient Roman theater, that provided a better view of the body. The first permanent anatomical theater was founded at the University of Padua in 1594. The second change is that professors now cut up the body themselves rather than following the medieval and early Renaissance method of reading from a classic medical text while servants,

An undated engraving depicting an interior view of a dissection lesson at the College of Surgery. (National Library of Medicine)

often surgeons or butchers, cut up the body and displayed the various organs to illustrate the points being made. Active dissection by the professor emphasized the evidence of the body over the explication of the text.

One long-standing issue in human dissection, exacerbated by the development of medical education, was finding a source for the bodies. One common source was the penal system, which regularly executed criminals. When a number of executions were scheduled in the city of Padua, the city government would stage them at intervals to suit the needs of the medical school. However, since most executed criminals were male, it was difficult to get female cadavers, particularly pregnant ones. Another important consideration was temperature. The prime dissection season was the winter, when the cold temperatures kept the body in better condition during the dissection, which might take several days.

The dissector needed a deft hand and strong stomach; Michelangelo was so sickened by the dissections he carried out in collaboration with the anatomist Realdo Colombo (c. 1510–1559) that he was unable to eat afterwards. The dissector worked with knives and scalpels for a variety of tasks, with a saw

for cutting bones, and sometimes, as when digging veins out of fat, with his fingers. Bellows and pipes inflated the lungs. A wicker basket next to the dissecting table received the body parts when the dissector was finished with them.

Not everyone approved of dissection. Cutting up dead bodies aroused the suspicion of some theologians and horror on the part of many common people, particularly the relatives of criminals and other people used for dissection. Champions of dissection depicted such opposition in gendered terms, with themselves as manly and those who opposed dissection or did not practice it as "little old women." However, dissection was increasingly popular among medical practitioners, particularly after the controversy over the great dissector Andreas Vesalius's attacks on Galen in his *Of the Fabric of the Human Body* (1543). With the collapse of ancient medical authorities, dissection became a process more oriented toward discovering new knowledge than toward restoring Galen's anatomical system.

Not all dissections were carried out in a university context. Some were forensic, usually carried out to determine if a dead person had been poisoned, and later in the early modern period, it became customary to perform autopsies of popes and kings. (Colombo dissected the corpse of Ignatius of Loyola, founder of the Jesuits, finding stones in the kidneys, lungs, and liver.) Colleges of physicians sponsored dissections as a form of continuing education, and scientific societies, including both the English Royal Society and the French Royal Academy of Sciences, held dissections. The city of Bologna even scheduled a public dissection during carnival week as a form of edifying entertainment.

Although Galen was condemned during the early modern period for extrapolating from animals to humans, animal dissection and vivisection was a common practice. Vivisectors such as Colombo claimed that only through the examination of the living animal, rather than the dead body, could actual life processes be understood. Paracelsus ridiculed dissection as an examination of the dead body to understand the living. Vivisection was considered particularly important for understanding gestation and the workings of the lungs and heart, and William Harvey's argument for the circulation of the blood relied on evidence from vivisection. Although the vivisection of human beings was forbidden throughout Europe and was carried out only under extreme secrecy, if at all, animal vivisection and experimentation was viewed as morally unproblematic until the late seventeenth century. Animals were asphyxiated in air pumps, and the lungs of a dog, the most popular animal for vivisection, were inflated and evacuated with a bellows to keep the dog alive after its ribs had been removed to allow its heart to be viewed. By the late sixteenth century, animals were also being dissected and vivisected to gain knowledge about the animals themselves, not just to extrapolate to human functioning.

See also Anatomy; University of Padua; Vesalius, Andreas.
Reference
French, Roger. *Dissection and Vivisection in the European Renaissance*. Aldershot, England: Ashgate, 1999.

Dynamics
See Force; Mechanics; Physics.

E

East Asian Science

In the medieval period, the flow of knowledge across the Eurasian continent was mostly east to west—a "grand titration" in which Chinese technological innovations, such as printing and gunpowder, made their way westward. Chinese influence on European science during this time was indirect, but it was detectable in the field of alchemy. By the time of the scientific revolution, Europeans remained interested in Chinese astronomical observations, which went back many centuries, but the flow of influence had mostly reversed itself. East Asia had become the area, outside Europe and its colonies, where European science had the most effect. This played out differently in different East Asian nations; in China the greatest impact was on astronomy, in Japan it was on medicine.

The focus on astronomy in China resulted from the missionary strategy of the Jesuits. In developed societies, the Jesuits usually followed a top-down model of conversion, beginning with the political and intellectual elite. They hoped to demonstrate the value of Western culture, including science, before beginning large-scale conversion. The publication of works of Aristotelian natural philosophy and Galenic medicine in Chinese had little effect; the Chinese had their own systems of natural philosophy and medicine to which

Western systems were not obviously superior. However, one avenue of influence open to the Jesuits was astronomy. Astronomy was important in the creation of calendars. Eclipses could be seen as bad omens for the ruling dynasty, and the Astronomical Bureau, which had a long tradition of hiring foreign astronomers, heretofore Indian and Muslim, was responsible for predicting them. Thus it was that early-seventeenth-century missionaries published astronomical works in Chinese, setting forth Western cosmological schemes and even reporting Galileo's telescopic discoveries. Along with astronomy, they introduced elements of Western mathematics, such as trigonometry and logarithms. A telescope was ceremonially presented to the emperor in 1634.

The condemnation of Copernicanism in 1616, however, meant that the missionaries' version of European science increasingly diverged from that of most European scientists. Like the Jesuits in Europe, the missionaries shifted from the Ptolemaic system to Tycho Brahe's picture of a stationary Earth, with the Sun rotating around it and the planets around the Sun. They modified this system with Keplerian elliptical orbits around the Earth for the Sun and Moon, but they shunned Copernicanism.

Although the Jesuits did not reap harvests

of converts, their astronomical proselytization was successful and had a significant influence on Chinese science. With the overthrow of the Ming dynasty by the Manchu Qing dynasty in 1644, the Jesuits gained the big prize in Chinese astronomy, the directorate of the Astronomical Bureau. The politically surefooted Johann Adam Schall von Bell (1591–1666) was appointed head of the bureau in 1645, quickly purging the Muslim staff. Thereafter, a Jesuit would head the Astronomical Bureau until news of the suppression of the order by the pope in 1773 reached China in 1774.

Jesuit cosmology aroused great interest among Chinese astronomers. Traditional Chinese astronomy was moribund and had little cosmological theory, and the Jesuits seemed to provide one. One astronomer, Wang Hsi-Shan (1628–1682), probably never met a Westerner, but learned the Tychonic system and techniques such as trigonometry from Jesuit astronomical manuals. He used this knowledge to introduce, for the first time in China, methods of predicting solar transits and planetary occultations. He also invented a new cosmology that was based on Tycho's, but with substantial modifications. The Jesuit writings led Chinese astronomers to be more interested in physical descriptions of the cosmos as opposed to algebraic techniques for predicting planetary movements.

However, Western astronomy never displaced or even seriously rivaled the prestige of the native tradition. To legitimate the study of Jesuit astronomy by the Chinese, late-seventeenth-century Chinese scholars even argued that this knowledge was originally Chinese, that it had been forgotten in its native land but was developed in the barbarian kingdoms of the West. Now it had returned to China. A common pattern of astronomical training in eighteenth-century China was to study the Western writings and then progress to the Chinese astronomical classics. Ultimately, the greatest impact of Western astronomy in China was the revitalization of the indigenous Chinese astronomical tradition.

The situation in Japan was different. The Japanese, who had adopted much of their culture from China and Korea, did not reject foreign knowledge simply because it was foreign. Thus the mid-sixteenth-century Jesuits in Japan were able to teach Western learning, mostly Aristotelian natural philosophy and Ptolemaic astronomy. The Japanese also adopted some European navigational and surgical techniques. However, after the expulsion of Catholic missionaries and the closing of the country in the early seventeenth century, the openings for scientific ideas to enter Japan were narrow. In 1630, Jesuit works in Chinese, which Chinese-literate Japanese scholars could read, were barred from the country along with other books relating to Christianity. But Jesuit ideas did continue to circulate. Shibukawa Harumi (1639–1715) employed Jesuit works in Chinese along with purely Chinese calendars to create Japan's first native calendar in 1684.

The main opening for Western science after the closing of the country was the Dutch colony. Dutch traders, the only Westerners allowed in Japan, were restricted to a small trading post in Nagasaki. They imported scientific and medical books, which attracted the attention of Japanese intellectuals, particularly physicians. Some Japanese people learned Dutch and translated Western works available in Dutch into Japanese for others to read; others were trained in Western surgery by European doctors attached to the Dutch colony. The shogunate's 1720 liberalization of the law barring foreign books from circulating eventually led to a widespread interest in *rangaku,* or "Dutch learning," in astronomy, cosmology, physics, and particularly medicine in the eighteenth century.

See also Jesuits.
References
Sivin, Nathan. *Science in Ancient China: Researches and Reflections.* Aldershot, England: Variorum, 1995.
Sugimoto, Masayoshi, and David L. Swain. *Science and Culture in Traditional Japan* A.D. *600–1854.* Cambridge: MIT Press, 1978.

Education

The early modern period saw a tremendous expansion of the educational system at every level. This was inspired by several developments: the need to indoctrinate youth in the new religious doctrines and practices of the Protestant and Catholic Reformations, the needs of burgeoning state officialdoms for men able to read and write in the vernacular and Latin, and the growth of humanistic learning. Although religion and humanistic concerns dominated the curriculum, the expansion of education also affected and was affected by the scientific revolution.

In the sixteenth century, humanist schools, which often positioned themselves as teaching a more practical curriculum than did the universities, emphasized rhetoric and moral instruction more than science. Their rejection of the Aristotelian-Galenist orthodoxy of the universities often meant the rejection of natural philosophy itself. Practical mathematics and science was taught, especially in Italy, only in trade schools for businessmen and artisans, which existed separately from the humanistic schools.

As the humanistic curriculum broadened and technical education became more prestigious, the lack of educational interest in science outside medicine and Aristotelian natural philosophy began to change late in the sixteenth century with the founding of institutions such as London's Gresham College. Dozens of early modern scientists and natural philosophers, from the most illustrious—such as Galileo Galilei, who gave private mathematics lessons as well as teaching at the University of Padua—to the most obscure, made a living for at least part of their careers as teachers, tutors, or schoolmasters. The expansion of scientific education led to a tremendous demand for materials on natural philosophy and science in a form that could be easily taught. The publication of scientific textbooks was a burgeoning field. Much of the popularity of Ramism, with its charts and mechanical formulae for organizing a subject, was based on the belief that it made subjects easily digestible for students. Other rhetorical styles, such as the mysterious, symbolic, and beautiful language of classical alchemy, could not be coherently presented to a roomful of adolescent boys. The Ramist-influenced schoolmaster Andreas Libavius therefore reformed chemical language in his textbook *Alchemia* (1597), making it more prosaic.

However significant the change in curricular structure, the substance of the science that was taught changed only slowly. The excellent Jesuit schools, leading institutions within the Catholic world and the envy of Protestants such as Francis Bacon, were limited both by their humanistic emphasis and by the Society's adherence to Aristotelianism. For the first five years, the curriculum was basically humanistic, after which it was principally devoted to theology and Aristotelian philosophy, including natural philosophy. Mathematics, including applied fields such as perspective, mechanics, and architecture, was a subsidiary subject. A pupil of the Jesuits, such as René Descartes, who attended the Jesuit College of La Flèche in Anjou from 1606 to 1614, got an excellent education in what was becoming an obsolete natural-philosophical system. (The name "college" can be misleading; Descartes was ten years old in 1606, and the colleges were more like present-day high schools.) The curriculum of many Protestant academies followed that of the Jesuits, except in theology, and indeed, many Protestant schools used the excellent Jesuit textbooks, such as the mathematics text of Christoph Clavius. Another Catholic teaching order, the Oratorians, was introduced into France in 1611 by a future cardinal and ally of Descartes, Pierre de Berulle (1575–1629). The French Oratorians, following the example of the Italian order founded in 1575 by St. Philip Neri (1515–1595), were more innovative in natural philosophy than were the Jesuits, boasting several distinguished Cartesians, such as Nicholas Malebranche, in their ranks.

What little education beyond basic literacy that was available to women focused either on traditional humanism or on household skills. One of the few advocates for a scientific education for women was the French alchemist Marie Le Jars de Gournay (1565–1645), who debated the Dutch humanist Anna Maria Van Schurman (1607–1678), an advocate of humanist and religious education for women. This debate was remote from reality, as the vast majority of even upper-class women had access to neither scientific nor humanist education.

A variety of projects for educational reform in the seventeenth century made more room in the curriculum for science. Some of these projects were millenarian in nature, designed in the hopes that a proper education could lead to a spiritual rebirth and the restoration of human rule over nature that had been lost with the Fall of Adam. The Czech millenarian educational reformer John Amos Comenius (1592–1670) thought education the key to universal reform. Comenius, one of the few male educational reformers to advocate the education of women in advanced subjects, worked from a central European tradition of encyclopedism—the creation of elaborate works setting forth the whole of knowledge on a systematic basis—which he called *pansophy,* or "all-wisdom."

Comenius's impact on European education was enhanced by the fact that the defeat of Czech Protestantism in the Thirty Years War sent him traveling through many countries, including Germany, Britain, Hungary, and the Dutch Republic. Working through the Hartlib circle, he inspired a host of educational reform projects in England during the Civil War of the mid-seventeenth century and the period of Puritan rule that followed. Some of these projects emphasized the importance of Copernican astronomy, alchemy, natural magic, and astrology, as opposed to Aristotelian natural philosophy. A member of the Hartlib circle, Sir William Petty (1623–1687), published a plan for vocational, technical, and scientific education titled *The Advice of W. P. to Samuel Hartlib, for the Advancement of Some Particular Parts of Learning*

(1648). Petty's Baconian-inspired plan would have provided a universal education in craft skills, including such disciplines auxiliary to the sciences as lensmaking and gardening, and an advanced institution, including a laboratory and an anatomical theater, for further studies. These plans had little real impact on English education in the period, as leading new philosophers such as John Wilkins defended the traditional curriculum.

Schools in the later seventeenth century saw more emphasis on practical and scientific skills. In England, Dissenting Academies were set up after the Restoration of Charles II (r. 1660–1685) in 1660 to educate those youths excluded from Oxford and Cambridge, admission to which was restricted to members of the Church of England. The Dissenting Academy curricula had more emphasis on science, although the humanistic and religious elements remained strong. The most influential educational theorist of the late seventeenth century was John Locke, author of *Some Thoughts Concerning Education* (1693). Locke's picture of the young child as a blank slate, his emphasis on the senses as the only way of acquiring knowledge, and his utilitarian approach to the curriculum led him to make a distinct place for science alongside the humanities in the education of the male social elite.

See also Gresham College; Universities.
References
Bowen, James. *A History of Western Education*. Vol. 3, *The Modern West: Europe and the New World*. London: Methuen, 1981.
Debus, Allen G., ed. *Science and Education in the Seventeenth Century: The Webster-Ward Debate*. New York: American Elsevier, 1970.

Elements

See Matter.

Embryology

The process by which new living beings are formed remained in some senses mysterious in the early modern period. Paracelsus in particular was almost terrified by the creative powers of semen, believing that it had the power of creating monstrous life nearly anywhere it was deposited, or even if it was retained within a man's body. He thought intestinal worms were the result of semen deposited there in an act of sodomy. The only healthy options for an adult male were marriage, where the semen would be deposited in the womb, the proper place for the creation of a human, or self-castration. Although he did not perform it himself, Paracelsus highly exalted self-castration as an act pleasing to God, arguing that God had placed the male genitals outside the body to facilitate castration. On the other hand, Paracelsus and other alchemists did view seminal fluid as necessary for one of the supreme alchemical works, the creation of a homunculus, an artificial man produced through alchemy, without passing through the female. This was often viewed as the height of blasphemous alchemical arrogance.

In the mainstream of embryological thought, the coming together of a male and female was regarded as usually necessary for the creation of a living being, although many thought the lower orders of animals could be created by spontaneous generation. The main divide was between those who believed that the living being develops out of something else, the epigenicists, and those who believed that the living being is already present at the act of generation, the preformationists. Preformationists originally believed that the living being is carried in the egg, and that the male seed serves to quicken it. In this "ovist" view, the egg has the advantage of being spherical, the most perfect shape (even the irregularly shaped eggs of birds and reptiles have spherical yolks), but it has the disadvantage of being carried in the inferior vessel, the female. The early epigeneticists, following Aristotle, the greatest and most influential embryologist before the scientific revolution, believed that the substance of the living being is provided by the female (Aristotle believed

it to be the menstrual blood). In this view, the form of the particular species of living being is provided by the male, but the individual creature itself is created in the process of conception, rather than being preformed.

Research into embryological development during the scientific revolution originated with Ulisse Aldrovandi, who followed the example of Aristotle in opening the eggs of chickens at different stages in their incubation to examine the developmental process. The greatest embryological researcher of the scientific revolution was William Harvey. His *On the Generation of Animals* (1651), based on elaborate dissections and studies of hen's eggs and of pregnant does donated to Harvey by King Charles I (r. 1625–1649), established the principle that all animals proceed from eggs, denying spontaneous generation. Despite his emphasis on the egg, Harvey was an epigeneticist. Like many others, he believed that the semen does not even have to come into contact with the egg, but that it lends an energizing essence merely by its proximity. A traditionalist in natural philosophy, Harvey shunned the mechanical philosophy, but his contemporaries René Descartes and Pierre Gassendi attempted unsuccessfully to apply it to fetal development.

In addition to enabling much closer study of the developing embryo, the microscope revolutionized approaches to the question of generation through the discovery of the sperm. The semen, or, as it was originally called, the *animalcules,* or "little animals," had previously been thought to be a coagulated form of milk or blood. Antoni van Leeuwenhoek and his rival claimant for the title of discoverer of the sperm, the natural philosopher Nicholas Hartsoeker (1656–1725), both believed that the sperm, not the egg, carries the preformed living being. Hartsoeker's illustration of the tiny human curled up in the head of the sperm has become famous, although he himself put it forward somewhat tentatively. "Spermism" had the advantage of placing the creative power in the male (although the testicles were not thought to be the most reputable area for human origins, either). However, it had the disadvantage of being terribly wasteful, given the millions of sperm in an ejaculation, only one of which, Leeuwenhoek claimed, would develop into a living being. Leeuwenhoek asserted that this wastefulness is merely analogous to the wastefulness of the myriad of seeds a plant generates, only a few of which grow to adult plants. Whether the sperm even played a role in reproduction was also a matter of controversy, as some claimed the sperm has nothing to do with reproduction but is the product of the putrefaction of the semen.

The microscope also greatly advanced the traditional ovist view of preformation, which located the preformed individual in the egg. Now it was at least theoretically possible to identify the tiny eggs from which those beings born alive are generated within their mothers' wombs. Nicolaus Steno suggested in 1667 that the "testes" of women (thought to be analogous to the male testicles) were the same organs as the "ovaries" of egg-laying creatures. The eggs were not identified immediately, but in 1672, the Dutch physician Reinier de Graaf (1641–1673) mistakenly identified as eggs what are now known as the Graafian follicles, also spherical. Ovism was endorsed by leading microscopic anatomists Marcello Malpighi and Jan Swammerdam, as well as by the French Cartesian philosopher Nicholas Malebranche, even though Descartes's mechanical theory had been epigenicist. Malebranche's *Search after Truth* (1674) sets forth a theory of encasement, wherein each egg is described as containing the preformed bodies of all its descendants, encased one inside the other like Russian dolls to an unimaginable degree of smallness. Malebranche supported this theory less because he had strong feelings about generation than because of the heavy emphasis on divine causation in his philosophical system: The creator of the "Russian dolls" could only be God. Preformationist theories, whether spermist or ovist, also had the advantage of seeming modern, without the Aristotelian associations of epigenesis, and

they dominated early-eighteenth-century embryological thought.

> *See also* Leeuwenhoek, Antoni van; Sexual Difference.
> *References*
> Needham, Joseph. *A History of Embryology.* 2d ed., revised with the assistance of Arthur Hughes. Cambridge: Cambridge University Press, 1959.
> Pinto-Correia, Clara. *The Ovary of Eve: Egg and Sperm and Preformation.* Chicago and London: University of Chicago Press, 1997.

Epicureanism

The ancient philosophy of Epicurus (341–270 B.C.) offered both exciting intellectual possibilities and fearsome religious dangers to early modern natural philosophers. Epicureanism was mostly known not through the fragmentary writings of Epicurus himself but through other ancient Greeks and Romans, notably the Latin Epicurean poet Lucretius (c. 96–c. 55 B.C.), who wrote the epic work *Of the Nature of Things.* Epicurean physics, which proclaimed the existence of both atoms and a vacuum, attracted many who sought an alternative to Aristotelianism. But Epicurus was also associated with a hedonistic ethics that claimed pleasure as the sole good, and even worse, denied that the gods care for the world—a stance that early moderns defined as "atheism." The Epicurean universe ran without any providential guidance, whether from the pagan deities of Epicurus's own culture or from the Christian God. Epicurean materialism also meant the denial of any spiritual reality, describing both the gods and the human soul as material, a doctrine shocking to early modern Christians.

The man who tried to solve the Epicurean problem was the seventeenth-century French priest Pierre Gassendi. Gassendi "baptized Epicurus," providing a version of Epicureanism that was compatible with Christianity. For example, for ancient Epicureans the atoms were eternal, neither created nor destructible. Gassendi harmonized Epicureanism with the Christian doctrine of creation by claiming that God had created the atoms. He denied that atoms are infinite in number or that the human soul is mortal.

Gassendi succeeded in demonstrating that Epicurean atomism could be reconciled with Christianity, but in so doing he did not entirely purge Epicurus of his dangerous reputation. Epicurean atomism fed into the mainstream of the mechanical philosophy, but scoffers at religion continued to identify themselves as Epicureans, and Christian apologists continued to denounce Epicurus as a blasphemous atheist. Robert Boyle, while accepting some aspects of Epicurean atomism, attacked "Epicurean" atheism. Isaac Newton, whose atomism was influenced by Epicureanism, tried to defend Epicurus from the Epicureans, remarking to his disciple David Gregory (1659–1708), "The philosophy of Epicurus and Lucretius is true and old, but was wrongly interpreted by the ancients as atheism" (Turnbull 1961, p. 335).

See also Atomism; Gassendi, Pierre.
References
Osler, Margaret J., ed. *Atoms, Pneuma, and Tranquillity: Epicurean and Stoic Themes in European Thought.* Cambridge: Cambridge University Press, 1991.
Turnbull, H. W., ed. *The Correspondence of Isaac Newton.* Vol. 3. Cambridge: For the Royal Society at the University Press, 1961.

Epigenesis
See Embryology.

Experiments
Although experiments were not an invention of the scientific revolution, they did attain a more central place in scientific practice and ideology during the early modern period. Experiments had played a role in Aristotelian natural philosophy and Galenic medicine, but few experiments were carried out during the Middle Ages or the Renaissance. This began to change during the scientific revolution, particularly in the early seventeenth century. An emphasis on experiments is sometimes identified with Baconianism; indeed, Francis Bacon frequently used the term "experiment" and was later identified as the founder of experimental science. He was arguably killed by an experiment, a chill caught by stuffing a goose with snow to see if it could be preserved. Generally, though, Bacon was more interested in the close observation of nature than in experiments per se. The difference between "experiment" in the root sense of experience and "experiment" in the modern scientific sense of the creation of an artificial situation designed to study scientific principles held to apply in all situations causes confusion in the study of early modern science, and the modern meaning of the word emerged only gradually.

Galileo Galilei, another person often identified as an originator of experimentalism, frequently employed experimental arguments in his writings, but scholars still debate which of these highly abstract experiments he had actually performed and which were "thought experiments"—imagined but not performed. It is certain, however, that he did not perform the famous experiment of dropping a light and a heavy weight off the Leaning Tower of Pisa to show that they fell at the same speed. This experiment was actually performed later by an Aristotelian who discovered that the heavy weight did indeed fall faster. Galileo's emphasis on mathematical abstraction and his efforts to consider motion under ideal conditions were not congenial to a thoroughgoing experimentalism.

The first work to systematically relate the performance of experiments to natural philosophical theory was William Gilbert's *On the Magnet* (1600), although many of the experiments it describes were not new. *On the Magnet* was followed by a boom in dramatic experiments, carried out in many fields and by supporters of all the major natural-philosophical traditions. The Aristotelian William Harvey experimented by tying off the veins in his arm with ligatures to ascertain the circulation of the blood, and the Aristotelian Jesuit natural philosophers were also active experimenters. Johannes Baptista van Helmont experimented to demonstrate his alchemical theories, and Galileo's disciple Evangelista Torricelli (1608–1647) went beyond his master as an experimentalist, carrying out the famous and well-publicized "Torricellian experiment." In this experiment, Torricelli took a glass tube about two yards long and sealed at one end, filled it with mercury, covered the open end with a finger, and inserted the tube, open end downward, into a container of mercury. The mercury in the tube sank to a level about 30 inches higher than the mercury in the container. This demonstrated the pressure of the air that sustained the column of mercury and also provided experimental evidence for the existence of a vacuum, in the gap between the top of the column and the top of the tube. Blaise Pascal followed Torricelli in carrying out barometric experiments, and Otto von Guericke (1602–1686) carried out another famous experiment, creating a vacuum between two hemispheres of brass and demonstrating that teams of horses could not pull them apart.

Not all seventeenth-century natural philosophers were enthusiastic experimenters. René Descartes was dubious about the value of experiment in achieving certainty. For him, experiment had to be subordinated to theory, and true certainty—the goal of natural philosophy—could be attained only by reason. There were Cartesian experimentalists—the most famous being the spectacular showman Jacques Rohault (1618–1672)—but in the Cartesian tradition, experiments were usually performed to demonstrate truths already arrived at through reason rather than to find new knowledge.

By the late seventeenth century, experiments were a central part of the program of scientific societies. In mid-seventeenth-century Florence, members of the Accademia del Cimento, the "Academy of Experiment," carried on a variety of experiments, although they did not relate their experiments to a theory of nature. Dramatic experiments involving expensive equipment and preferably spectacular effects were prized in court society, like that of the Medici family, who sponsored the Accademia del Cimento. Natural-philosophical theorizing was not as entertaining. After the Accademia del Cimento closed down in 1667, the most convinced experimentalists of the late seventeenth century were the natural philosophers of the English Royal Society, who often referred to what they were doing as "experimental philosophy." Robert Boyle's famous experiments with the air pump were revered by English natural philosophers. Isaac Newton's experiments with a prism, which established that white light is a mixture of colors, were also classic examples of experimental procedure. Society members also enjoyed the services of the most gifted designer and conductor of experiments, Robert Hooke. As curator of experiments, Hooke had the grueling task of providing several experiments, preferably dramatic ones, for each of the society's meetings. Hooke's own discussions of the method of natural-philosophical investigation, which he called "Philosophical Algebra," incorporated experimentalism into the Baconian tradition. Boyle also attempted to identify an experimental method. He and other English natural philosophers considered experimental investigations as superior to theoretical disputes because they saw them as less contentious (a position that aroused the ire of Thomas Hobbes), and also as embodying a more humble and respectful attitude toward knowledge. The English, Boyle prominent among them, developed the genre of the "experimental account" to enable those who did not themselves witness the experiment to become "virtual witnesses." Since experiments, particularly those requiring such expensive and finicky devices as air pumps, could not be replicated on demand, it was necessary that experimental narratives convince readers of the truth of the experiment. This required a great deal of detail, careful descriptions of procedure, and a specific narrative of an event, including its setting at a particular place and time.

See also Accademia del Cimento; Hooke, Robert; Royal Society.
References
Gillispie, Charles Coulston. *The Edge of Objectivity: An Essay in the History of Scientific Ideas.* Princeton: Princeton University Press, 1960.
Hall, Marie B. *Promoting Experimental Learning: Experiment and the Royal Society, 1660–1727.* Cambridge: Cambridge University Press, 1991.
Shaffer, Simon, and Steven Shapin. *Leviathan and the Air-Pump: Hobbes, Boyle, and the Experimental Life.* Princeton: Princeton University Press, 1985.

Exploration, Discovery, and Colonization

The early modern period saw a tremendous increase in European involvement with and knowledge of the "outside world," whether previously unknown, like the Americas, or previously known only from a distance, like India and East Asia. This new knowledge transformed European science, gradually making it clear to all that the books of the ancients were not complete repositories of natural knowledge.

Exploration revealed the falsity of much traditional European knowledge. Aristotle's belief that the equatorial zone of the Earth's surface is too hot for human habitation was thoroughly disproved by Iberian sailors and the African and American peoples they encountered. Belief in the traditional monstrous races thought since classical times to be living outside Europe, such as the people with dog's heads, declined in the early modern period as more was learned about the world. Columbus's first letter from the New World specifically mentioned the lack of monstrosities. Exploration also put new problems dramatically on the intellectual agenda, perhaps most importantly those of navigation and cartography. Astronomers were faced with the southern sky.

Asia and Africa had been known to the

ancients, but the Americas presented a more fundamental challenge. Once it was realized that the Americas were not an extension of Asia, the basic intellectual problem became the ignorance of the New World in both classical writings and the Bible. Attempts were made to solve this problem by grafting classical and biblical knowledge onto the Americas. The Amazon River was given its name by Spanish explorers who believed it was inhabited by women warriors of the kind that ancient Greek writers had described. Some suggested that the Native Americans were the ten lost tribes of Israel, and even though that was never a majority opinion among European thinkers, nearly all agreed on the necessity of fitting them into the biblical framework of descent from the sons of Noah. Some argued that America could be found in the Bible, or that various places in the New World were named after biblical figures.

But Europeans gradually became aware that the Americas were not dealt with specifically by either the classics or the Bible, and that neither were terribly helpful on the further reaches of Asia, either. By 1570, the classic ancient work on geography, that by Claudius Ptolemy (A.D. 90–168), was openly referred to as possessing merely historical interest. European expansion became the standard example given by those who argued that modern civilization had advanced beyond the classical world. That would-be intellectual revolutionary, Francis Bacon, specifically compared himself to Columbus, and the frontispiece of his *The Great Instauration* (1620) shows a ship sailing past the Pillars of Hercules, the western limit of the Mediterranean and a symbol of the limits of ancient knowledge.

The impact, intellectual and otherwise, of the New World discoveries was initially greatest by far in Spain, and much Spanish writing on the subject was translated into other European languages. Spain possessed a monopoly over the New World, however ineffective that monopoly soon became, and thus it received the greatest flow of information thence. Spanish investigators dominated New World natural history, producing a long series of works, of which the earliest notable example is Gonzalo Fernández de Oviedo's (1478–1557) *Natural History of the Indies* (1547). The most complex debates on New World peoples and subjects in the sixteenth century were carried on in Aristotelian terms in the very conservative Spanish universities, particularly the University of Salamanca. The Spanish debate between Bartolomé de Las Casas (1474–1556) and Juan Ginés de Sepúlveda (1490–c. 1573) on the status of the Native Americans was the last major European intellectual debate to be framed entirely in Aristotelian terms.

The disciplines most transformed by the new knowledge were those of natural history, botany, and zoology. New natural knowledge came to Europe in a flood, particularly from the only recently encountered lands of the West. The properties of all kinds of previously unknown plants and animals had to be worked out with little help from the ancients. Europeans gained much knowledge from indigenous informants, ranging from the tribespeople of South America to the learned Ayurvedic Hindu physicians of Portuguese Goa.

The Europeans were particularly interested in the medical properties of new plants, and many of the scientific explorers and knowledge gatherers sent from Europe were physicians, such as García d'Orta and Francisco Hernández. These investigators were sometimes carried away with enthusiasm over the medical potential of foreign plants. For example, tobacco, a New World plant, was hailed as a wonder drug. It was claimed to cure migraine, "cold stomach," kidney pain, hysteria, gout, toothache, worms, scabies, nettles, burns, wounds from poisoned arrows, and gunshots. Tobacco was also viewed as psychologically beneficial in calming the mind and making men more attractive to women.

The myriad of new plant and animal species encountered eventually made much of classical natural history obsolete. Amerigo Vespucci (c. 1454–1512), whose pamphlets on his encounters in the New World were more influential than Columbus's writings in shaping the first European perceptions, specifically pointed out the abundance of new species not found in the works of the ancient Roman natural historian Pliny the Elder (A.D. 23–79). These new species lacked the traditional symbolic and allegorical associations of animals and plants already known to Europeans, and thus they could be discussed in a much more empirical way.

In European colonial possessions, the creation of an exhaustive natural history both inventoried a colony's natural resources and made a textual claim on it. The production of such natural histories, however, required a substantial commitment of resources, both to the collection of knowledge and to its publication in book form. Hernández's ambitious and state-sponsored natural history of Mexico was never published in its complete form, and the proposed natural history of Virginia that Thomas Harriot was involved in never got past the initial stages because of poor financing and the general shakiness of the English position in Virginia. Georg Markgraf's (1610–1643) *Natural History of Brazil* (1648) was more successful. Markgraf, one of the first astronomers to study the stars of the Southern Hemisphere, worked under the sponsorship of the Dutch during their short-lived rule in Brazil. Ironically, his work appeared after the Portuguese rule had been restored, too late to benefit his Dutch sponsors.

The tradition of European investigation of the natural history of foreign lands continued throughout the early modern period, although by the late seventeenth century, the predominantly medical interest of many of the early naturalists had faded in favor of the less use-oriented view of people such as Georg Eberhard Rumph and Anna Maria Sibylla Merian. Curios, dried plants, and stuffed animal specimens from foreign lands became the subjects of an international trade, mostly centered in

Things from far away were often felt to have medicinal value. This illustration of men gathering cinnamon bark in India appeared in a 1579 medical book by the French surgeon Ambroise Paré. (National Library of Medicine)

Amsterdam and appealing to European collectors and museums.

Most of these natural historians and other scientists accompanied vessels whose primary mission was nonscientific. But by the late seventeenth century, the English government was backing a few voyages whose primary mission was the gathering of scientific knowledge. For example, the voyages of Edmond Halley to the South Atlantic from 1698 to 1700—one of the few cases of a scientist actually captaining a vessel—were aimed at the study of the southern sky and magnetic variation for navigational purposes. Also, William Dampier (1652–1715), not as original a scientist as Halley but a seasoned scientific observer, went on a voyage to the Pacific in 1698 with primarily scientific ends.

European attitudes toward indigenous scientific knowledge varied. Specifically local knowledge of geography, natural history, and local diseases was valued everywhere. The importation and cultivation of new crops—tobacco, tomatoes, corn, and potatoes—required some appropriation of indigenous knowledge. The calendars of the Mayans and Aztecs aroused curiosity and respect from European chronologists. But non-European knowledge in the more theoretical and abstract areas of natural philosophy received little study. The natural-philosophical theories and observations of "civil" peoples—those who had cities and a social hierarchy, such as the Chinese, Japanese, and Muslims—were sometimes investigated by Europeans, although all believed European knowledge to be superior.

The Jesuits in China hoped that their Aristotelian science would provide a wedge into the world of the Chinese elite, eventually leading to their conversion to Catholicism, but this proved fruitless, as the Chinese, although intrigued, preferred their own natural-philosophical traditions. However, the Chinese and other non-Western people possessing developed cartographic traditions did incorporate new geographical information derived from the Europeans onto their maps.

As Europeans established permanent settlements in the New World, the institutions and practices of European intellectual life went with them. Institutions such as the University of Mexico City or Harvard University mostly taught an Aristotelian curriculum, adapting to new intellectual currents after some time lag relative to Europe. But colonial institutions and groups could not compete with their European counterparts as generators of new knowledge. The most important natural philosopher from the British colonies in America, George Starkey, was educated at Harvard and found fellow alchemical workers in New England, but in order to pursue his studies he had to relocate to England. Intellectuals based in the colonies could participate in European scientific discourse only from a distance, as Rumph from the East Indies and New Englanders such as the physician and alchemist John Winthrop (1606–1676) and the minister Cotton Mather (1663–1728), both fellows of the Royal Society, did from America.

See also Capitalism; Cartography; Hernández, Francisco; Jesuits; Merian, Anna Maria Sibylla; Natural History; Navigation; Orta, García d'; Race; Rumph, Georg Eberhard.

References
Dickenson, Victoria. *Drawn from Life: Science and Art in the Portrayal of the New World.* Toronto: University of Toronto Press, 1998.
Grafton, Anthony, with April Shelford and Nancy Siraisi. *New Worlds, Ancient Texts: The Power of Tradition and the Shock of Discovery.* Cambridge: Belknap Press of Harvard University Press, 1992.
Storey, William K., ed. *Scientific Aspects of European Expansion.* An Expanding World, vol. 6. Aldershot, England: Variorum, 1996.

F

Fermat, Pierre de (1601–1665)

A French lawyer, Pierre de Fermat was one of the most innovative pure mathematicians of the seventeenth century and the greatest amateur mathematician of all time. From a middle-class family of merchants and lawyers, Fermat received a degree in law from the University of Orléans in 1631. He settled down in Toulouse as a lawyer and member of the local *parlement,* or law court. Fermat's interest in mathematics can be traced to the 1620s, but because of his reluctance to publish, his work was known only to a few associates. In 1636, his friend and fellow Toulouse lawyer Pierre de Carcavi (1600–1684) moved to Paris and introduced Fermat's work to Marin Mersenne, who disseminated it to his correspondents in the scientific world.

Fermat shares the glory of having invented analytic geometry, the representation of geometrical curves by algebraic equations, with René Descartes, who despised him. The controversy between Fermat and Descartes began in 1637 with Fermat's criticism of Descartes's *Optics* (1637), and it quickly spread to Descartes's attacks on Fermat's mathematical methods, such as that for deriving tangents. Descartes was concerned with his priority in the invention of analytic geometry, and he seems to have convinced himself that Fermat had plagiarized him.

In a correspondence with Blaise Pascal in 1654, Fermat originated probability theory; Pascals' father, Etienne (1588–1651), had been involved in the Fermat-Descartes controversy. Fermat left it to Pascal to follow up, as he himself was concerned principally with number theory by this time. In number theory, Fermat stands alone in the seventeenth century, both for the brilliance of his contributions and for the lack of interest among other leading mathematicians, including Pascal and Christiaan Huygens, both of whom Fermat tried to interest. Fermat's approach to mathematics was that of a problem solver rather than a system builder, and much of his number theory originated in the study of the ancient problems of the Greek mathematician Diophantus (3rd c. A.D.). He tried to drum up interest in number theory through a series of challenge problems addressed to the mathematical community of France, England, and the Netherlands. This led to a controversy with the English mathematician John Wallis, who lacked Fermat's interest in number theory and saw his problems as basically trivial. After Fermat, number theory lay quiescent until the eighteenth century.

Fermat's best known contribution to number theory is known as "Fermat's last theorem," even though it was not his last theorem chronologically speaking. Fermat's last theorem is the claim that there is no integral solution for the equation $x^n + y^n = zn$ for n greater than 2. Fermat claimed he had a proof of this theorem, but he did not write it down and it has never been discovered. This was typical of Fermat, whose reluctance to reveal his methods or the intermediate steps of his arguments was one reason he did not publish. (Some of his mathematical works were published posthumously, edited by his eldest son, Clement-Samuel Fermat.) Most contemporary mathematicians believe that any proof Fermat had must have been invalid and was possibly based on the assumption that the proofs he had devised for third and fourth powers could be extended to all higher powers. The theorem itself, which inspired centuries of productive mathematical work, has recently been proved by Andrew Wiles using means unknown to Fermat (Singh 1997).

As a natural philosopher, Fermat was an Aristotelian and suspicious of the application of mathematics to physical questions. In optics, Fermat is remembered for "Fermat's principle," which states that light always follows the shortest possible path. This was based on the Aristotelian principle that nature in general operates along the shortest path. From this principle, Fermat was able to deduce Descartes's sine law of refraction.

See also Cycloid; Mathematics; Probability.
References
Mahoney, Michael Sean. *The Mathematical Career of Pierre de Fermat, 1601–1665.* 2d ed. Princeton: Princeton University Press, 1994.
Singh, Simon. *Fermat's Enigma: The Epic Quest to Solve the World's Greatest Mathematical Problem.* New York: Walker, 1997.

Flamsteed, John (1646–1719)

The first royal astronomer of England, John Flamsteed, was one of the finest star mappers of the scientific revolution. From a merchant family in the north of England, Flamsteed did not attend university because of poor health, but he taught himself astronomy. He received an M.A. from Cambridge by royal command

in 1674, and shortly after he took orders in the Church of England. He attracted the notice of Sir Jonas Moore (1617–1679), mathematician and surveyor-general of the Royal Ordinance, who proposed to set him up in an observatory in Chelsea.

Instead, after Flamsteed demonstrated that existing astronomical data could not support schemes to discover longitude by lunar observation, he was appointed astronomer royal with a stipend of £100 a year. He was based in the new Royal Observatory at Greenwich, into which he sunk a great deal of his own money. Flamsteed was an ideal choice for the position because he believed that the highest task of astronomy is the patient accumulation of precise observations rather than the creation of cosmological theory or astrological prediction. His hero was Tycho Brahe. Obsessed with the precision of his instruments, Flamsteed introduced several improvements and innovations, notably the large mural arc. He also invented new observational techniques, such as the technique for finding the point of the vernal equinox.

Flamsteed's emphasis on precise observation and his disdain for astronomical theory led to conflict with Edmond Halley and with Isaac Newton, who was one of the Royal Society's visitors, or supervisors, to the observatory. These men were eager for Flamsteed to publish his data, which they regarded as valuable evidence for the theory of universal gravitation, and they regarded Flamsteed as a civil servant whose observations were public property. But the independent and pugnacious Flamsteed wanted the catalog to be as perfect as possible, and since he had spent his own money, he regarded the data as his own property.

In 1709, Newton had Flamsteed, a fellow of the Royal Society since 1677, kicked out for nonpayment of dues. In 1712, Halley published an abbreviated edition of the Greenwich data, arousing Flamsteed's wrath. Flamsteed bought up and burned 300 of the 400 copies printed. In any case, the affair motivated him to finally prepare *Historia*

Coelestis Britannica, posthumously published in three volumes by his widow, Margaret Flamsteed (c. 1670–1730), who had assisted with his work, and two of his former associates. Published in 1725, it was the finest star catalog to that time.

> *See also* Greenwich Observatory; Halley, Edmond.
> *Reference*
> Willmoth, Frances, ed. *Flamsteed's Stars: New Perspectives on the Life and Work of the First Astronomer Royal (1646–1719)*. Woodbridge, England: Boydell and Brewer, in association with the National Maritime Museum, 1997.

Fludd, Robert (1574–1637)

Robert Fludd, one of the last major occult philosophers of the scientific revolution, came from a prosperous family in Kent. He graduated with a B.A., an M.A., and an M.D. from Oxford and then traveled on the Continent. After some difficulty caused by his militant Paracelsianism, he was admitted to the London College of Physicians in 1609 and practiced medicine successfully in London. He was inspired by the Rosicrucian manifestos (although he complained that the Rosicrucians had not recruited him), and in 1616, he published the first part of his long-prepared work on the nature of the universe, a short piece called *Apology for the Brothers of the Rosy Cross*. This work was reissued in an expanded version the next year along with a theological work, *Theologico-Philosophical Tract,* and the first volume of Fludd's great work, *History of Both Cosmoses,* the title referring to the universe as a macrocosm and humanity as a microcosm.

Fludd's lavishly illustrated volumes set forth a unified occultism based on a mystical approach to mathematics, science, and medicine. He rejected both the ancient heritage of Aristotelian philosophy and Galenic medicine and the new mechanical philosophy. Believing the Book of Genesis to be the ultimate source of wisdom on nature, Fludd held that true philosophy had descended

Undated engraving of the occult philosopher Robert Fludd. His emphasis on the spiritual in his theories brought about harsh criticism from contemporaries such as Johannes Kepler and Marin Mersenne. (National Library of Medicine)

from Moses—to Plato and Pythagoras, incorporating Hermeticism and the Kabbalah. Among more recent thinkers, Fludd was influenced by William Gilbert and Giambattista della Porta.

The *History of Both Cosmoses* attracted the attention of Johannes Kepler, who denounced Fludd's work in an appendix to his *Harmonies of the World* (1619). Fludd replied to Kepler in his *Theater of Truth* (1621). The Kepler-Fludd debate illuminates the different developments of occult philosophy. Kepler, sympathetic to the idea of cosmic harmony, objected to Fludd's work because of its mysticism, its obscurity, and its attempt to describe the universe in terms of pictures rather than precise mathematical relations. Fludd believed Kepler's mathematics to be vulgar and trivial in the tradition of Euclid (fl. c. 300 B.C.), concerned with the outward appearance rather than the true inward essence of things.

This controversy was followed by another with the French mechanical philosopher Marin Mersenne, whose *Questions on Genesis* (1623) bitterly attacked Fludd and other magicians and alchemists as pantheists and *hereticomagi*. Fludd, who never feared controversy, responded by the end of the decade, defending his views of the relation of microcosm and macrocosm and the magical tradition generally. The conflict then drew in Mersenne's friend Pierre Gassendi, who, in one of the first printed discussions of Harvey's theory of the circulation of the blood, attacked Harvey's theory while Fludd defended it. Fludd did not fully accept Harvey's theory—he thought arterial and venous blood to be separate substances—and he approached the question from the viewpoint of his microcosm-macrocosm analogy rather than from anatomy: As the Sun circulates the spirit of life in the universe, so does the heart circulate the blood in the body. In England, Fludd was also involved in the controversy over the weapon salve, whose efficacy he defended. Fludd's controversies, which attracted international interest, helped create the division between mechanical science and magic.

See also Kepler, Johannes; Magic; Mersenne, Marin; Music; Weapon Salve.
Reference
Debus, Allen G. *The Chemical Philosophy: Paracelsian Science and Medicine in the Sixteenth and Seventeenth Centuries.* 2 vols. New York: Science History Publications, 1977.

Fontenelle, Bernard Le Bouvier de (1657–1757)

Bernard Fontenelle, the most influential popularizer of the scientific revolution, was at the heart of French scientific life for several decades. From a family of noble magistrates based in Rouen, Fontenelle arrived in Paris in the late 1680s, determined to make his way in the intellectual and literary circles of the capital. Welcome in several leading Paris salons, Fontenelle made his mark in 1686 with the publication of *Conversations on the Plurality of Worlds,* a series of imaginary dialogues between a male scientist and a young noblewoman in which the scientist expounds a Copernican and Cartesian theory of the world. This work, written with literary charm and grace, also covers a number of topics outside natural philosophy, as the two characters make an imaginary tour of the world by watching it turn from an immobile point in space. *Conversations* was frequently translated and reprinted well into the eighteenth century, and Fontenelle, who himself did no original scientific work, became a leading figure of Parisian science and intellectual life.

Accepted into the Royal Academy of Sciences in 1691, Fontenelle served as its secretary from 1697 to 1740, combining this position with that of secretary to the French Academy. He played an important role in the 1699 reorganization of the Royal Academy, and he was also elected a member of the Royal Society in 1733 and the Berlin Academy of Sciences in 1749. As secretary of the Royal Academy, Fontenelle originated and was responsible for bringing out its annual *History* and its collections of *Memoirs,* which he wrote and presented in a light, accessible style, describing various experiments and observations and relating them to the Cartesian system of the world. He died a few weeks before his 100th birthday.

See also Popularization of Science; Royal Academy of Sciences.
Reference
Sutton, Geoffrey V. *Science for a Polite Society: Gender, Culture, and the Demonstration of Enlightenment.* Boulder, CO: Westview Press, 1995.

Force

Like many physical terms, "force" took on increasing precision and clarity during the scientific revolution. It was used with a variety of meanings in the early seventeenth century, often by the same scientist. The Aristotelian tradition had treated force (in Latin, *vis*) as that which constrains bodies to move in violent motions against their natural inclination. Thus, when a rock is thrown, force is what causes the rock to move upward, and natural motion causes it to fall. Quantitative analysis of force was dominated by the mathematical theory of the lever, in which the problem was stated in terms of the amount of force needed to move an object of a given weight (the distinction between weight and mass emerged only at the end of the seventeenth century).

The Aristotelian distinction between motions continued in the work of Galileo Galilei, who preserved the distinction between natural and unnatural motion, identifying natural motion as the tendency of bodies to fall or to move uniformly in a circle, as do objects on the surface of the Earth or in the heavens. For Galileo, force remained that which causes bodies to move in a violent way. Thus his thinking about force remained dominated by the principle of the lever—that which causes a falling body to accelerate is not force in the strict sense, since falling is a natural motion.

Descartes abandoned the distinction between natural and unnatural motion and categorized force (a word he, like Galileo, used with a variety of different meanings) as that which causes change in a body's motion, whether that change be acceleration or deceleration, the most radical form of the latter being stopping a moving body. Since all motion in the world, with the exception of motion caused by the human will, is caused by bodies acting on each other in a mechanical way, the force of a body is its ability to affect other bodies, not the ability of a body to move itself. Descartes defined the force of a body as the quantity of its motion. "Force" as something independent that causes a body's motion would be an occult quality not reconcilable with Descartes's strict mechanical philosophy, and the Cartesian Chris-

tiaan Huygens came close to eliminating it entirely from his mechanics.

The new dynamical philosophies of Isaac Newton and Gottfried Wilhelm Leibniz involved new and much more precise and rigorous definitions and uses of the concept of force. In a 1686 article in the German journal *Acta Eruditorum* entitled "A Short Demonstration of a Memorable Error of Descartes," Leibniz disproved the identification of force with quantity of motion, substituting the formula force = mass × velocity with the formula force = mass × velocity^2. This is closer to the modern concept of kinetic energy than the modern concept of force. Leibniz distinguished this "living force" *(vis viva)* of a body in motion from the dead force of static situations, such as the force the weight of an unmoving body exerts on its support.

The concept of force that prevailed in mechanics, however, was not Leibniz's living force but Newton's, which was actually more similar to Leibniz's dead force. Newton's force, as stated in his second law of motion, was something outside a body that acts on it to produce changes in its motion. These changes are proportional to the force. Given knowledge of the mass of the body and its changes in velocity, the quantity of force can be expressed with mathematical precision, eventually expressed in the famous formula f = ma, or force = mass × acceleration. Thus "forces" include phenomena such as magnetism and gravity, as well as the physical interaction of bodies.

See also Gravity; Newton, Isaac; Physics.
Reference
Westfall, Richard S. *Force in Newton's Physics: The Science of Dynamics in the Seventeenth Century.* London: Macdonald, 1971.

Fossils

The nature and origin of fossils were hotly debated during the scientific revolution. The modern concept of a fossil—a piece of stone resembling part of a living thing—did not emerge all at once, and originally the

This turtle fossil from Turkey was a gift to Sir Hans Sloane, the greatest collector of the scientific revolution. (Courtesy of the Trustees of the Natural History Museum)

category of fossil included crystals and other unusually shaped stones. Georgius Agricola's *On the Nature of Fossils* (1546), for example, included all kinds of unusual stones. All of these objects were avidly collected as curiosities, or "jokes" of nature, and a fossil could be considered the same kind of object as a potato shaped like a human head.

What we now consider to be fossils began to emerge as a separate category in the seventeenth century. The major split in the study of these fossils was between those like Athanasius Kircher, who believed that fossils are formed from within the Earth by a "plastic virtue" that shapes stone into forms resembling those of living things, and those like Nicolaus Steno and Robert Hooke, who believed that fossils are the remains of actual living things. In this latter view, the living things do not have to be ancient; some argued that the eggs or seeds of animals and plants could be caught in stone and develop into a stone version of the original creature. This did not explain fossils of marine crea-

tures found on mountaintops many miles from the ocean or deep within solid rock, however.

The idea that fossils are in some way remnants of living things seemed easier to reconcile with mechanical philosophy than rival theories suggesting that they are shaped by the Earth itself or the influence of the stars. However, the former view introduced a number of problems related to biblical chronology. It was difficult to place all fossils into the short history of the Earth that was upheld by early modern biblical interpreters, such as Archbishop James Ussher (1581–1656), who granted only a few thousand years between Creation and the seventeenth century. Some fossils could be explained as remnants of Noah's Flood, but it was difficult to explain fossils of unknown creatures because early modern science lacked the concept of extinction. The problem of the nature of fossils was not solved during the scientific revolution, and late-seventeenth-century scholars such as the English physician and fellow of the Royal

Society Martin Lister (1639–1712) continued to uphold the theory that fossils are
formed by the Earth itself.

See also Geology.

Reference
Oldroyd, D. R. *Thinking about the Earth: A History
of Ideas in Geology.* Cambridge: Harvard
University Press, 1996.

G

Galenism

The dominant body of medical knowledge and medical philosophy in medieval and Renaissance Europe was identified with the ancient Greek physician Galen (A.D. 129–c. 199). Much of this tradition was not original to Galen but was characteristic of ancient physicians generally. Galen, however, was both an original medical thinker and the greatest system builder among ancient physicians, and the voluminous Galenic corpus, altered by generations of ancient, Arabic, and medieval commentators and systematizers, was the basis of the Renaissance medical school curriculum. Galen's belief that the true physician is also a knowledgeable philosopher was congenial to the European tradition of learned medicine, with its scorn for empirical practitioners who lacked theory.

Galenic anatomy was limited by the fact that the culture of the Roman Empire, where Galen worked, forbade human dissection. Galen dissected animals and extrapolated from them to humans. Galenic anatomy treated the heart, brain, and liver as the major organs of the human body, each holding a portion of the vital spirits. In this, it conflicted with the Aristotelian emphasis on the heart, and Aristotelian biology and Galenic medicine were to remain the chief rivals in the analysis of living things well into the sixteenth century.

Galen was influenced by Aristotle, however, and he analyzed the human body in terms of Aristotle's four elements: earth, air, fire, and water. Of these, only air is taken into the body directly in Galen's system. The other elements—earth, fire, and water—are represented by bodily fluids, or "humors"—fire by yellow bile, earth by black bile, and water by phlegm. The fourth humor is blood. The theory of the four humors did not originate with Galen; it can be traced back to Hippocrates (c. 460–377 B.C.). What Galen did was to systematically link the humors with the Aristotelian elements.

Medieval physicians went on to correlate each humor with a personality type that was believed to appear with the domination of that humor: phlegmatic personality for those dominated by phlegm, sanguine for those dominated by blood, melancholic for those dominated by black bile, and choleric for those dominated by yellow bile, also known as choler. The imbalance of humors was thought to cause disease. This scheme was widely known in the early modern period and could be exploited to comic effect, as in Ben Jonson's (1572–1637) plays *Every Man Out of His Humour* (1600) and *Every Man In His Humour* (1601).

Galen's pharmacology was based on the four qualities of heat, cold, dryness, and moisture. Individual drugs were thought to promote one or two of these qualities, curing diseases characterized by the opposite qualities. Thus a hot disease like a fever would be cured by drugs that had the quality of coldness. Alternatively, a fever could be attacked by letting blood, a warm substance, out of the body, thus cooling it. Galen and Galenists were zealous advocates of bloodletting.

Scholars of the Renaissance humanist movement wanted to read classical science in the original Greek, unencumbered with subsequent Arab or Latin commentators. The works of Galen were no exception. The publication of a massive complete edition of Galen's Greek writings in 1525 seemed to add luster to his reputation, and hundreds of editions and translations of Galen's treatises appeared in the sixteenth century. But Galen's authority was also increasingly challenged during the scientific revolution. The first challenge was that of Paracelsus, who had a particular dislike of Galen as a corrupt pagan. The subsequent development of Paracelsian "chemical medicine" was implicitly or explicitly anti-Galenist. In the long run, though, a more important attack came from a Galenist, Andreas Vesalius. Steeped in Galenic learning, Vesalius actually applied it to the dissection of human beings, discovering that many of Galen's extrapolations were invalid. His findings were published in *Of the Fabric of the Human Body* (1543).

The innovations of subsequent anatomists drew anatomy further and further from Galen, culminating in the Aristotelian William Harvey's discovery of the circulation of the blood, published as *On the Motion of the Heart* in 1628. Harvey's discovery discredited Galen's theory that the blood passes between the ventricles through tiny pores in the septum. Diehard Galenists attempted to defend Galenic anatomy—one even claiming that the difference between Galenic and Vesalian anatomy could be explained by the degeneration of modern bodies from the perfect bodies of the ancients Galen described. But Galen's work was increasingly obsolete. The

rising importance in medicine of surgeons and apothecaries, who did not have the investment in Galen that learned physicians did, also contributed to his decline. Increased use of quantitative and mechanical approaches to medicine also meant that Galen's analysis of heat and cold as qualities in themselves was no longer acceptable. Galenism went out of fashion, and by the late seventeenth century, Galen's works were no longer central to the medical school curriculum. The empirically oriented Hippocrates replaced Galen as the most admired ancient physician.

See also Medicine.
References
Porter, Roy. *The Greatest Benefit to Mankind: A Medical History of Humanity.* New York: W. W. Norton, 1998.
Temkin, Owsei. *Galenism: Rise and Decline of a Medical Philosophy.* Ithaca and London: Cornell University Press, 1973.

Galilei, Galileo (1564–1642)

Galileo Galilei, one of the greatest minds of the scientific revolution, also lived one of the most dramatic lives. The eldest child of a distinguished Florentine family of minor nobility, Galileo was exposed to science in his own family, as his father, Vincenzo Galilei (c. 1520–1591), performed significant experiments in musical science and wrote treatises on it. After entering the University of Pisa in 1581 as a medical student, Galileo discovered mathematics and promptly became enraptured. The ancient Greek mathematician Archimedes (c. 287–212 B.C.) became his intellectual hero. Galileo supplemented classes in natural philosophy at Pisa with private mathematical study in Florence. He left Pisa without a degree in 1585 and became a mathematics tutor in Florence, where he established the isochronal nature of the pendulum—the fact that the frequency of a pendulum is a constant. In 1589, his Archimedes-inspired work won him the mathematics chair at Pisa.

In 1592, Galileo became professor of mathematics at the University of Padua,

Europe's leading scientific university. He later described the Padua years as the best of his life. He established a friendship with the Aristotelian professor of natural philosophy Cesare Cremonini (1550–1631) and made connections with many leading citizens of the Republic of Venice, in whose territory Padua was situated. He also visited the famous Arsenal of Venice, the largest workshop in Europe. However, a Padua professor's salary was inadequate to support Galileo and his siblings, for whom he was responsible as the head of the family. Galileo was also frustrated by the low rank of mathematics in the intellectual hierarchy of the university.

Whatever the personal and financial stresses of the Padua years, they did not prevent this from being the most intellectually fruitful time of Galileo's life. He moved from a highly mathematical approach to knowledge to a greater interest in experiment. (The story of his dropping balls from the Leaning Tower of Pisa, however, is only a legend.) Although his attitude to Aristotelian logic was fairly positive, Galileo was a convinced and vociferous opponent of Aristotelian physics. At Padua, he began to elaborate a non-Aristotelian approach to the problems of moving bodies. His most famous result was the law of falling bodies: the distance covered by a falling body varies with the square of the time of the fall. Galileo introduced into mechanics the idea of uniform acceleration and made the first steps toward formulating the principle of inertia. He had a workshop attached to his house and developed a horse-powered pump and a geometrical instrument known as the "military compass." Although astronomy was not a major interest for him at this time, Galileo was clearly a discreet Copernican. He began corresponding with Johannes Kepler in 1597. The two would never meet and their intellectual and personal styles were vastly different, but as the foremost champions of Copernicanism, they would be linked as allies.

What catapulted this obscure, if well-respected, university professor to Italian and European fame was his work with the telescope. Galileo did not claim to have invented the telescope; rather, he first heard of it as a device invented in the Netherlands. He designed his own from the information he was able to gather, and the resulting telescope was superior to the contemporary Dutch telescopes. With his telescope Galileo observed the moons of Jupiter—the first known satellites of a planet other than the Earth—as well as the mountains of the Moon, the phases of Venus, and the composition of the Milky Way out of innumerable stars. These epochal discoveries were announced in *Nuncius Sidereus* (The Starry Messenger) in 1610.

Galileo offered his telescope to the city of Venice, but he was thinking of moving to the court of the grand duke of Tuscany in Florence. Not only was Tuscany Galileo's native land, but the patronage of a prince was also more prestigious than that of a republic. The Tuscan court also offered the opportunity to concentrate on scientific work rather than teaching. Galileo gave Jupiter's moons the name "the Medicean stars" after the ruling dynasty of Tuscany—a brilliant stroke to win the duke's favor that secured Galileo's appointment as court mathematician. He insisted that he be given the title not merely of "mathematician," but of "philosopher" as well. Since the actual physical nature of the universe was the province of natural philosophers, Galileo as a philosopher could make claims about the nature of the universe that he could not make as a mathematician.

Galileo's move to the Tuscan court angered the Venetians, who wanted to keep him at Padua. However, it aided Galileo in addressing a wide audience outside the universities, particularly since he published in Italian rather than Latin. Galileo proved himself a skillful courtier and a merciless debater who made many enemies. The court was an arena where he could demonstrate his expertise in natural philosophy. For instance, his anti-Aristotelian *Discourse on Floating Bodies* (1612) originated in a conversation at the duke's table. The sphere in which he now operated was Roman as well as Florentine, and in 1611, Galileo was admitted to the Accademia dei Lincei, a Roman scientific society under the patronage of Federico Cesi (1585–1630).

It was from Rome that Galileo faced what would prove to be the greatest challenge of his career, that of the Catholic Church's condemnation of Copernicanism. His pro-Copernican *Letters on Sunspots* (1613), arguing that sunspots are indeed spots on the Sun, led to a bitter dispute with the Jesuit astronomer Christoph Scheiner (1573–1650). Although Scheiner later changed his opinion, at the time he believed sunspots to be celestial bodies between the Sun and the Earth. Galileo's relations with the Jesuits had up to this time been not particularly hostile, but the order, and Scheiner in particular, became his bitter enemies. On a general level, church authorities took an increasingly repressive attitude toward Copernicanism and toward Galileo as its principal Catholic champion, putting Copernicus's *On the Revolution of the Celestial Spheres* (1543) on the Index of Forbidden Books in 1616. Unlike Kepler and Newton, among others, Galileo always pursued science as something separate from religion. He argued that Copernicanism was a purely philosophical concept having no bearing on theology, but he was defeated.

Galileo was a powerful and well-connected man, though, and his works were not specifically condemned. Instead, on February 26, 1616, Cardinal Robert Bellarmine (1542–1621) privately admonished Galileo not to teach Copernicanism. Although Galileo obeyed the admonition, later in the same year he got into a bitter dispute over comets with another Jesuit, Orazio Grassi (1583–1654). Galileo, ironically, explained comets in basically Aristotelian terms as atmospheric phenomena in his later work titled *Assayer* (1623), but despite this incorrect theory *Assayer* remains a great work of scientific argument.

Despite his enormous importance in the history of astronomy, Galileo was in many ways not an astronomer at all. He was not concerned with the precise observations and elaborate calculations required to predict the courses of the stars—activities that were central to the discipline of astronomy as then conceived. The most significant work he wrote on astronomy after *The Starry Messenger* was *Dialogue on the Two Chief Systems of the World* (1632). Galileo's identification of the "two chief systems" as the Copernican system and the Ptolemaic Earth-centered system was disingenuous, as Ptolemaicism was increasingly obsolete, by now supplanted by Tycho Brahe's system, with a Sun rotating around the Earth and the other planets around the Sun. Galileo despised Tycho's system as a feeble compromise, and the fact that he deemed it unworthy even of mention was the ultimate condemnation.

The argument in the *Dialogue,* written in sparkling Italian, was between two fictional characters, the Aristotelian-Ptolemaic Simplicio and the Copernican Salviati, the latter representing Galileo himself. A third character, Sagredo, acts as the balance between these two, eventually siding with Salviati. In this dialogue, Salviati demonstrates the falsity of objections to the idea of the Earth's motions, which he uses to explain the tides. Galileo was very proud of this argument, regarding it as superior to the idea that the tides are caused by an attraction exerted by the Moon. Salviati got the better of the argument, but the *Dialogue* led to Galileo's trial and his condemnation to house arrest at his villa outside Florence for the rest of his life.

The last years of Galileo's life were filled with personal tragedy—his beloved illegitimate daughter, the Franciscan nun Sister Maria Celeste, died in 1634—and, with ironies he probably did not relish, the courtier was now under house arrest, and the man who had seen farther than any before him had lost his sight. He did acquire two new scientific disciples, the physicist Evangelista Torricelli (1608–1647) and Vincenzo Viviani (1622–1703), his earliest biographer. He continued his work on the pendulum, designing a workable pendulum clock.

Galileo's *Discourse on Two New Sciences* (1638) was the summation of his lifelong work on physics. Most of the work was in the form of a conversation between the same three characters as in the *Dialogue*. One of the "new sciences" they discuss is that of the strength of materials. The second is dynamics, the study of matter in motion, drawing on the work Galileo had done at Padua. In a crowning irony, this last great masterpiece of Galileo, the glory of Italian science, could not be published in Italy and had to be smuggled to Leiden in the Dutch Republic.

> See also Accademia dei Lincei; Astronomy;
> Courts; Mechanics; Physics; Telescopes; Trial of
> Galileo.
>
> *References*
> Biagioli, Mario. *Galileo, Courtier: The Practice of*
> *Science in the Culture of Absolutism.* Chicago:
> University of Chicago Press, 1993.
> Geymonat, Ludovico. *Galileo Galilei: A Biography*
> *and Inquiry into His Philosophy of Science.*
> Translated from the Italian with additional
> notes and appendix by Stillman Drake. New
> York: McGraw-Hill, 1965.
> Shea, William R. *Galileo's Intellectual Revolution.*
> London: Macmillan, 1972.

Gassendi, Pierre (1592–1655)

The great rival of Descartes, Pierre Gassendi, presented Epicureanism in terms of seventeenth-century mechanical philosophy, reviving the ancient doctrine of atomism and asserting the existence of a void. Influenced by humanists and skeptics, he conceived a dislike for Aristotle while he was a student at the College of Aix-en-Provence from 1604 to 1611. Gassendi was ordained a priest in 1616, after receiving a doctorate from the papal university at Avignon. He received a chair of philosophy at Aix-en-Provence in 1616, a position which, ironically, required him to teach Aristotelianism. When the Jesuits took over the college in 1622, Gassendi lost his academic position. He divided the rest of his life between Paris, where he held a brief appointment at the College Royale from 1645 to 1646, and the Provence home of Nicolas-Claude Fabri de

Peiresc, his good friend and patron, and Digne, where he was a canon of the cathedral of the town. Gassendi corresponded with Marin Mersenne and Galileo Galilei.

Gassendi's natural philosophy combined astronomical observations with humanist allegiance to ancient texts. His most notable astronomical achievement was the only accurate observation of the transit of Mercury, in 1631, which he believed confirmed Copernicanism. Combining the roles of Catholic priest and a zealous Copernican, Gassendi was shocked by the condemnation of Galileo and afterwards publicly advocated the Tychonic system. His central humanist intellectual project, beginning in 1624, was the creation of a Christian Epicureanism to supplant Aristotelianism. This was a difficult challenge, since Epicurus was identified with materialism, atheism, and immorality. Gassendi emphasized God's power to create the universe any way He pleased, but despite his claims that Epicureanism was compatible with Christianity, the church viewed his philosophy with grave suspicion.

Gassendi also asserted that the empirical examination of sense-data and the construction of probable explanations were superior to Aristotelian dogmatism and claims to know the inner essences of things. He opposed magical and Platonic philosophies, as in his controversy with the English magician Robert Fludd, and he also criticized Cartesianism, contributing skeptical objections to Descartes's *Meditations* (1641). In both controversies, Gassendi was involved at the behest of Mersenne. Gassendi was one of the earliest scientific biographers, writing lives of Peiresc and Tycho Brahe and shorter studies of Nicolaus Copernicus and the fifteenth-century astronomers Georg von Peurbach (1423–1461) and Johann Müller, known as Regiomontanus (1436–1476).

Although much of Gassendi's work was not published in his lifetime, his complete works, including many of his surviving manuscripts, were printed in 1658, soon after his death. Gassendism, mostly expressed in difficult Latin rather than Descartes's elegant French, would be eclipsed in France by Cartesianism in the second half of the seventeenth century, but it had a great influence in England, particularly after the publication of Walter Charleton's *Physiologia Epicuro-Gassendo-Charletoniana* in 1654.

See also Atomism; Epicureanism; Peiresc,
    Nicolas-Claude Fabri de.
References
Brundell, Barry. *Pierre Gassendi: From Aristotelianism
    to a New Natural Philosophy.* Dordrecht, the
    Netherlands: Reidel, 1987.
Lennon, Thomas M. *The Battle of the Gods and
    Giants: The Legacies of Descartes and Gassendi,
    1655–1715.* Princeton: Princeton University
    Press, 1993.

Gender
See Sexual Difference; Women.

Geography

During the early modern period, geography was transformed both by the humanistic revival of the classical heritage and by the flood of new geographic information introduced to Europe as a result of the voyages of exploration and conquest. European governments also desired more accurate knowledge of their own territories, in Europe and abroad.

The humanist revival focused on the geographical classics of the ancient Greeks, notably the treatises of Claudius Ptolemy (A.D. 90–168) and Strabo (c. 63–c. 25 B.C.). These treatises, of which Strabo's was more inclined toward historical and social description and Ptolemy's more inclined toward mathematics, were translated into Latin early in the fifteenth century. They provided a more sophisticated way of thinking about spatial locations on Earth, including Ptolemy's use of a system of coordinates of latitude and longitude. Humanistic study of ancient texts also raised geographical questions about the location of ancient places. The Bible also remained a source for geographic knowledge and questions.

In the late fifteenth and sixteenth century, Europeans vastly expanded their geographic knowledge in voyages that circumnavigated Africa, encountered and explored the islands of the Caribbean and the coasts of the Americas, and braved the broad Pacific and freezing Arctic Oceans. The voyages of Christopher Columbus (1451–1506), Amerigo Vespucci (c. 1454–1512), Vasco da Gama (c. 1460–1524), Ferdinand Magellan (c. 1480–1521), and many others certainly had an effect on geography, but it was not to convince Europeans that the world is round. All educated Europeans thought of the world as roughly spherical, in accordance with Aristotelian natural philosophy. What the new information did was to present irrefutable evidence that the geographical knowledge of the ancients was incomplete, and in some cases, such as Aristotle's assertion that the tropics would be uninhabitable due to their extreme heat, simply wrong. (Indeed, the voyages of Columbus, who had some knowledge of humanist geography, were premised on Ptolemy's severe underestimation of the size of the Earth, which led Columbus to believe that he could get to the East much more quickly than he actually could.)

Ancient and medieval travelers' fantastic accounts of such phenomena as the people with dogs' heads supposedly living in central Asia also diminished in credibility with the reception of this new knowledge. The ancients' ignorance of the Americas became a standard trope of those who asserted the superiority of modern to ancient knowledge, most memorably expressed in the frontispiece to Francis Bacon's *The Great Instauration* (1620); it shows Columbus sailing beyond the "Pillars of Hercules," the western limit of the Mediterranean and a symbol of the limits of ancient knowledge, into the Atlantic—precisely analogous, in Bacon's view, to his own project of intellectual renewal.

Early modern history is full of the accounts of travelers and explorers. In the English tradition, these are most notably collected in the work of Richard Hakluyt (c.

1552–1616), *The Principall Navigations* (1589). Certainly the global geographic awareness of early modern Europeans was increasing rapidly; the problem with this flood of information was that it was difficult to distinguish good information from bad. In addition to the repetition of ancient and medieval myths, there were problems caused by the cultural narrowness of many early modern European travelers, who evaluated other civilizations and regions only in terms of their likeness or unlikeness to Europe.

Whether good, bad, or indifferent, all this new information clearly demanded new geographic treatises and textbooks rather than merely improved editions of Ptolemy and Strabo. The German Sebastian Münster (1489–1552) produced a number of geographical works, culminating in *Cosmographia* (1544), which attempted to present worldwide geographical knowledge in text and map form and in the German language. Indeed, geography was one science that had an extensive reading public in the vernacular languages. Another vernacular treatise was that of the Englishman Nathanael Carpenter (1589–1628). His *Geographie Delineated Forth in Two Bookes* (1625) took a more formal and mathematical approach than Münster's work.

There were also Latin treatises, often intended as university textbooks. In 1611, the German Bartholomew Keckermann (c. 1571–c. 1609) published his *System of Geography,* which followed Ptolemy in its mathematical emphasis. Keckermann's widely circulated book influenced the less humanist and more theoretical and explicitly Cartesian work of the German physician Bernhard Varen, known by the latinized name Bernhardus Varenius (1622–1650). His *Geographia Generalis* (1650) was one of the few seventeenth-century geographical treatises to discuss Copernicanism, which Varenius viewed as at least equally possible as competing systems. *Geographia Generalis* was frequently reprinted. Isaac Newton brought out an expanded edition in 1672 for use by his Cambridge students; with Newtonian

explanations and arguments substituted for Cartesian ones, the book would continue to be reissued into the eighteenth century.

The division between the Strabonian tradition of description and the Ptolemaic mathematical tradition was not resolved in early modern thought. The disciples of Ptolemy tended to place geography with astronomy as a mathematical science, while the successors of Strabo tended to place geography with history as a human science. There was also a distinction between "universal" and "special" geography—universal geography dealing with the globe as a whole and special geography dealing with the physical and human characteristics of particular locations. Special geography, sometimes described as "cosmography," was practiced in untheoretical compendia—encyclopedic works such as Peter Heylyn's (1600–1662) *Cosmography in Four Books* (1652), which was frequently reprinted into the eighteenth century. These works often presented geography principally as an aid to the understanding of history.

Chorography, or the study of the particular features of specific areas of the Earth, was also practiced in a series of locally and regionally based descriptive treatises, such as Robert Plot's (1640–1696) books on the natural histories of Oxfordshire and Staffordshire. A more politicized variant inventoried the resources of a particular area for the central government, as in the survey of Ireland undertaken by the English regime of Oliver Cromwell (r. 1653–1658) in the 1650s. Similarly, the Royal Academy of Sciences undertook a mapping and geographic definition of France.

What all of these genres had in common was empiricism. Universal and theoretical geography was declining in the late seventeenth century. There were some efforts to handle information regarding the Earth in a quantitative way, most notably in Edmond Halley's work on wind, tides, and magnetic declination, but these findings were not incorporated into the geographical literature. By the early eighteenth century, geography was dominated by cheap surveys and reprints of seventeenth-century books.

See also Cartography; Exploration, Discovery, and Colonization.
References
Bowen, Margarita. *Empiricism and Geographical Thought: From Francis Bacon to Alexander von Humboldt.* Cambridge and New York: Cambridge University Press, 1981.
Livingstone, David N. *The Geographical Tradition: Episodes in the History of a Contested Enterprise.* Oxford: Blackwell, 1993.

Geology

The Earth and its minerals were the subject of several different disciplines in the early modern period. Something resembling the modern science of geology emerged only toward the end of the scientific revolution, and the word itself did not come into common use until the late eighteenth century.

The overall framework of knowledge about the Earth during the Renaissance was that of Aristotelian natural philosophy, in which the Earth is a spherical body, composed of the heavy elements earth and water and existing at the center of the universe. ("Earth" has two meanings here: one the world, the terrestrial globe, and the other one of the four elements.) Volcanoes and earthquakes, topics of great interest, were viewed as explosions of subterranean gases, in effect as gigantic belches.

One Aristotelian claim, however—that the Earth had existed from eternity—had to be rejected as incompatible with Christianity. For Christians, the history of the Earth was that recounted in Genesis, including the seven-day Creation, the Fall, and the Flood of Noah. Chronological study of the Bible in the early modern period led to the conclusion that the Earth was about 6,000 years old. (Medieval thinkers, less literal in their biblical interpretation, had been willing to consider the possibility that the Earth had existed for millions of years.) Many believed that the Earth was decaying and growing old and senile, in preparation for the Apocalypse.

Other bodies of knowledge dealing with the Earth and its minerals included the science of mining, which often viewed the Earth as a fruitful womb in which minerals were born. Minerals, particularly odd or striking items like jewels, gems, and fossils, were also collected and classified by natural historians. Jewels and gems were widely believed to have occult and healing powers, and fossils were the object of great curiosity and several conflicting theories.

The displacement of the Earth from the center of the universe by Copernican astronomy destroyed the Aristotelian physical picture of the Earth as composed of earth and water at rest. The idea of a central fire in the Earth gained popularity in the seventeenth century, particularly as it could be identified with the Christian hell. The most notable geologist of the mid-seventeenth century was Athanasius Kircher, who actually had himself lowered into the crater of Vesuvius. Although Kircher was Aristotelian and anti-Copernican, his *The Subterranean World* (1665) presented a picture of the world with a central fire, for which volcanoes on the surface are outlets. He also brought the phenomenon of the Earth's increasing heat at lower levels, known to miners, to the attention of natural philosophers. Kircher thought of the Earth as analogous to a living body, of which stones are the bones. In this view, the circulation of fires and fluids within the Earth keep it healthy.

All this was anathema to the mechanical philosophy of Kircher's contemporary René Descartes, who in his *Principles of Philosophy* (1644) theorized that the Earth and other planets had originally been suns. The buildup of sunspots, which Descartes believed to be solid bodies, had cooled these suns, dimmed their light, and given them a solid crust, although a fiery center remained. Descartes attempted to explain various features of the Earth, such as mountains, through the mechanical movements of corpuscles, of which he thought the Earth was composed. Like many early modern people, Descartes

*Athanasius Kircher believed the Earth is filled with fire that makes its way out through volcanoes, as depicted in this
illustration from his* The Subterranean World *(1665).*

explained mountains not by an upthrust of
the peak, but by a large-scale collapse that
formed the slopes. Of course, he also denied
the occult and healing properties of jewels,
which Kircher supported but which could
not be explained mechanically. (Skepticism
on this matter was not universal. Robert
Boyle continued to believe in the medical
virtues of some jewels and crystals.)
Mechanical philosophers also denied the possibility that minerals are replenished in the
fruitful womb of the Earth—the kind of
metaphor they despised.

The formative stage for the discipline of
geology was the late seventeenth century,
beginning with the work of Nicolaus Steno.
Steno's study of Tuscany, *Prodromus* (1669),
was the first study of a specific area's geological stratification. Steno formulated the prin-

ciple that the layers of rock in an area, or
"strata," serve to recount its changes over
time, with the lower levels representing a
more distant past. In his view, strata that contain marine fossils must at one time have been
seafloors. Steno, whose natural philosophy
was basically Cartesian, concentrated on
Tuscany, where he was then resident, rather
than producing a grand theory of the Earth's
development like Descartes's. *Prodromus* was
translated into English soon after its publication, and Steno's idea of strata would have
immense influence on subsequent geologists.
The discussion of crystals in *Prodromus* was
also influential. Steno contended that they do
not grow from within, like living things, but
by accreting matter from outside.

Assertions about the history of the Earth
had to fit into the framework provided by the

Book of Genesis. The short lifespan that early modern Christians ascribed to the Earth biased them in favor of geological explanations based on sudden and dramatic catastrophes, rather than the slow unfolding of processes like erosion. Robert Hooke proposed a terrestrial history based on earthquakes caused by the wanderings of the Earth relative to the poles. In his *Posthumous Works* (1705), Hooke invoked the ancient pagan writers Plato and Ovid to support his thesis, but the theory was rejected, partially because it lacked a biblical basis. In *Sacred Theory of the Earth* (1681 and 1689), the English clergyman Thomas Burnet (c. 1635–1715) described in Cartesian terms the events in Earth history recounted in Genesis. Burnet gave major credit to Noah's Flood in explaining geological changes such as the emergence of mountains.

The furor over Burnet's work, which some claimed subordinated the Bible to man-made natural philosophy, contributed to the publication of several more books on geological history in late-seventeenth-century England. John Ray's *Three Physico-Theological Discourses* (1693) emphasized the benevolence of God and the usefulness of mountains, which the rather pessimistic Burnet had disliked. William Whiston (1667–1752) published *A New Theory of the Earth* (1696), applying Newtonianism rather than Burnet's Cartesianism to the scriptural account and suggesting that the Earth had originated as a fragment of the Sun knocked off by a comet. On a more practical level, in a paper presented to the Royal Society in 1683 the physician and natural historian Martin Lister (1639–1712) proposed the creation of geological maps of the strata in different localities. Another physician, John Woodward (c. 1665–1728), who published *Essay Toward a Natural History of the Earth* (1695), carried on investigations of pits dug in England for mines or wells and distributed lists of geological questions to natural philosophers in other countries, claiming that the division of Earth into strata was universal.

See also Fossils; Hooke, Robert; Steno, Nicolaus.
References
Gohau, Gabriel. *A History of Geology.* Revised and translated by Albert V. Carozzi and Marguerite Carozzi. New Brunswick, NJ: Rutgers University Press, 1990.
Oldroyd, D. R. *Thinking about the Earth: A History of Ideas in Geology.* Cambridge: Harvard University Press, 1996.

Geometry
See Cycloid; Mathematics.

Gilbert, William (1544–1603)
William Gilbert, English physician and natural philosopher, wrote an influential tract on the magnet that was one of the first works of systematic experimental investigation connected to scientific theory. Born in Colchester, he received a B.A., an M.A., and an M.D. from Cambridge and eventually moved to London, where he was a successful physician to the city's elite. In 1600, he became both president of the London College of Physicians and physician to Queen Elizabeth I (r. 1558–1603).

Gilbert's *On the Magnet and Magnetic Bodies and the Great Magnet, the Earth; a New Physics* (1600) was the culmination of several years work on magnets, building on the work of the English navigator Robert Norman. Gilbert examined the tendency of magnets to point in a direction other than true north—a tendency called "the magnetic variation"—and to vary from the horizontal when suspended by their centers—a tendency called "the dip." Norman and previous investigators of these subjects had been primarily interested in navigational questions, writing in English for an audience of navigators and sailors. Taking their work, Gilbert applied it to natural philosophy, writing in Latin for an international audience of scientists. His work was read and discussed by leading Continental scientists such as Galileo Galilei, Johannes Kepler, and Giambattista della Porta, and later by René Descartes.

As a natural philosopher, Gilbert was influenced by several traditions, including Aristotelianism and Platonism. His explanation for the behavior of the magnet was to identify the Earth itself as a vast magnet—the contribution for which he is best known today. Gilbert believed that the magnetic north is identical to the geographical north, and he theorized that the magnetic variation is caused by irregularities in the Earth's surface. He tried to use this theory to argue for the Earth's rotation, in support of Copernicanism, claiming that the Earth rotates because of its magnetic properties. Gilbert did not analyze magnetism in mechanical terms, but conceived of the Earth as alive and magnetism as a force analogous to that exerted by living bodies. *On the Magnet* also included the first use of the term "electricity," building on the Greek term for amber, whose attractive properties were already known. Gilbert identified other electric substances as well. A second, posthumously published work, *A New Philosophy of the World below the Moon* (1651), was compiled from Gilbert's papers by his brother William Gilbert of Melford. It includes discussions of physics and meteorology, but it was not as influential as *On the Magnet*.

See also Compasses.
Reference
Roller, Duane H. D. *The De Magnete of William Gilbert*. Amsterdam: Hertzberger, 1959.

God

Nearly all early modern scientists, particularly in the seventeenth century, saw what they were doing in relation to God. The general trend of the scientific revolution was away from the Aristotelian and medieval scholastic conception of God as primarily the highest good, an idea tracing back to the ancient Greeks, and toward the more specifically Jewish-Christian idea of God as creator, ruler, and lawgiver of the universe. How God ruled, though, remained subject to debate.

Nicolaus Copernicus argued that astronomy is a path to God, and that his design for the cosmos, more elegant than Ptolemy's, was therefore more worthy of God. Johannes Kepler, like many who were influenced by the magical tradition, saw God as having left marks of design everywhere in the universe, giving it a rational structure based on cosmic harmonies and symbolism. The task of the natural philosopher, then, is to solve the riddles and decipher the clues of God's existence everywhere. Galileo's God was also a creator, but in Galileo's view, the structure of the created universe is more geometrical and less musical and harmonious than in Kepler's. For both Kepler and Galileo, God operates through mathematics, but Kepler's God was more like an artist or musician, and Galileo's more like an architect or engineer.

The mechanical philosophers also believed in a divine creator, but they emphasized God's separation from Creation. René Descartes spoke of God's power and the rationality with which the universe is imbued, while his philosophical rival Pierre Gassendi emphasized the essential arbitrariness of God's will in creating the universe. Descartes did not believe that God's creative power is intrinsically limited by what we call rationality. He made the very radical assertion that God could have created a universe in which, for example, 2 + 1 would not equal 3. God had, however, chosen to be bound by rationality. This argument was radical, and arguably heretical, because it implied that God has the power to annihilate even himself.

This notion was Descartes's version of an old Scholastic distinction between God's "absolute" power to create the universe in whatever fashion he chose, and God's "ordained" power, which was limited once the universe had been created. For Descartes, once God set up rational mechanical laws, the universe ran itself without the need for further divine intervention. The possibility that the mechanical philosophy, and particularly its Cartesian variant, denied God's providential action in the world led some, such as the English Platonist Henry More (1614–1687), to attack it as "atheistic." "Atheism" in the seventeenth-century sense was not a denial of God's existence, but a denial of God's providential care for humanity and the universe.

By the late seventeenth century, the extreme positions on the relation of God to the scientifically knowable cosmos were represented by Blaise Pascal and Baruch Spinoza. Pascal accepted the mechanical philosophy's separation of God from creation, going so far as to deny that science is a way of knowing God at all. In this view, God is not known through study of the laws of nature but rather is manifest in the very miracles that violate those laws. Spinoza, on the other hand, saw God as completely manifest in nature; he frequently referred to God as "God or Nature," and he viewed miracles as nonexistent. He attained an unenviable reputation as an atheist.

The most prominent figure in the broad intermediate range between Spinoza and Pascal was Isaac Newton. Newton's God stood outside the cosmos, functioning as its maintainer. Rather than taking the purely mechanical or "clockwork universe" approach identified with Descartes, in which God simply created the world and left it to run itself, Newton saw God as actively maintaining the universe, both in providing the force of gravity, which is not inherent in material things themselves, and in occasional interventions to prevent the universe from running down, such as limiting the cumulative effect of irregularities in planetary orbits. In this view, space and time themselves do not exist outside of God but rather are aspects of God's Being. Newton's God is also a unity; Newton denied the doctrine of the Trinity in favor of Arianism, the belief that the Son was a created being rather than an aspect of God. He kept this heretical belief very quiet, however.

Newton's great rival, Gottfried Wilhelm Leibniz, viewed Newton's position as an insult to God, in that the need for occasional divine intervention means that the universe is not perfect. Leibniz laid great stress on the perfection of creation, defining perfection as the combination of the maximum diversity of phenomena with the minimum complexity of laws. Since God is perfect, the universe he had created has to be perfect as well. The difference between Newton's theological position and Leibniz's was memorably set forth in the Leibniz-Clarke correspondence.

See also Boyle Lectures; Leibniz-Clarke Correspondence; Natural Theology; Religion and Science.

References

Burtt, E. A. *The Metaphysical Foundations of Modern Physical Science.* 2d ed., rev. Garden City, NY: Doubleday, 1954.

Funkenstein, Amos. *Theology and the Scientific Imagination from the Middle Ages to the Seventeenth Century.* Princeton: Princeton University Press, 1986.

Osler, Margaret J. *Divine Will and the Mechanical Philosophy: Gassendi and Descartes on Contingency and Necessity in the Created World.* Cambridge: Cambridge University Press, 1994.

Gravity

The attraction between physical bodies and the Earth received several explanations in the course of the scientific revolution, culminating in Isaac Newton's formulation of the inverse square law in *Mathematical Principles of Natural Philosophy* (1687). In the Aristotelian physics of medieval and Renaissance Europe, all bodies seek their natural place. In this view, the natural place of heavy bodies is the center of the universe, the Earth to which they fall when dropped. Since this Earth-seeking behavior is inherent to the elements of earth and water, objects fall more quickly if they contain more earth. The other elements, air and fire, seek their natural places by rising. The stars and planets are not subject to these same forces, as their natural place is the heavens, where they already are. Thus, the Copernican displacement of the Earth from the center of the universe was a direct challenge to Aristotelian gravitational theory, which was perhaps the most important academic objection to Copernicanism in the sixteenth century. Nicolaus Copernicus attempted to replace the Aristotelian notion of gravity with a theory that holds that objects tend naturally to the most perfect shape, the sphere. Copernicus's gravity remained a local force drawing objects to each celestial body, rather than a universal force binding the planets and stars to each other.

The most popular non-Aristotelian theories of the forces that hold the universe together in the early seventeenth century were magnetic theories, particularly after the publication of William Gilbert's *On the Magnet* in 1600. Gilbert claimed that the rotation of the Earth can be explained magnetically. Johannes Kepler, influenced by Gilbert, invoked a magnetic force emanating from the Sun to explain the behavior of the planets. Galileo Galilei, who also employed magnetic explanations, attacked the Aristotelian theory of terrestrial gravity, demonstrating that objects fall at the same rate regardless of the amount of earth they contain (although he was far from the first to do so). Galileo continued to maintain the distinction between the terrestrial realm of gravity and the realm of the skies, which he supposed was subject to different physical principles.

For his part, René Descartes did not admit this distinction or the possibility of a nonmechanical magnetic influence. He put forth the most influential gravitational theory before Newton's. Descartes and subsequent Cartesians believed that gravity is caused by the vortices that they saw everywhere in nature. These vortices force objects to their center, as whirlpools force objects to theirs. Although this tradition dominated on the Continent, the lingering influence of Gilbert's magnetical philosophy in England, particularly on Christopher Wren and Robert Hooke, made English natural philosophers more likely to analyze planetary and cometary motion in terms of nonmechanical attractive forces. Hooke, along with the Italian astronomer Giovanni Alfonso Borelli (1608–1679), had argued that the motions of heavenly bodies can be analyzed as a combination of centrifugal force and a constant attraction to a body at the center of the orbit.

The idea that attractive forces between

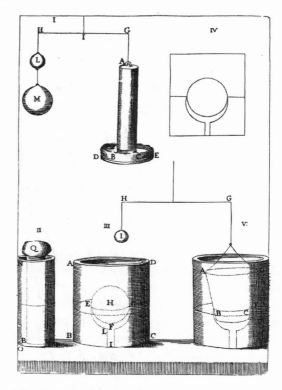

This illustration depicts an experiment by the Accademia del Cimento to prove that levity, the opposite of gravity, is not a positive quality. (Saggi di Naturali Esperienze, 1667)

two bodies operate on the basis of an inverse square of the distance between them was not unique to Newton. Kepler had already demonstrated an inverse-square relationship in considering the diminution of light from a luminous surface. Hooke formulated an inverse-square law of gravitational attraction in a letter to Newton in 1680. The publication of Newton's own work on gravity, which dated from his *annus mirabilis,* or "year of wonder," in 1665–1666, was precipitated by a conversation between Hooke, Wren, and Edmond Halley after a Royal Society meeting on January 14, 1684. The three agreed that a force that diminishes in inverse proportion to the square of the distance would explain planetary orbits, but none of them had the mathematical skills to demonstrate this. Visiting Newton in August, Halley asked what path a planet would take if it was assumed that gravity diminishes as the square of the distance, and Newton replied, "An ellipse." Their conversation re-awoke Newton's interest in the subject, the first fruits being a short article on the movement of bodies Newton sent to London that winter.

Halley's subsequent prodding resulted in the publication of *Mathematical Principles of Natural Philosophy*, which set forth Newton's law of gravitation. Newton stated that between any two bodies in the universe there exists a force directly proportional to the product of the masses of the two bodies and inversely proportional to the square of their distance. Unlike Hooke, who mistakenly accused Newton of having stolen the inverse-square law, Newton demonstrated the theory mathematically, systematically working out its implications for astronomy and physics. He also demonstrated that the gravitational attraction of a large spherical body, such as the Sun or a planet, can be mathematically analyzed as if it were exerted from a point at its center. Newton's gravitation is *universal*; that is, it explains not only the orbits of planets and their satellites but also the fall of dropped objects (he claimed that his work on gravity was inspired by a falling apple) and the tides of the Earth.

Newton did not suggest a cause for gravitation or a mechanism by which it operates. He seems to have conceived of gravity as a nonmaterial force permeating the universe, a view that had affinities with the ideas of alchemists, magnetic philosophers, and ancient Stoics. Like magnetism, Newtonian gravity challenged the mechanical philosophy. One of the prime tenets of the mechanical philosophy, whether atomistic or Cartesian, is that bodies can only affect other bodies if they are touching. The ability of one body to affect another at a distance without intermediary bodies seemed to open the door for the return of occult qualities, such as the celestial influences of the astrologers—a point made by mechanical philosophers on the European Continent,

mostly Cartesians such as Gottfried Wilhelm Leibniz.

See also Force; Newton, Isaac.
References
Cohen, I. Bernard. *The Birth of a New Physics.* New York: Anchor Books, 1960.
Hall, A. Rupert. *The Scientific Revolution, 1500–1800: The Formation of the Modern Scientific Attitude.* London: Longmans, 1954.

Great Chain of Being

In the early modern period, one way of understanding the relationship between different levels of existence was the idea of the great chain of being or the ladder of life. This idea described the universe as a hierarchical continuum that spans, without really including, the two endpoints of nonexistence and God. This theory had both Platonic and Aristotelian roots. Plato had promoted the idea of a hierarchy of forms, and Aristotle proclaimed the existence of natural continua, emphasizing intermediate forms like the marine life that seemed partway between plants and animals. The lack of a rigid distinction between the living and the not-living in Aristotelian thought also contributed to the picture of the continuous universe. In the fully developed theory, the hierarchy of created things stretches from rocks and stones through worms and vermin to beasts and eventually to the angels that surround God's throne, carefully arranged in hierarchies. Humanity is located around the middle of this sequence, between the brute beasts and the angels.

The chain of being was sometimes expanded into the "principle of plenitude," a theory that originated with the late antique Neoplatonists. This principle held that all things that can possibly exist must exist. This was explained by the perfection of God's power and benevolence, which cannot be limited by the lack of existence of a thing and cannot deny the blessing of existence to a thing that can exist. The principle of plenitude was sometimes thought to require the infinity of the universe, or at least the necessity for a universe very much larger than that of Aristotelian and Ptolemaic cosmology.

Giordano Bruno used the principle of plenitude to argue for an infinite universe, and Gottfried Wilhelm Leibniz's belief in the chain of being was an important part of his philosophical and scientific system. In order for the universe to be the best possible created by God, it has to include everything that can possibly exist, requiring a multitude of inhabited worlds. Belief in plenitude also caused Leibniz to deny the existence of a vacuum. Not every seventeenth-century natural philosopher believed in plenitude or the chain of being; René Descartes denied it as an infringement on God's freedom not to create. But the idea of plenitude, incorporated into natural theology, went on to be very popular in the eighteenth century.

Reference
Lovejoy, Arthur O. *The Great Chain of Being: A Study of the History of an Idea.* Cambridge: Harvard University Press, 1936.

Greenwich Observatory

Greenwich Observatory was founded in 1675 by King Charles II (r. 1660–1685) of Great Britain to establish more accurate tables of stellar and planetary positions for use by navigators. The leading spirits in its establishment were Sir Jonas Moore (1617–1679), the Surveyor of the Ordinance, Christopher Wren, who suggested Greenwich as the site and along with Moore and Robert Hooke helped design the building, and John Flamsteed, who was appointed royal astronomer at an annual salary of £100. The Greenwich Observatory was not subordinated to the Royal Society, which had only vague rights of supervision; this ambiguous relationship would lead to conflict between Flamsteed and the society later.

Like many of Charles II's ventures, the observatory was underfunded, particularly in comparison with the contemporary Paris Observatory. Indeed, Flamsteed, who served

as royal astronomer for 45 years, had to pay for many of the instruments out of his own pocket. The most notable of Flamsteed's instruments at Greenwich was a mural arc of 140 degrees, constructed at a cost of £120 and over a year's work. Greenwich was also the base for the construction of Flamsteed's great star catalogue. After his death in 1719, Greenwich came under the control of Edmond Halley, who succeeded Flamsteed as royal astronomer. There was a crisis at the observatory after Flamsteed's death, when many of his instruments were taken out by his executors. Halley replaced the instruments with the aid of a government grant, and he served until his death in 1742. The needs of navigation rather than pure science remained paramount at Greenwich.

See also Flamsteed, John; Halley, Edmond.
Reference
Wolf, A., with the cooperation of F. Dannemann and A. Armitage. *A History of Science, Technology, and Philosophy in the Sixteenth and Seventeenth Centuries.* 2d ed., prepared by Douglas McKie. London: Allen and Unwin, 1950.

Gregorian Reform of the Calendar

The Gregorian calendar used today originated in the Rome of Pope Gregory XIII (pope, 1572–1585) as the culmination of a long-standing frustration with the increasing inaccuracy of the Julian calendar, which had slipped ten days behind the actual year. Like previous abortive medieval and Renaissance calendar reform projects, Gregory XIII's reform was intimately bound up with religious concerns because of the need to find the correct date for the celebration of religious holidays such as Easter.

The Gregorian commission was assembled some time in the mid-1570s. It included astronomers such as Christoph Clavius, scholars of Greek and Arabic astronomical texts, and clerics. Influenced by the work of the physician and astronomer Luigi Giglio, known by the Latin form of his name, Aloisius Lilius (1510–1576), the commission founded its improved version of the Julian calendar on the average rather than the real motions of the Sun and Moon. The new calendar dropped ten days from the current reckoning to bring the date back into harmony with the seasons. It abolished the leap day on years ending in 00 unless the first two digits of the year are divisible by four—thus 1600 and 2000 are leap years. It also introduced a new method for calculating the date of Easter, on which the religious calendar depended.

The Gregorian calendar was announced and enjoined upon Catholics in a papal bull dated February 24, 1582. Although all European astronomers agreed on the need for calendar reform, reception of the Gregorian calendar was mixed and did not always follow the Protestant-Catholic divide. Johannes Kepler's teacher, Michael Maestlin (1550–1631), attacked the new calendar in a debate with Clavius, but another Lutheran, Tycho Brahe, enthusiastically adopted it. Protestant governments were another matter. The Protestant states of the Holy Roman Empire and Denmark did not adopt the Gregorian calendar until 1699; England, not until 1752; and Sweden, not until 1753.

See also Clavius, Christoph; Papacy.
Reference
Coyne, G. V., S.J., M. A. Hoskin, and O. Pedersen, eds. *Gregorian Reform of the Calendar: Proceedings of the Vatican Conference to Commemorate Its 400th Anniversary, 1582-1982.* Vatican City: Specolo Vaticana, 1983.

Gresham College

Although ultimately a failure, the London college founded by the financier Sir Thomas Gresham (1519–1579) played an important role in disseminating scientific culture to middle-class Londoners. The college, an institution for offering public lectures in English rather than granting degrees, was actually founded on the death of Gresham's widow in 1596, but Gresham's will had been confirmed by parliamentary statute in 1581, which made it very difficult for the college trustees to modify any aspect of it. The first few decades of the college were its most successful, as its leading personality was Henry Briggs (1561–1630), the Gresham Professor of Geometry (one of seven Gresham professorships) until 1620, when he left to be the first

An exterior view of Gresham College, where the first meetings of the Royal Society were held. (Library of Medicine)

Savilian Professor of Geometry at Oxford. Briggs and two of his protégés, Edmund Gunter (1581–1626), Gresham Professor of Astronomy from 1619 to 1626, and Henry Gellibrand (1597–1636), Gresham Professor of Astronomy from 1626 to 1636, emphasized useful knowledge for navigators and businessmen and made the college an early center for disseminating the use of logarithms for astronomical and navigational calculations.

The college began to decline in the 1630s, when the government of Charles I (r. 1625–1649) treated it as a source of patronage. Its decline continued throughout the English Civil War of the 1640s, as successive English governments imposed political litmus tests on potential professors. The building itself was used as a storehouse, a prison, and a garrison. Although the faculty of the 1650s contained some stars, such as Sir William Petty (1623–1687) and Christopher Wren, no effort was made to restore the institution. After the Restoration of 1660, Gresham College became the headquarters of the Royal Society, but as an institution it was shut down completely after the Great Fire of 1666, when the London government took over the building.

After its restoration in 1673, Gresham was an institution characterized by the lackadaisical performance of its professors. Even Robert Hooke, Gresham Professor of Geometry from 1665 to 1703 and one of Gresham's more conscientious professors, lectured to very small audiences. In an effort to put Gresham on a sounder financial footing, the trustees planned to demolish the building, replace it with a smaller one, and rent out the remaining land. The Royal Society, however, blocked this move, fearing it would lose its space in the building. The college exists in London to the present day.

See also Logarithms; Royal Society.

Reference
Adamson, J. R. "The Administration of Gresham College and Its Fluctuating Fortunes as a Scientific Institution in the Seventeenth Century." *History of Education* 9 (no. 1, March 1980): 13–25.

Grew, Nehemiah (1641–1712)

Along with Marcello Malpighi, Nehemiah Grew founded plant anatomy. The son of a Puritan clergyman, Grew was educated at Cambridge. In 1671, he took an M.D. at the University of Leiden by examination, without attending classes. His first botanical work, *The Anatomy of Vegetables Begun* (1672), was followed by highly detailed studies of particular parts of plants such as trunks, leaves, and flowers, eventually collected in *The Anatomy of Plants* (1682). Grew examined the structures of plants through observation and microscopic and chemical analysis. Inspired to study plants by contemporary researches in animal anatomy, he analyzed many plant functions in terms of analogies to animal functions. He was the first to analyze the sexual organs of plants.

The Royal Society was impressed by the manuscript of *The Anatomy of Vegetables Begun* and published the book. It was at the Royal Society that Robert Hooke introduced Grew to the microscope. With John Wilkins taking the lead, the society also made Grew an offer in hopes of luring him from Coventry, where he was practicing medicine, to London. Grew became one of the few scientific researchers that the Royal Society attempted to support financially; he was appointed curator of the anatomy of plants in 1672. Grew's *Idea of a Phytological History* (1673) set forth an ambitious program of botanical research to be supported by the society. His annual salary of £50 was supposed to be paid by ten members of the society who would pay £5 apiece. But several members did not pay, and so Grew supported himself in London by lecturing at Gresham College and working as secretary of the society from 1677 to 1680. Most of his secretarial labors were devoted to an inventory of the society's collections, published and widely distributed in 1681 as *Musaeum Regalis Societatis*. After publishing *The Anatomy of Plants,* Grew abandoned scientific research in favor of a London medical practice.

See also Microscopes; Royal Society.

References

Fournier, Marian. *The Fabric of Life: Microscopy in the Seventeenth Century.* Baltimore: Johns Hopkins University Press, 1996.

Hunter, Michael. *Establishing the New Science: The Experience of the Early Royal Society.* Woodbridge, England: Boydell, 1989.

Halley, Edmond (1656–1742)

Edmond Halley is best known for his prediction of the return of the comet of 1682, now called Halley's comet. The leading astronomer and one of the most versatile natural philosophers of his time, for decades Halley occupied a central role in British science and its relation to government and traveled more extensively in the pursuit of scientific knowledge than any other major early modern scientist. Born into a wealthy family of London businesspeople, Halley attended Oxford, although he did not take a degree. At the age of 19, he submitted to the Royal Society the first of many papers, a discussion of methods for computing planetary orbits.

Halley's first major scientific achievement was the creation of a chart of the stars of the Southern Hemisphere, for which he established an observatory on the island of St. Helena in 1677. The expedition was financed by the royal treasury, with passage provided by the East India Company at the behest of King Charles II (r. 1660–1685) in an effort to improve navigation in the Southern Hemisphere. Halley's chart was the best made of the southern sky to that point and the first stellar chart based on telescopic observation. The points of reference were based on Tycho Brahe's, but Halley believed that they could be recalculated when more accurate charts

became available. Upon his return to England, Halley was elected a fellow of the Royal Society, and shortly afterwards he was sent as the society's unofficial emissary to the Danzig astronomer Johannes Hevelius. He would serve the Royal Society in a variety of positions throughout his career, and in 1729 he was also admitted to its French rival, the Royal Academy of Sciences.

Halley was essential to the publication of Newton's *Mathematical Principles of Natural Philosophy* in 1687. After a 1684 conversation with Robert Hooke and Christopher Wren, Halley visited Newton at Cambridge. He asked Newton what he thought of an attraction between celestial bodies based on an inverse-square law. On discovering that Newton had already calculated that such orbits would be elliptical, Halley encouraged him to publish the results. As clerk to the Royal Society, Halley saw the finished work through to press, personally bore the financial responsibility, and contributed a Latin ode to Newton.

Much of Halley's work was concerned, directly or indirectly, with navigation and sailing, and particularly with the problem of determining the longitude. Halley led three Atlantic voyages on the *Paramore* from 1698 to 1701 to map the variations of the Earth's magnetic field. These were the first ocean voyages exclusively dedicated to scientific purposes. Halley's map of the variations, published in 1701, was the first to use "isogonic" lines connecting positions that had the same variation. He worked on tides, applying Newtonian physics to the problem, and produced the first survey of the tidal patterns of the English Channel. Halley also charted trade winds.

Halley's contributions to astronomy were immense and varied. He discovered the movement of the stars relative to each other and the acceleration of the Moon over long periods of time. He invented a method of using the transits of Venus across the Sun as a means of determining the distance from the Earth to the Sun, although the next transit was in 1761 and he did not live long enough to actually apply them. Halley's *Synopsis of Cometary Astronomy,* first published in 1705 and revised in subsequent editions, set forth the modern theory of cometary orbits of the Sun in elliptical paths, and predicted that the comet of 1682 would return in 1758. Halley originally worked harmoniously with the other major English astronomer, the quick-tempered John Flamsteed, but the two quarreled beginning in the early 1680s, most bitterly over Halley's publication of Flamsteed's star chart in 1712. In 1691, Halley was a candidate for the Savilian Chair of Astronomy at Oxford, but despite a recommendation from the Royal Society, he was rejected in favor of the Scotsman David Gregory (1659–1708). This may have been caused by doubts about Halley's religious orthodoxy.

Whatever the cause of the earlier rejection, Halley was appointed Savilian Professor of Geometry in 1703. As a mathematician, he was skilled if not creative. His best known mathematical accomplishment was his edition of the Greek text of Apollonius (c. 262–c. 190 B.C.), the *Conics,* a work requiring competence in Arabic as well as Greek, in which Halley collaborated with Gregory. In 1720, following Flamsteed's death, Halley succeeded his old enemy as royal astronomer. Although he replaced the instruments Flam-

Edmond Halley's extensive work in the field of science is overshadowed in the modern mind by his accurate prediction of the return of a comet in 1682. (National Library of Medicine)

steed's heirs had removed from the Greenwich Observatory with more modern ones, his tenure as royal astronomer was not distinguished. Halley continued to pursue his scientific interests throughout his life, supporting the efforts of John Harrison (1693–1776) to produce the first marine clock accurate enough for navigation.

See also Astronomy; Comets; Flamsteed, John; Greenwich Observatory; Navigation; Newton, Isaac; Royal Society.

References
Cook, Alan. *Edmond Halley: Charting the Heavens and the Seas.* Oxford: Clarendon Press, 1998.
Wolf, A., with the cooperation of F. Dannemann and A. Armitage. *A History of Science, Technology, and Philosophy in the Sixteenth and Seventeenth Centuries.* 2d ed., prepared by Douglas McKie. London: Allen and Unwin, 1950.

Harriot, Thomas (1560–1621)

Thomas Harriot was an innovative scientist and mathematician who had less impact than he might have had because he left the vast majority of his scientific work in manuscript form. Of obscure background, Harriot graduated from Oxford with a B.A. in 1580. He made his way to London, where he became involved with a circle of men interested in exploration, navigation, and cartography. Harriot acquired the patronage of Sir Walter Ralegh (1554–1618); he conducted navigational classes for Ralegh's sea captains and applied his mathematical knowledge for the improvement of navigational technique. In 1585, he journeyed to Ralegh's first ill-fated colony in Virginia, returning with the rest of the colonists to publish the first printed description of Virginia, *A Briefe and True Report of the New Found Land of Virginia* (1588), which included discussion of the natural history and geography of the colony. (Like many in Ralegh's circle, Harriot was a keen smoker of the American plant, tobacco. He became the first recorded smoker to die of cancer.)

Shortly after his return from Virginia, Harriot added to Ralegh's patronage that of Ralegh's friend Sir Henry Percy, ninth earl of Northumberland (1564–1632). Immensely rich, the "Wizard Earl" became the greatest patron of mathematicians and magicians in England, paying Harriot a particularly generous pension. Harriot's association with Ralegh and Northumberland and his friendship with John Dee contributed to his bad reputation as a magician and an atheist. He narrowly avoided being caught up in the disgrace and imprisonment suffered by Ralegh and Northumberland in 1603 after the death of Queen Elizabeth I (r. 1558–1603) and the accession of King James I (r. 1603–1625), suffering only a brief period of imprisonment.

Beginning in the 1590s, Harriot broadened his interests to include mathematics, physics, optics, ballistics, and alchemy. He was interested in the quantitative properties of various substances, measuring their specific gravities and refractive indices, a subject on which he corresponded briefly with Johannes Kepler. Harriot began using telescopes, which he fashioned himself, at around the same time as Galileo Galilei, and he was one of the first astronomers after Galileo to observe the moons of Jupiter and sunspots. In optics, Harriot discovered the sine law of refraction that would later be independently rediscovered by René Descartes, and he was an early supporter of atomism. In his will, Harriot directed his literary executor, Nathaniel Torporley (1564–1632), to edit and publish his mathematical and scientific manuscripts, but Torporley failed in this task, and the only portion of Harriot's mathematical work published was *Practice of the Art of Analysis* (1631), on the solution of algebraic equations. Posthumously, Harriot became a cause for dispute between the English mathematician John Wallis and French mathematicians over whether Descartes had plagiarized Harriot's methods in his *Geometry*.

See also Telescopes.
Reference
Shirley, John W. *Thomas Harriot: A Biography.* Oxford: Clarendon Press, 1983.

Hartlib Circle

Samuel Hartlib (c. 1600–1662), born in Prussian Poland of mixed German and English descent, was at the center of a network of scientific correspondence and joint projects extending throughout Protestant Europe and as far away as North America. He attended Cambridge University in the mid-1620s, and after a brief return home he settled permanently in England in 1628. In England, Hartlib sought patronage from the leading Puritan families. He became involved in projects to restore learning and unite the Protestant churches in a millenarian context, preparing for the Second Coming of Jesus Christ and his reign on Earth. Hartlib was also inspired by Francis Bacon and the Czech educator John Amos Comenius (1592–1670). He and his close friend and collaborator, the Scottish minister John Dury (1596–1680), saw themselves at the head of an international brotherhood to advance true learning and godly religion. Hartlib sought to improve human technological ability in areas such as mining and agriculture, sponsoring the activities of the engineer and metallurgist Gabriel Plattes.

Hartlib's heyday occurred with the English Civil War and Interregnum (1640–1660), when his friends among the Puritan gentry seized control of the country. He sought government backing for new educational and scientific institutions and for the reform of old ones, emphasizing natural and religious knowledge over humanistic studies, and he spearheaded other projects for social and intellectual reform. His A Description of the Famous Kingdome of Macaria (1641) described an ideal state where applied natural knowledge leads to a better and more prosperous life for its citizens. The Hartlib circle of the 1640s and 1650s included such natural philosophers as George Starkey, Robert Boyle, and Sir William Petty (1623–1687). Hartlib himself was appointed to a position at Oxford to superintend an Office of Address for the advancement of learning.

This office, however, was never actually instituted. This became true of many of Hartlib's projects as the English regime in the 1650s lost interest in radical reform and the conservatism of the educational establishment reasserted itself. After the Restoration of Charles II (r. 1660–1685) in 1660, members of the Hartlib circle such as Boyle and Henry Oldenburg, who began his career as a scientific intelligencer by writing letters to Hartlib, were instrumental in the new Royal Society. Hartlib himself, tainted by associations with the Puritan regime, became a forgotten man and died in poverty and obscurity in 1662.

See also Correspondence; Millenarianism; Puritanism and Science.

Reference
Webster, Charles, ed. *Samuel Hartlib and the Advancement of Learning.* London: Cambridge University Press, 1970.

Harvey, William (1578–1657)

The discoverer of the circulation of the blood, William Harvey, was the most renowned anatomist of the seventeenth century. He received his B.A. from Gonville and Caius College of Cambridge University in 1597, but his intellectually formative period was spent at the University of Padua, where he received an M.D. in 1602. On his return to England, Harvey married the daughter of a prominent London physician and made a successful medical career, serving as the physician of Saint Bartholomew's Hospital from 1609 to 1643 and as physician to Kings James I (r. 1603–1625) and Charles I (r. 1625–1649). Harvey became a fellow of the London College of Physicians in 1607 and was appointed Lumleian Lecturer in anatomy and surgery to the college in 1615.

Harvey's interest in anatomy took an Aristotelian as opposed to a Galenic form. He was influenced by a revival of interest in Aristotle's works of natural history at Padua, led by Harvey's instructor, the anatomist Fabricius of Acquapendente (1533–1619). Rather than maintaining the focus on human anatomy characteristic of medical Galenists, who dissected only those animals that seemed relatively close to humans, Fabricius's anatomy, influenced by the Aristotelian natural philosophy, was concerned with discovering the operation of the organs across a number of species. Harvey's studies of the heart through both dissection and vivisection, described in some of the surviving notes for his Lumleian lectures, led him to discover the circulation of the blood. His vivisection of animals, including animals remote from humans such as reptiles, enabled him to realize that the force of the heart is in the contraction part of the heartbeat, the systole. Estimating the amount of blood that would be ejected from the heart at each contraction, Harvey concluded that the amount of blood expelled in a short time would exceed the total amount of blood in the body. He also noted that the valves of the veins, described by Fabricius, indicate that the flow of blood in the veins can only be to the heart. He employed these arguments, along with experiments based on the behavior of blood in the arm when circulation is cut off by a tourniquet, in his *On the Motion of the Heart* (1628), which presents the full theory of the circulation.

The other area that absorbed Harvey's energy was embryology. His *Of Generation*, which like *On the Motion of the Heart* was based on numerous dissections, was finished in the late 1630s, but the English Civil War delayed its publication until 1651. In this work, Harvey argued that all living things, including those born alive, originate in eggs or seeds, also indicating some skepticism about spontaneous generation. His explanations of generative processes continued to be Aristotelian and nonmechanical, and he gave nonmaterial processes an essential role in conception. Harvey was an epigeneticist who denied that living beings are preformed in the seed.

Besides the specific content of his science, Harvey's importance lies in the centrality of experiment, dissection, and demonstration to his arguments. As one of the first major innovations in natural philosophy to be put forth and accepted largely on the basis of experimental evidence, Harvey's theory of circulation helped create the prestige of experiment and experimental science in the seventeenth century. This was particularly true in England, where Harvey was invoked along with Francis Bacon and William Gilbert to identify the experimental tradition as peculiarly English.

But Harvey was critical of Bacon, and he despised Paracelsianism. Although Harvey treated Bacon medically, and his use of experiment could be seen as Baconian, he had little respect for Bacon's philosophical works and is alleged to have described Bacon as writing philosophy "like a lord chancellor," that is, without dealing with specifics. His Aristotelianism is also manifest in that he was much less religious in his presentation than were other anatomists in this period, who often described their anatomical claims as revealing the divine glory through God's handiwork. Harvey, following

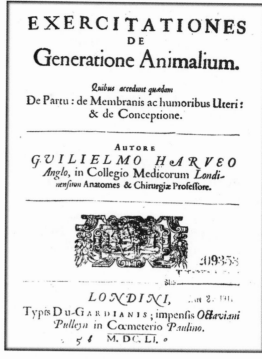

William Harvey, *best known as the discoverer of the circulation of the blood, published this work on animal reproduction* (Of Generation) *near the end of his life, in 1651. (National Library of Medicine)*

Aristotle, did not usually speak of anatomy in religious terms. His own religious position was that of a conservative member of the Church of England, as befit an associate of Charles I, who allowed Harvey to dissect deer from the royal parks when he was studying embryology.

Harvey was not politically active, but during the English Civil War he took the side of the king. Parliament's victory put him somewhat in the shade, but he remained a leading figure in the London College of Physicians to his death. He left the college a great deal, including an endowment for a Harveian Lectureship in experimental philosophy and his landed estate. The library of the college that Harvey financed and many of his manuscripts were destroyed in the Great Fire of London in 1666.

See also Circulation of the Blood; London College of Physicians; Physiology.

References

Bylebyl, Jerome J., ed. *William Harvey and His Age: The Professional and Social Context of the Discovery of the Circulation*. Baltimore: Johns Hopkins University Press, 1979.

French, Roger. *William Harvey's Natural Philosophy*. Cambridge and New York: Cambridge University Press, 1994.

Keynes, Geoffrey. *The Life of William Harvey*. Oxford: Clarendon Press, 1966.

Helmont, Francis Mercury van (1614–1698)

The youngest son of Johannes Baptista van Helmont, Francis Mercury van Helmont played an important role in seventeenth-century natural philosophy by introducing and spreading Kabbalistic and heterodox ideas. Giving his child the name "Mercury," a name of profound significance to any alchemist, may indicate that Johannes Baptista thought this son would have a special destiny. In any event, Francis Mercury was the only son in the family to outlive his father, and he edited and published the elder van Helmont's posthumous *Ortus Medicinae* in 1648.

After serving as a diplomat and adviser for minor German courts, van Helmont settled in 1651 in the small principality of Sulzbach, where he worked with the translator of Kabbalistic texts, Christian Knorr von Rosenroth (1636–1689). Becoming an ardent propagandist for a Christianized version of the Lurianic Kabbalah, van Helmont spread various doctrines derived from it such as reincarnation and universal salvation—heresies that landed him in the prisons of the Inquisition from late 1661 to early 1663. He shared his father's alchemical interests, setting up a laboratory at Sulzbach and publishing a collection of 153 *Chemical Aphorisms* in 1668. A well-traveled individual, van Helmont spent several years in England in the household of Anne Conway and her husband and was the friend and intellectual colleague of John Locke and Gottfried Wilhelm Leibniz. He saw through to press Conway's *The Principles of the Most Ancient and Modern Philosophy* (1690). Van Helmont's own *Spirit of Diseases* (1692) restated his father's ideas about the spiritual nature of disease. A collection of remedies attributed to van Helmont, *Medicina Experimentalis Helmontiana* (1704), was largely traditional and even somewhat Galenic in its approach.

See also Conway, Anne; Helmont, Johannes Baptista van; Kabbalah.

References

Coudert, Allison P. *The Impact of the Kabbalah in the Seventeenth Century: The Life and Thought of Francis Mercury van Helmont (1614–1698)*. Leiden, the Netherlands: Brill, 1999.

Helmont, Johannes Baptista van (1579–1644)

The most influential chemist between Paracelsus and Robert Boyle, Johannes Baptista van Helmont fundamentally reshaped the Paracelsian inheritance, creating a system of "chemical philosophy" that was influential, particularly in medical circles, throughout the seventeenth century. Van Helmont was born into a landed gentry family in the Spanish Netherlands (modern Belgium), and his inherited wealth, augmented through marriage to a woman of the same class in 1609, gave him independence to pursue his studies. He attended the University of Louvain but was disgusted by what he saw as the pretension and disguised ignorance of academic life, earning but refusing to accept an M.A. He attended lectures on philosophy given by the Jesuits outside the university but found little to satisfy in them either, and his thoroughly reciprocated dislike of Jesuits would prove lifelong. He gave lectures on surgery at Louvain at the age of 17, taking an M.D. there in 1599. He then spent some years traveling in France, Switzerland, Italy, and England, turning down an invitation to settle at the court of Rudolf II, among other offers from German princes. He settled in his native land, living in seclusion to study medicine and dispense free care to the poor.

Van Helmont was obsessed by the worthlessness of Galenic medicine and Aristotelian

This posthumous edition of Johannes Baptista van Helmont's complete works shows him with his son Francis Mercury, the editor of his works. The display of coats of arms emphasizes the van Helmont family's aristocratic status. (National Library of Medicine)

philosophy. He preferred Paracelsianism, although he was by no means a slavish follower of Paracelsus. He attacked what he viewed as Paracelsian materialism and eventually rejected its fundamental analogy of the body as microcosm and the universe as macrocosm. A practicing alchemist who believed that he had transmuted mercury to gold, van

Helmont claimed that chemical knowledge is the key to understanding the universe. Although he avoided the mystifications and secretiveness of traditional alchemical writings and put more emphasis on quantitative analysis than was common in the alchemical or Paracelsian traditions, van Helmont was highly suspicious of both mathematics and reason as ways of understanding the universe. Reason, he believed, is a deceiver, originating in the Fall of humanity from Eden; in its place, he put forth a philosophy of experiment and observation as superior to one of logical deduction. In this view, biological processes need to be understood in chemical terms, and for starters, van Helmont provided the most complete description of digestion as the dissolution of food in the stomach by an acid, breaking with the Galenic theory of heat as the means of digestion.

Van Helmont invented the term "gas," from the Greek for chaos, and he described 15 different gases. The concept of gas emerged from van Helmont's vitalist matter theory, which despite his proclaimed anti-Aristotelianism was strongly influenced by the Aristotelian distinction between form and substance. Van Helmont believed that the most fundamental and undifferentiated matter is water, and that water in its natural state is purely inert matter without any spiritual element, or "seed," causing it to take a specific form. Gases, then, are water vapors charged with "seeds." In an experiment only possible for a wealthy man, he burned 62 pounds of coal in a closed vessel, claiming that only 1 pound of ash remained and that the rest had turned to gas.

Expanding on a Paracelsian idea, van Helmont asserted that living things are governed by *archei*—nonmaterial, living, gaseous principles that control biological processes and are arrayed within each individual body in a hierarchy under the control of a supreme *archeus*. What appear to be interactions between material things are in fact interactions of their *archei*. Diseases, for example, are caused by the invasion of the body and attacks on the supreme *archeus* by flawed or unbalanced *archei*. So diseases are specific, positive entities rather than negative states of absence of health. Van Helmont denied Galenic humoral theory, particularly the idea that diseases can be caused by *catarrh,* the displacement of phlegm downwards.

Van Helmont was drawn out of his reclusiveness by the controversy over the weapon salve, an ointment that some claimed could cure wounds by being applied to the weapon that caused them. Van Helmont's writings on the weapon salve and spas aroused the suspicion of the Jesuits and other Catholic authorities, worried about the spread of magic and Paracelsianism. The medical faculty of the University of Louvain denounced his work in 1623, and in 1625 the Spanish Inquisition, which had jurisdiction over the Southern Netherlands, denounced 27 propositions drawn from Helmont's work as heretical, magical, and crypto-Protestant. He spent much of the next two decades under various forms of house arrest, and he published nothing between 1624 and 1642. Despite his use of magical concepts, van Helmont became a friend and correspondent of the arch antimagician Marin Mersenne after treating him in 1630. Most of van Helmont's work was published posthumously by his son, Francis Mercury van Helmont.

See also Alchemy; Chemistry; Helmont, Francis Mercury van; Paracelsianism; Weapon Salve.

References

Debus, Allen G. *The Chemical Philosophy: Paracelsian Science and Medicine in the Sixteenth and Seventeenth Centuries.* 2 vols. New York: Science History Publications, 1977.

Pagel, Walter. *Joan Baptista van Helmont: Reformer of Science and Medicine.* Cambridge: Cambridge University Press, 1982.

Herbals

One of the most common forms of scientific writing in early modern Europe was the herbal, a list of plants with illustrations and descriptions. Nearly all herbals focused on medicinal uses of plants, but beyond this there was a wide variety. The classical herbal of the ancient Greek physician Dioscorides (A.D. c. 40–c. 90) was reprinted and translated into both Latin and vernacular languages. Pier Andrea Mattioli's (1501–1577) commentary on Dioscorides, first published in 1544, went through over 50 editions. Other herbals were specific to a particular region, in Europe or beyond. Astrological herbals, such as the popular *Physicall Directory* (1649) of the Englishman Nicholas Culpeper (1616–1654), explained the correlation of each plant with a particular planet. Large and elaborate herbals were confined to libraries, whereas small, portable herbals could be carried on a walk in the fields to identify and gather plants. Entries on individual plants were usually listed alphabetically or by the medical conditions they alleviated rather than by botanical similarities.

The first decades of printing in the late fifteenth century saw the publication of a number of ancient and medieval herbals that had been circulating in manuscript form. These early herbals included fabulous and mythological plants, such as the tree of the knowledge of good and evil from the Book of Genesis. Some herbs were described as having magical powers, such as the ability of wormwood held in the hand to make a journey easy. The most innovative early herbal was *Living Images of Herbs* (1530), published in Strasbourg and created by the partnership of the botanist Otto Brunfels (1489–1534) and the woodcut artist Heinz Weiditz, a disciple of Albrecht Dürer (1471–1528). The illustrations were the most accurate and lifelike to date because Weiditz copied from living plants rather than from previous illustrations, as had been the earlier practice.

Other notable herbals were created in Germany at this time by Jerome Bock (1498–1554) and Leonhard Fuchs (1501–1556), whose well-illustrated herbal included descriptions of American plants such as the pumpkin. There were also a myriad of European plants remaining to be recorded, and the Fleming Charles de l'Ecluse

(1526–1609), author of *Very Rare Plants* (1601), added hundreds of plant types to the stock of botanical knowledge, mostly from the Iberian Peninsula and Hungary. New World and Asian herbals, often incorporating indigenous knowledge about the medicinal uses of plants, came from Spanish and Portuguese writers. Two Aztecs composed a Latin manuscript herbal of Mexican plants.

Although the tradition of small herbals—written in vernacular languages, focusing on medical uses, and affordable for a wide range of people—remained lively, the seventeenth century also saw the rise of "super-herbals," containing descriptions of thousands of plants or even aiming at universal coverage. These herbals, often written by physicians, had drifted from their medical roots to a more purely botanical approach. The works of the early-seventeenth-century French-Protestant Bauhin brothers (Jean, 1541–1613, and Gaspard, 1560–1624) were the first attempts at universal herbals, and they remained influential into the eighteenth century. The massive botanical works of John Ray and the *Memoirs of Plants* produced by the French Royal Academy of Sciences in the late seventeenth century had clearly transcended the genre, but their origins lay in the humble herbals of the Renaissance.

See also Apothecaries and Pharmacology; Botany; Illustration.
Reference
Arber, Agnes. *Herbals, Their Origin and Evolution: A Chapter in the History of Botany, 1470–1670.* 2d ed. Cambridge: Cambridge University Press, 1938.

Hermeticism

The *Corpus Hermeticum* is a collection of ancient pagan Greek texts influenced by astrology, Gnosticism, and Neoplatonism and produced in Egypt around the second century A.D. They emphasize knowledge of the divine through mystical contemplation. During the Renaissance, they were generally thought to be much older and were ascribed to a legendary sage contemporary with or preceding Moses called Hermes Trismegistus (Thrice-Great.) The Hermetic writings were introduced into the West around 1460, part of the recovery of Greek literature in the Italian Renaissance. They were translated into Latin shortly thereafter by the Florentine Platonic philosopher Marsilio Ficino (1433–1499) and were circulated widely in both print and manuscript form. Also thought to be authored by Hermes was a lost Greek text that survived in a Latin translation, *Asclepius,* describing magical procedures, allegedly used by Egyptian priests, for animating statues of the gods by drawing down celestial powers.

Hermeticism was received in two chief ways in early modern Europe—as a system of natural theology and as a way to legitimize the practice of magic. As a system of natural theology, Hermeticism was assimilated to the idea of the *prisca theologica,* or ancient wisdom, handed down directly by God through a series of Gentile sages paralleling the Hebrew prophets and now surviving only in fragments. Things Egyptian were often associated with the *prisca theologica.* It was known in Renaissance Europe that Egypt was the oldest ancient civilization, and that the ancient Greeks themselves had viewed Egypt as a source of mystical wisdom. Zoroaster, Pythagoras (c. 580–c. 500 B.C.), Plato (fl. 428–389 B.C.), and other subsequent sages were identified as disciples and followers of Hermes.

Hermetic theology was also believed to be perfectly compatible with Christianity. It was often claimed that Hermes derived his knowledge from Moses and prophesied the birth of Christ, and thus Hermetic wisdom could be seen as an expression of Christianity. (Hermes was conspicuously portrayed as a proto-Christian sage in the pavement of the Dome of the Cathedral of Siena, laid down in the late fifteenth century.) Hermetic natural religion was put forward as an alternative to Catholicism and Protestantism that would end the sharp religious conflicts of the Reformation era.

Hermeticism was also central to the early modern revival of high magic. Except for the *Asclepius,* the Hermetic writings contained little magic, but in the Middle Ages Hermes Trismegistus had been known as a magician, and a number of magical and alchemical works originally produced in Greek, Arabic, and Latin had been ascribed to him or mentioned his name with honor. Egypt was also associated with magic through the story of Pharaoh's magicians in the Book of Exodus. Like the ancients, early modern people ascribed mystical meanings to Egyptian hieroglyphics, which had not yet been deciphered, and Ficino asserted that Hermes had invented them.

As a respected and holy sage, the Hermes Trismegistus of the Renaissance could legitimize magic for the would-be Christian magician. Ficino's system of natural magic, based on the manipulation of celestial influences and the innate properties of natural substances, drew on Hermes along with other ancient Neoplatonic sources. Hermetic high magic was consciously elitist, holding itself far distant from the popular magical practices of witches, cunning folk, and "the vulgar," as well as from diabolic magic. Some combined Hermeticism with other magical traditions, notably the Kabbalah, as did the highly eclectic Renaissance philosopher Giovanni Pico della Mirandola (1463–1494).

A number of significant figures in the scientific revolution were influenced by Hermeticism. It provided an alternative natural philosophy to Aristotelianism, and anti-Aristotelian natural philosophers such as Paracelsus drew on the Hermetic writings. Nicolaus Copernicus quoted the *Corpus Hermeticum,* and Hermetic reverence for the Sun as a divine being may have helped inspire his heliocentrism. John Dee combined interest in Hermes with active magical practice and sustained attention to science and mathematics. Giordano Bruno, who unlike most Hermeticists took Hermeticism so far as to oppose Christianity, made heliocentrism and the infinity of the universe central to his magical and religious system, hoping to revive a version of "Egyptian" religion derived from the Hermetic texts. Magicians in the Hermetic tradition hoped to use the powers of nature to attain their goals, promoting a more active approach to the universe than did Aristotelians. For some scientists, Hermeticism was also important as a negative quality that they defined themselves against; mostly notably anti-Hermetic was Marin Mersenne, who tirelessly promoted the mechanical philosophy as an antidote for Hermetic and other magic, which he saw as a pagan belief in nature's power to act without God.

The true dating of the *Corpus Hermeticum* was revealed in 1614 by a French Protestant humanistic scholar, Isaac Casaubon (1559–1614), based on philological arguments concerning the vocabulary and Greek forms in the text. Although some, like Isaac Newton, continued to believe in the *prisca theologica,* and seventeenth-century magicians like Robert Fludd and Athanasius Kircher continued to draw upon Hermes, Casaubon's discovery did contribute to the decline of Hermeticism and magic generally in European culture.

See also Magic.
References
Merkel, Ingrid, and Allen G. Debus, eds. *Hermeticism and the Renaissance: Intellectual History and the Occult in Early Modern Europe.* Washington, DC: Folger Shakespeare Library, 1988.
Yates, Frances Amelia. *Giordano Bruno and the Hermetic Tradition.* London: Routledge and Kegan Paul, 1964.

Hernández, Francisco (c. 1517–1587)

The most important natural historian of Mexico in the sixteenth century, Francisco Hernández received an M.D. from the University of Alcalá and had a successful career in the Spanish medical community. (Like many Spanish physicians, he may have been the descendant of converted Jews.) He reached the exalted position of physician to

the chamber of the king in the late 1560s. Hernández was already an experienced medical botanist, and in 1570 the king of Spain, Philip II (r. 1556–1598), ordered him to go to Mexico to study its natural history with a view to understanding the medical uses of plants there. He was also appointed *protomédico,* or head of the medical profession, for Mexico.

Hernández arrived in Mexico in 1571, staying until 1577. While there, he traveled extensively, gathering specimens, illustrations, and descriptions of plants. He consulted with Native American healers about Mexican plants and their medicinal uses, and he also studied the terrible epidemic of *cocolitzli* (possibly typhus) that killed half the Mexican population in 1577. Hernández learned Nahuatl, the language of the native population there, and translated some of his materials into it for their use.

Hernández's return to Spain brought disappointment, as the bulk of his work remained unpublished. Abridged versions were published in Mexico in 1579 and 1615. The first European publication was sponsored by Federico Cesi (1585–1630) and his Accademia dei Lincei in Rome in 1628, and a fuller version was published as *A Treasury of Medical Things from New Spain* in 1648. Unfortunately, many of Hernández's manuscripts were destroyed in a fire in 1671, but his work remains an important source on those Mexican plants and animals that became extinct following the Spanish conquest.

See also Accademia dei Lincei; Botany; Exploration, Discovery, and Colonization; Natural History.
Reference
Goodman, David. "The Scientific Revolution in Spain and Portugal." In Roy Porter and Mikulas Teich, eds., *The Scientific Revolution in National Context.* Cambridge: Cambridge University Press, 1992.

Hevelius, Johannes (1611–1687)

One of the most accurate and prolific astronomers of the late seventeenth century,

Undated engraving of Johannes Hevelius, one of the most accurate and prolific astronomers of the late seventeenth century. (National Library of Medicine)

Johannes Hevel, known as Hevelius, was also the last major astronomer to use instruments without telescopic sights and to work at his own private observatory, largely independent of governments. From a wealthy merchant family of Danzig, Poland, Hevelius visited England and France as a young man and studied law at the University of Leiden without taking a degree. He set up an observatory called Stellaeburg in his home town and began making observations with the help of his wife, Catherina Elisabeth (1647–1693). He received the patronage of Louis XIV (r. 1643–1715) and three successive kings of Poland. Hevelius also played a leading role in the Danzig city government.

As an astronomer, Hevelius discovered several comets and the first variable star to be described, which he called Mira. He observed transits of Venus and Mercury, and he mapped the Moon, discovering its "libration," or oscillation. He set forth his results and described his instruments in elaborate folios that are some of the most visually

impressive astronomy books of the scientific revolution. Hevelius was one of the first foreigners admitted to the Royal Society, in 1664, and he made many contributions to *Philosophical Transactions*. His good relations with the society were marred by a dispute with Robert Hooke over telescopic sights, which Hooke supported and Hevelius disdained. The Royal Society sent the young Edmond Halley to Danzig with telescopic instruments in 1679 to examine Hevelius's instruments and make observations alongside Hevelius and his staff. Halley found Hevelius's observations highly accurate, although he himself continued to prefer telescopic sights. Two months after Halley's departure, a fire in Stellaeburg destroyed all of Hevelius's instruments. He continued observing but became an increasingly archaic figure. His extensive star catalog, *Prodromus Astronomica,* was published in 1690 by Catherina Elisabeth. It became a standard astronomical source.

See also Astronomy; Observatories.
Reference
MacPike, E. F. *Hevelius, Flamsteed, and Halley: Three Contemporary Astronomers and Their Mutual Relations.* London: Taylor and Francis, 1937.

Hobbes, Thomas (1588–1679)

Thomas Hobbes is today chiefly known for his political philosophy, but he was also a natural philosopher and mathematician. His political philosophy can be seen as an early attempt to derive political philosophy from science rather than religion, law, or humanistic learning. Educated at Magdalen College, Oxford, Hobbes entered the household of William Cavendish (1555–1626) in 1608. Hobbes's connection with the Cavendish family would last, at intervals, for the rest of his life.

Hobbes worked briefly as Francis Bacon's secretary in the 1620s, but his interest in natural philosophy developed through the non-Baconian path of geometry, which seemed to offer certainty where rhetoric offered only persuasion. Hobbes became familiar with the mechanical philosophy when traveling in France and Italy as tutor to Cavendish's grandson William, the third earl of Devonshire (1617–1684), from 1634 to 1636. He met Galileo Galilei and made the acquaintance of Marin Mersenne's circle of natural philosophers, including René Descartes, Kenelm Digby, and Pierre Gassendi. At Mersenne's prompting, Hobbes contributed a set of objections to Descartes's Meditations (1641). He distrusted Cartesian dualism, holding that the mind, which affects material things, must itself be material. Hobbes and Descartes also disagreed on optical questions. Both attempted to mechanize optics, but Hobbes refused to accept Descartes's distinction between the motion of light and the direction of that motion. Hobbes's Of Body (1655) was one of the first published mechanical philosophies.

A Royalist in the Civil War and mathematics tutor to the future Charles II (r. 1660–1685) from 1646 to 1648, Hobbes spent the 1640s in Paris. The influence of the mechanical philosophy on his political thought is apparent in his classic, Leviathan (1651). As the mechanical philosophy reduced the universe to matter and motion, so Hobbes's work reduced politics to individuals and their desire to survive. Leviathan's anti-Catholicism made Paris dangerous for Hobbes, and he returned to England. There his materialism and his denial of free will gave Hobbes a reputation as an atheist that would cling to him the rest of his life.

In 1655, Hobbes got into a nasty and prolonged mathematical dispute with John Wallis over his claims to have squared the circle, which Wallis scorned. Wallis unquestionably had the best of the mathematics—Hobbes's geometry-dominated mathematics was increasingly out of date—but the vituperation of their quarrel had more to do with religious differences. Hobbes thought Wallis was a revolutionary Puritan, and Wallis thought Hobbes was an atheist. Hobbes's bad reputation kept him out of the Royal Society following the Restoration, although his old pupil, now king, protected him from serious harm. Hobbes's mechanical philosophy, which like Descartes's was based on a plenum, was opposed to the increasingly fashionable atomism of the Royal Society. Hobbes's controversy with Robert Boyle in the early 1660s over Boyle's air-pump experiments did as much damage to his reputation as a natural philosopher as the controversy with Wallis did to his reputation as a mathematician, at least in Britain. However, he continued to publish on scientific subjects and was revered as one of Europe's greatest natural philosophers by Continental scientists, including the young Leibniz.

See also Mechanical Philosophy; Politics and Science.
Reference

Sorell, Tom, ed. *The Cambridge Companion to Hobbes.* Cambridge: Cambridge University Press, 1996.

Hooke, Robert (1635–1703)

Robert Hooke was one of the first professional scientists outside the universities and the medical profession and one of the most inventive and versatile thinkers in the scientific revolution. He was from a poor background, but while a student at Oxford, he attracted the interest of the anatomist Thomas Willis (1621–1675) and Robert Boyle, both of whom employed him as a lab assistant. Boyle and Hooke formed a close patron-client relationship. Hooke constructed the air pump that Boyle used to conduct his famous experiments. But Hooke's social inferiority to other British scientists would plague him throughout his life. Hooke was not initially made a fellow at the founding of the Royal Society in 1662. Instead, through Boyle's patronage he was hired as an employee of the society with the title of curator of experiments. He was made a fellow of the society in 1663.

From his society-provided lodgings at Gresham College (he was appointed Gresham Professor of Geometry in 1665, an appointment he kept for the rest of his life), Hooke dominated the society's experimental program in its early years, assuming the very heavy burden of work the society laid on him. The list of inventions and improvements in instrument design that can be credited to "the ingenious Mr. Hooke" is enormous. He improved air pumps, barometers, clocks, telescopes, and microscopes, among others. He invented the universal joint, still known as "Hooke's joint," and the iris diaphragm for telescopes. He also discussed technical problems with Thomas Newcomen (1664–1729), the originator of the "Newcomen engine," an early steam engine. Hooke's technical genius in performing experiments and designing instruments was reinforced by his connections with London's instrument makers. His whole approach to science was technical and mechanical, and his work does not show the theological interests of his contemporaries Boyle and Isaac Newton. Hooke believed in the importance of up-to-date instrumentation, and he campaigned against the Danzig astronomer Johannes Hevelius's refusal to use telescopic sights on his astronomical instruments.

One of Hooke's earliest publications was a collection of microscopic observations called *Micrographia* (1665). This elaborate and beautifully illustrated work, dedicated to King Charles II (r. 1660–1685), is a semiofficial statement of the Royal Society's program, ranging over many scientific subjects. It includes the first biological use of the word and concept of "cell," which appears in a discussion of the structure of cork, and it also includes a discussion of combustion as analogous to breathing. Hooke was also the first to demonstrate the rotation of Jupiter, and he devised "Hooke's law," relating the pressure on a spring to its deformation. Springs were a particular area of Hooke's interest because of his extensive and ingenious work on timepieces. He was also an important geologist, insisting that fossils represent actual early life, as opposed to being jokes of nature, and putting forth a controversial theory claiming that the Earth had shifted its axes. His published *Cutlerian Lectures,* the products of a professorship of mechanics endowed for him by Sir John Cutler, were among the classic works of seventeenth-century mechanics.

Hooke remained a professional technician and mechanic rather than a gentlemanly amateur natural philosopher like Boyle. As such, he put great stock in the ownership of his ideas, and he was involved in several disputes with those he claimed took credit for his ideas or distributed them improperly. This problem was compounded by Hooke's versatility; he would set forth a myriad of ideas, without always fully developing them or following up on them. He accused Henry Oldenburg of depriving him of the credit for inventing the balance spring watch by leaking Hooke's idea and giving the credit to Christiaan Huygens. Hooke's most lengthy feud was with Isaac Newton, whom he charged with stealing from him the idea for the inverse-square law of gravitation. Although not a mathematical ignoramus, Hooke did not have the mathematical skills to understand Newton's achievement, and the charge was inaccurate. As might be expected of any dispute involving Hooke and Newton, two of the most gifted grudge-holders of the scientific revolution, the feud was long and ugly, and it still tends to divide scholars into Newtonites and Hookeists. Newton took up the presidency of the Royal Society only on Hooke's death.

Hooke's social situation improved when he succeeded his old enemy, Oldenburg, as secretary to the Royal Society in 1677. He held the post until 1682. He also received a medical degree by order of the archbishop of Canterbury in 1691. (Although Hooke had no formal medical training, there was a certain appropriateness to this, as he was sickly and one of the great medicine-takers of the seventeenth century.) Hooke's life was centered in London, where he was an active participant in social and coffee-house life. Unlike most English scientists, he never traveled to the Continent. His intellectual world was also insular; for instance, he displayed little interest in Cartesian theory. The important influences on his natural philosophy were English, particularly the "magnetical philosophy" of William Gilbert and the work of Francis Bacon, which Hooke tried to develop into a "Philosophical Algebra," or method for making discoveries.

Hooke's chief source of financial support was not his scientific activities but his extensive business as a London surveyor and architect. Like his close friend Christopher Wren, he was in great demand after the Great Fire of London in 1666. Hooke, like many early modern scientists a lifelong bachelor, died a wealthy man. In a final irony that he would not have relished, his *Posthumous Works* appeared two years after his death, with a dedication to Isaac Newton.

> *See also* Air Pumps; Boyle, Robert; Clocks and Watches; Experiments; Geology; Gravity; Newton, Isaac; Royal Society; Technology and Engineering.
> **References**
> Espinasse, Margaret. *Robert Hooke*. London: William Heinemann, 1956.
> Hunter, Michael, and Simon Schaffer, eds. *Robert Hooke: New Studies*. Woodbridge, England: Boydell, 1989.

Humanism

Humanism, the intellectual movement associated with the Renaissance, also played an important role in the genesis of the scientific revolution. The movement originated in Italy in the fourteenth century and spread to the rest of Europe in the fifteenth century. Humanists studied the texts of ancient writers, pagan and Christian, rather than focusing on the disciplines of logic, natural philosophy, and abstract theology that dominated the medieval university curriculum. Humanists were not anti-Christian, although originally they studied "human" texts rather than the Bible. Humanists emphasized original texts rather than commentaries, and whole works rather than selected statements taken out of context, as had been the Scholastic method.

Humanists were largely responsible for the revival of the knowledge of ancient Greek in the West, as well as for promoting a Latin style based on ancient Roman rather than the Latin of medieval writers. Humanistic study was accompanied by efforts to search out surviving manuscripts of ancient Greek and Latin works from libraries and to make them public, eventually through print. In the sixteenth century, humanism became a great educational movement at the university- and lower-school level, setting the curriculum for European schools until the nineteenth century. Most early modern natural philosophers had some humanistic education, and humanistic Latin was an important medium of scientific communication through the seventeenth century.

An anachronistic reading of the twentieth-century divide between the "two cultures" of the sciences and the humanities has led some to describe the early modern humanists as antiscience. This is not true. Although science never matched rhetoric or ethics as a concern for humanists, a number of the texts they studied and published relate to issues of natural philosophy. The humanists wanted to create improved texts of the Greek authorities in natural philosophy—the greatest, of course, being Aristotle, several of whose works were rediscovered in this period. Humanists often opposed, or simply disregarded, the Aristotelianism of the schools, which had surrounded the original Aristotle with a thick crust of Arabic and Latin commentaries. But they were usually more positive about Aristotle himself and sometimes also his ancient Greek commentators, such as Alexander of Aphrodisias (fl. c. 200 A.D.). Humanists studied other ancient Greek scientific authorities as well, including Ptolemy (A.D. 90–168) in astronomy, astrology, and geography; Euclid (fl. c. 300 B.C.), Archimedes (c. 287–212 B.C.), and Apollonius (c. 262–c. 190 B.C.) in mathematics; Strabo (c. 63–c. 25 B.C.) in geography; Theophrastus (c. 372–c. 287 B.C.) and Dioscorides (A.D. c. 40–c. 90) in botany; Hermes Trismegistus in natural theology; and

CÆLIQVE VIAS ET SIDERA MONSTRAT

THE SPHERE of M. MANILIVS made An English Poem. by Edward Sherburne, Esq

NATVRÆ VNIVERSITAS

VNIVERSITATIS INTERPRE

This frontispiece to an English translation of an ancient Latin poem about astronomy reconciles science and humanism by showing the Muse of Astronomy, Urania, looking through a telescope. (M. Manlius, The Sphere, *London, 1673)*

Galen (A.D. 129–c. 199) and Hippocrates (c. 460–377 B.C.) in medicine. Although Roman writers were not accorded the same intellectual authority as Greeks, there were also important natural-philosophical aspects to the study of Pliny the Elder (A.D. 23–79) in natural history, the Stoic philosopher Seneca (c. 4 B.C.–A.D. 65), the epic poets Lucretius (c. 96–c. 55 B.C.) and Ovid (43 B.C.–A.D. 17), the engineer Sextus Julius Frontinus (A.D. c. 35–103), and the medical encyclopedist Celsus (2nd c. A.D.).

The reverence of humanists for Greek and to a lesser extent Roman authorities in natural science was exceeded only by their contempt for medieval Latin and to an even greater extent Arab writers. Such writers were *barbari*—barbarians. For mainstream humanists, study of the Arab writers such as Avicenna (980–1037) was valued only as a way of understanding the Greeks.

By the sixteenth century, much of the humanistic project revolved around textual editing, especially the recovery of the exact meaning of ancient texts in their ancient contexts. This included mathematical, scientific, and medical texts. Given all the corruptions introduced by the copying and recopying of manuscripts over the centuries, and given the differences between different manuscript traditions, humanist editors had to use natural-philosophical knowledge to reconstruct the wording and meaning of the texts. Thus, in reconstructing these texts, humanists made contributions to natural-philosophical knowledge. For example, Andreas Vesalius was restoring as much as challenging the text of Galen, some of which he had edited, when he published the results of his dissections in *Of the Fabric of the Human Body* (1543). The Italian mathematical humanist Federigo Commandino (1509–1575) translated many Greek mathematical works into elegant Latin, greatly contributing to the advance of mathematics. Many of the humanist botanical works were concerned with identifying the plants described by the ancient botanists, and sometimes botanists apologized for the inclusion of plants unknown to the ancients!

Humanists revived various non-Aristotelian schools of ancient philosophy as vital intellectual options. So rich and diverse was the ancient tradition that there was little interest before René Descartes in rejecting it totally. Galileo Galilei flaunted his contempt for Aristotelian physics but invoked Archimedes (c. 287–212 B.C.) as an alternative classical Greek hero and role model. There was a long-standing humanist interest in Plato, the vast majority of whose writings had not been available in the medieval West. Epicureanism, most notably represented in Lucretius's frequently translated and reprinted poem, *On the Nature of Things,* and Stoicism also attracted interest. Pierre Gassendi revived Epicurean atomist physics in a Christianized form.

Then Gassendi's contemporary and rival in mechanical philosophy, Descartes, consciously broke with the humanist tradition. Although he was the recipient of a fine humanist education from a Jesuit school, Descartes rejected the humanist focus on the classical texts and produced a new model of the universe that he claimed was not substantially indebted to the ancients. He also published works in French rather than Latin, aiming at a broad audience, including women, whom humanism mostly excluded. Thomas Hobbes, after spending much of his career as a humanist, also broke with what he considered to be the overly rhetorical and insufficiently mechanical ancient tradition in both his natural and political philosophy.

As the humanists recovered more knowledge of the particular context in which ancient thinkers worked, the limitations of these thinkers became more evident. Aristotle, for example, was not the medieval "master of those who know" but a particular man working in a particular time. The study of antiquity itself became more "scientific" during this period, with triumphs of interpretation, such as the 1614 work of Isaac Casaubon (1559–1614), demonstrating from linguistic evidence that the writings of Hermes Trismegistus date from the first Christian centuries rather than from the time of Moses. Greater awareness of the context of ancient intellectual activity was the basis of the late-seventeenth-century "quarrel of the ancients and moderns," most furious in England and France, in which the moderns attacked ancient authority not because they were ignorant or contemptuous of the ancients, but because they had a better grasp of the limitations of ancient thought. The rise of Cartesian and Newtonian natural philosophy allowed late-seventeenth-century scientists to claim that they had gone beyond the ancients, and that ancient natural-philosophical writings were no longer relevant to science. Latin was also declining as a medium for scientific communication.

The distinction between science and humanism at the end of the scientific revolution should not be exaggerated. The first Boyle lecturer, who employed Newtonian science to support Christianity, was Richard Bentley (1662–1742), a leading modern and England's greatest classical philologist. Even in the early eighteenth century, Edmond Halley's principal achievement as Savilian Professor of Geometry at Oxford was an edition of the Greek text of Apollonius's *Conics*.

See also Aristotelianism; Epicureanism; Galenism; Hermeticism; Platonism; Stoicism.
References
Gaukroger, Stephen, ed. *The Uses of Antiquity: The Scientific Revolution and the Classical Tradition.* Dordrecht, the Netherlands: Kluwer Academic Publishers, 1991.
Mandrou, Robert. *From Humanism to Science, 1480–1700.* Translated by Brian Pearce. N.p: Penguin Books, 1978.

Humors
See Galenism.

Huygens, Christiaan (1629–1695)
Christiaan Huygens, born into a wealthy and cultivated Dutch family of high civil servants,

became the most celebrated scientist between Galileo and Newton, making significant and original contributions to physics, optics, mathematics, clockmaking, and astronomy. He was a great international figure who corresponded widely, spent much of his career in Paris, and had significant German and English connections. Through his father, secretary to the stadtholder, or leader, of the Dutch Republic, the young Huygens met René Descartes and corresponded with Marin Mersenne.

Huygens was educated at home until he attended the University of Leiden in 1645, and he inherited enough wealth to free him for a life devoted to science. His earliest publications were in the field of mathematics, applying algebraic analysis to geometrical problems. He also had a lifelong interest in optical problems, telescopes, and microscopes. The Netherlands was a center of lens-grinding, and Huygens, along with his brother Constantijn, made excellent telescopes with improved eyepieces. Using one of these telescopes, in 1655 Christiaan was the first to observe the moon of Saturn called Titan. Huygens was also the first to identify Saturn as a ringed planet, solving the problem of the strange appendages to the planet, which had mystified astronomers since Galileo.

Like many Dutch thinkers, Huygens was Cartesian in his natural philosophy, but he was not dogmatically so; he disproved, for instance, Descartes's theories of impact. In mechanics, Huygens applied a mathematical approach, stemming from the work of Galileo, to a basically Cartesian universe. He invented a new theory of impact, and he produced the modern mathematical formula for centrifugal force.

Huygens's mechanics were always aimed at practical application. In 1657, he published *Horologium,* applying the pendulum principle to clocks. Unaware of Galileo's previous design, Huygens invented the pendulum clock, hoping to solve the longitude problem. Unfortunately, pendulum clocks proved too unreliable aboard ship for this to work.

Huygens continued to work on the problem of the marine clock for the rest of his life.

Huygens became one of the first foreign members of the Royal Society in 1663. He lived in Paris from 1666 to 1681, drawing a salary as a member of the Royal Academy of Sciences. (Huygens and the Italian astronomer Gian Domenico Cassini were the first foreign members of the Royal Academy.) While in Paris, he published further work on pendulum theory in *Horologium Oscillatorum* (1673), in which he demonstrated the method of creating isochronous pendulums through the pendulum describing a cycloid curve. This book was dedicated to Louis XIV (r. 1643–1715), who at the time of its publication had invaded the Dutch Republic, Huygens's native country, and was occupying much of it! After a visit to The Hague in 1681, Huygens decided not to return to France, where the government was increasing its persecution of Protestants. He spent the rest of his life at The Hague, taking a trip to England in 1689, when he met Newton.

Huygens's reactions to Newton's theories were hostile, although not totally dismissive, and his later works, such as *Treatise on Light* (1690), were defenses of pure Cartesian mechanism against Newton's theory of gravity and its reintroduction of what Huygens considered to be occult qualities. Huygens devised a Cartesian wave theory of light in part to refute Newton as well as to explain the peculiar double refraction of Icelandic spar. He also wrote a work of scientific popularization, published posthumously as *Kosmotheoros* (1698), setting forth a Copernican universe powered by Cartesian vortices with the addition of Newton's gravitational formulas.

See also Cartesianism; Clocks and Watches; Mechanics; Navigation; Physics; Royal Academy of Sciences.

References

Bell, A. E. *Christian Huygens and the Development of Science in the Seventeenth Century.* London: Arnold, 1947.

Van Berkel, Klaas, Albert Van Helden, and Lodewijk Palm, eds. *A History of Science in the Netherlands: Survey, Themes, and Reference.* Leiden, the Netherlands: Brill, 1999.

Illustration

The scientific revolution saw changes in illustration toward more precise images of nature and away from allegorical representations. Techniques derived from art, such as perspective, were applied to the representation of natural phenomena. Scientific illustration was also transformed by printing.

Printing made possible the replication of scientific images in a way that had not been possible in the manuscript era. Uniform illustrations were impossible when each had to be made by hand, particularly when copies were made from copies. Print made possible a standardized image that could in theory be reproduced many times without variation. In botany, for example, some of the medieval manuscript illustrators had produced remarkably accurate and beautiful drawings of plants. But each had to be created separately, and none were precisely identical.

The permanence and replicability made possible by print also had drawbacks. Illustrations, then as now, were among the most expensive elements of a printed book; cheaper editions of expensive texts often omitted them altogether. Illustrations were usually created by specialized artisans who worked with the author and printer, although some scientific authors, such as Johannes Hevelius and Anna Maria Sibylla Merian,

were skilled engravers and made their own illustrations. Once created, illustrations often lived a life of their own, not always connected to the original text. As a book went from edition to edition with abridgments, expansions, and translations, illustrations were often moved from their original placement in the text. This was a particularly significant problem for botanical texts, in which the same illustration could thus be attached to written descriptions of altogether different plants. Illustrations, particularly woodcuts, were vulnerable to degeneration after repeated imprints and could also be pirated or simply reused and attached to texts to which they had little intellectual relevance. Since the process of producing a new woodcut was expensive, woodcuts tended to be used over and over—one set of botanical illustrations was used for 200 years. The other major form of reproduction, engraving, produced a finer and more detailed image but degenerated even faster. It was also difficult to combine engraved images with printed text on the same page, meaning that a group of engravings were often segregated on a page of their own. Although the ideal was to take illustrations directly from nature, in practice particularly important images tended to serve as models for later illustrations.

The illustrations least meant to be taken as

ASPARAGVS   Sparġen.

This graceful illustration of an asparagus plant is part of a trend toward greater realism in sixteenth-century botanical illustration. (Fuchs, De Historia Stirpiun, 1542)

descriptions of actual phenomena were the elaborately allegorical and emblematic frontispieces to major treatises of natural philosophy. These compositions, which often require a great deal of decoding, were designed to advance an intellectual and cultural position through the manipulation of a language of symbols. Johannes Kepler actually used the frontispiece of his *Rudolphine Tables* (1627) to complain about poor funding; the frontispiece shows the double-headed eagle of the imperial house of Hapsburg dropping coins from its beak onto astronomers, but only a few coins reach Kepler, who is writing on a tablecloth because he is too poor to buy paper. Some new apparatuses of natural philosophy were included in frontispieces as emblems. For example, the telescope featured in the frontispiece to Athanasius Kircher's *The Great Art of Light and Shadow*

(1646) serves as an emblem for sense as the true source of knowledge. (Jesuits like Kircher were particularly fond of emblems.) The air pump also became an emblem, this time for the new science itself: The frontispiece to Thomas Sprat's (1635–1713) *History of the Royal Society* (1667) prominently features an air pump as an emblem of the experimental science that Sprat's history promotes. Old and new emblems could be combined in what seems to us to be incongruous ways. For instance, to argue that using the air pump in natural investigation can be a pious activity, one might have an illustration in which an air pump is worked by cherubs.

Just as elaborate were some of the emblematic illustrations used in alchemical texts, embodying the complex set of alchemical metaphors in pictures whose direct reference to chemical or natural phenomena and procedures could only be decoded by the knowing eye. The god Saturn urinating into a bowl could represent the amalgamation of the star regulus of antimony ("Saturne's pisse") with mercury. This allusive style of illustration reached its apogee in the elaborately illustrated works of the English physician and magician Robert Fludd, who viewed pictorialization as a way of understanding the cosmos.

Other illustrations claimed to be accurate representations of natural objects. Perhaps the most famous example of the early scientific revolution is the illustrations to Andreas Vesalius's *Of the Fabric of the Human Body* (1543), which had as much influence as the text and were also reproduced separately for the benefit of students of anatomy unable to afford Vesalius's treatise. Similarly, the great herbals of the early sixteenth century influenced the development of botany through the detail of their representations of plants. The practical use of botany for medicine meant that accuracy was more important in depicting plants than animals, and accurate botanical illustrations preceded accurate zoological illustrations by some time.

Many representations of animals still concentrated on their symbolic functions in the tradition of the medieval bestiaries. On maps, for example, animals resident in a particular area denoted that area, with bears, coyotes, and deer, for example, often denoting America. Animals were usually depicted as motionless, reflecting the fact that many of them had not originally been drawn from life, but from stuffed specimens. This was particularly the case with the exotic fauna of the New World because most printed images were produced by illustrators living and working in Europe.

Early modern illustrators also differed from artists in that they usually depicted individual plants and animals isolated from their natural context. Despite their proclaimed realism, these illustrations often represent not an actual thing, but an idealized version, emphasizing those aspects fitting the natural-philosophical argument being made. Even more idealized were abstract diagrams representing assertions about physics. These grew even more popular after the analytic geometry of René Descartes showed how to represent equations as curves.

New technological devices also had to be visually represented to communicate their workings. Fifteenth- and sixteenth-century engineers developed techniques for representing complex machines, such as the "exploded" drawing, which represents the parts separately, and the cutaway drawing, in which part of the surface is removed to reveal the inner workings. Experiments were also depicted in illustrations. These ranged from those that claimed to give a picture of the actual experiment to highly idealized representations meant to strip away all extraneous elements to display the physical principles involved. Isaac Newton's illustrations of his famous prism experiments in *Opticks* (1704) are simply geometrical drawings of the rays of light.

The discernment of new scientific objects through the sense enhancements provided by the telescope and microscope posed new challenges for illustrators. Illustrations were necessary in order to provide the reader who lacked access to such devices a direct experience of the wonders they revealed. This required close collaboration between the investigator and the artist, or better yet, that the investigator and the artist be the same person, as in the case of Hevelius, whose self-published lunar maps and other astronomical illustrations were of unsurpassed quality. In microscopy, the standard was set by Robert Hooke's *Micrographia* (1665). Despite the importance of Hooke's text, much of *Micrographia*'s initial popularity was based on the numerous, excellent, and often bizarre illustrations produced under Hooke's supervision, such as that of the compound eyes of a housefly.

See also Art; Printing.
References
Dickenson, Victoria. *Drawn from Life: Science and Art in the Portrayal of the New World.* Toronto: University of Toronto Press, 1998.
Hunter, Andrew, ed. *Thornton and Tully's Scientific Books, Libraries, and Collectors: A Study of Bibliography and the Book Trade in Relation to the History of Science.* 4th ed. Brookfield, VT: Ashgate, 2000.
Mazzolini, Renato G., ed. *Non-Verbal Communication in Science Prior to 1900.* Florence: L. S. Olschki, 1993.

Impetus
See Force; Medieval Science.

Indian Science
Although Europeans had many contacts with India in the early modern period, the two scientific traditions had surprisingly little impact on each other. Like much of Indian culture, the major contribution of the Indian scientific and mathematical tradition had already reached Europe through the intermediary of Islamic civilization. This contribution was the so-called Arabic numerals, actually invented in India. They reached Europe in the Middle Ages and slowly overcame their chief rivals, Roman numerals and the primitive

abacus of Europe. The sixteenth century was decisive for this transformation; the last arithmetic book in Roman numerals was published in 1514. The Indian numerals enabled faster and more accurate calculation and also made it easier to deal with much larger numbers, a necessity for scientific advance.

The circumnavigation of Africa by the Portuguese sailor Vasco da Gama (c. 1460–1524) in 1498 enabled direct European contact with the science of India, without the Islamic intermediary. This had the most impact in the field of medicine and medical botany, particularly with the work of the Portuguese physicians García d'Orta and Christovão da Costa (c. 1515–1580) on the drugs and remedies of India. Their works incorporated much Indian knowledge of plants and diseases, particularly from the Hindu Ayurvedic physicians of the Portuguese colony of Goa. An Italian visitor to Goa, Filippo Sassetti (1540–1588), also translated an Ayurvedic compendium, the *Nighantu,* from Sanskrit to Italian. In the seventeenth century, the Dutch governor of Malabar, Henricus Van Rheede Van Draakenstein (c. 1637–1691), employed Indian physicians in the preparation of his 12-volume botanical compilation, *The Garden of Malabar in India,* published in Amsterdam from 1678 to 1703. Indians were also interested in European medicine, and several European physicians practiced medicine among Indians.

Outside medicine, contacts between Europeans and Indians were mainly in the field of technology. Indian princes employed Europeans as gunners, engineers, and jewelers, and Indian painters adopted European techniques of mathematical perspective. But contacts in science remained few. The most distinguished European natural philosopher to visit India was the French Gassendist, Francois Bernier (1620–1688), the author of a very popular travel book called *Memoirs of the Grand Mogul's Empire,* which went through many editions and translations in Europe.

While in India, Bernier translated works of Gassendi and Descartes into Persian for the benefit of his patron Danishmand Khan, but these works had little influence on Indian thought.

See also Orta, García d'.

References
De Figuerido, John M. "Ayurvedic Medicine in Goa According to the European Sources in the Sixteenth and Seventeenth Centuries." *Bulletin of the History of Medicine* 58 (1984): 225–235.
Qaisar, Ahsan Jan. *The Indian Response to European Culture and Technology (A.D. 1498–1707).* Delhi: Oxford University Press, 1982.

Inertia

See Force; Mechanics; Physics.

Infinity

Medieval scholars had discussed infinity mainly as an attribute of God, but during the scientific revolution, Europe's natural philosophers and mathematicians also grew accustomed to the idea of the infinity of the universe. In cosmology, the Aristotelian-Ptolemaic picture of a universe enclosed within the sphere of the fixed stars was replaced by one without clear limits in any direction. In mathematics, a concept of an infinite geometrical space was developing.

The idea of an infinite universe was radical, associated mainly with ancient atomists and Epicureans. The traditional limited universe, enclosed by a vast sphere of the fixed stars, was not challenged even by Copernicus, who simply replaced the Earth with the Sun as the center of the universe. However, the idea of an infinite universe was soon put forth by Copernicans, notably Thomas Digges (c. 1545–1595) and Giordano Bruno. Digges, author of *Perfit Description of the Caelestiall Orbes According to the Mose Aunciene Doctrine of the Pythagoreans Lately Revived by Copernicus and by Geometricall Demonstrations Approved* (1579), like Copernicus believed in the celestial spheres, but he substituted an infinite space with stars for the sphere of the fixed stars, an "infinite Orbe immovable." Digges had relatively little influence, however, and the most zealous public champion of the infinite universe was Bruno, author of the forthrightly titled dialogue *The Infinite Universe and Worlds* (1584). Unlike Copernicus and Digges, Bruno pictured an Epicurean-influenced, infinite universe with no center, populated by an infinite number of stars and worlds like ours. Bruno's universe was also homogenous in that its qualities and the principles of its behavior were everywhere the same.

The idea of an infinite universe may not have gained much from its association with the heretic Bruno. Although endorsed by William Gilbert, it was denied by Johannes Kepler, who pointed out that its homogeneity was refuted by the uneven distribution of the stars. Kepler's idea of a mathematically ordered and harmonious universe was hardly compatible with its infinity. The universe of René Descartes was indefinite in that it could not be assigned limits, but it was not properly infinite—a quality Descartes reserved for God. The Cambridge Platonist Henry More (1614–1687) neatly combined infinite space with an infinite God by making space an attribute of God.

The question of the infinity of the universe was a concern primarily limited to those natural philosophers interested in vast cosmological questions. Many practicing scientists, such as Galileo Galilei, got along quite well by largely ignoring it. While cosmologists were wrestling with the notion of spatial infinity, geometers were approaching the problem from another direction. The French mathematician Girard Desargues (1591–1661), founder of projective geometry, was the first geometer to make the notion of infinity central to his geometry, defining lines and planes as infinitely extended. He also described parallel lines as meeting at infinity. This was followed up by Blaise Pascal, who famously spoke of the terrifying quality of infinite space.

Isaac Newton knew Desargues's work, as well as More's, and he mathematized More's infinity of space and its relation to God. Newtonian infinite space was absolute, independent of the relations between the bodies it contained. It was also compatible, for Newtonians, with a finite amount of matter. Thus, by the late seventeenth century, the notion of an infinite universe could be held in different ways. Newton's rival Gottfried Wilhelm Leibniz, a passionate opponent of the idea of absolute space, deduced an infinite universe from the fact that God would create the fullest universe possible—the "best of all possible worlds," combining the greatest diversity of phenomena with the maximum simplicity. For Leibniz, in contradiction to the Newtonians, an infinite universe meant infinite matter as well as infinite space since he denied the possibility of a vacuum. In the eighteenth century, Newtonians too would accept the infinity of matter.

See also Bruno, Giordano; Leibniz, Gottfried Wilhelm; Newton, Isaac.

References

Field, J. V. *The Invention of Infinity: Mathematics and Art in the Renaissance.* Oxford: Oxford University Press, 1997.

Koyre, Alexander. *From the Closed World to the Infinite Universe.* Baltimore: Johns Hopkins University Press, 1957.

Instruments

See Barometers; Clocks and Watches; Microscopes; Technology and Engineering; Telescopes; Thermometers.

J

Jesuits

The Society of Jesus, a Roman Catholic religious order founded by Ignatius of Loyola (1495–1556) and recognized by the pope in 1540, was not originally intended to be heavily involved in science. But along with discipline, the Jesuits emphasized education. Every member had a university education, and the Jesuits saw the excellence of their own schools and their intellectual prestige as weapons to convert Protestants. In the seventeenth century they emerged as the scientific elite among Catholic religious orders. The Jesuit college in Rome, the Collegio Romano, was a leading Catholic scientific institution, and Jesuits, notably Christoph Clavius, were in charge of the important task of making astronomical calculations for the pope. Although no other Catholic order was as hated by Protestants, particularly in England, many Protestants sent their sons to Jesuit schools, and Protestant schools used Jesuit textbooks on nonreligious subjects, such as Clavius's excellent mathematical textbooks.

From the late sixteenth century on, Jesuit schools were known for mathematics. Jesuit educators trained many important scientists and mathematicians in the Catholic world, such as René Descartes, and Jesuit scholars were leaders in mathematical sciences such as optics and astronomy. Jesuits were also lead-ers in experimentation, contributing to optics, electric science, and magnetism. The Jesuit Francesco Maria Grimaldi (1618–1663) performed important experiments on the diffraction of light and gave one of the first accurate accounts of it. The German Jesuit polymath Athanasius Kircher and another Jesuit, the Italian Niccolò Cabeo (1586–1650), were leading students of magnetism. Cabeo was the first to describe electrical repulsion.

The Society of Jesus was, and is, a global order, represented in the seventeenth century in every inhabited continent save Australia. The Jesuit José de Acosta was a leading natural historian of South America, and the Jesuit Matteo Ricci (1552–1610) introduced European science to China. In the mid-seventeenth century, Kircher in Rome was at the center of a vast worldwide web of Jesuit scientific correspondence. Jesuit missionaries sent information to the French Royal Academy of Sciences. Quinine, a South American import, was known in Europe as "Jesuit's bark" because of the zeal of Jesuits in promoting its medicinal use.

Nonetheless, there were limits on Jesuit intellectual endeavor. Jesuits were required not only to defend Catholic doctrine in matters scientific (as in all other areas) but to believe it. Ignatius specifically required

Jesuits to believe that white is black if that was the church's decree, so believing that the Sun goes around the Earth was not much of a challenge to a good Jesuit. Jesuits such as the astronomer and sunspot expert Christoph Scheiner (1573–1650) led the opposition to Galileo Galilei's Copernicanism (although Galileo had enjoyed good relations with the Jesuits early in his career). Although there was initially some intellectual pluralism among Jesuits, following a decree by Superior General of the Order Claudio Aquaviva (1543–1615) in 1611, all Jesuits were required to defend the authority of Aristotle in philosophy. During and after the trial of Galileo in 1633, Jesuits were the leading defenders of Aristotelian natural philosophy and Earth-centered astronomy (Tychonic rather than Ptolemaic) in the Catholic Church.

Early modern Jesuits always subordinated natural philosophy to religious ends, and their science was dominated by the search for signs and emblems of God in the created universe. Kircher, for example, was as fascinated by stones formed in the shape of the cross or other religious symbols and personalities as he was by the properties of magnets. Like the defense of Aristotelianism, this emblematic approach required the rejection of the mechanical philosophies put forth by the Catholics René Descartes and Pierre Gassendi. Jesuit scientists generally did not broaden their individual discoveries into general arguments on natural philosophy, and they were willing to put forth and analyze a number of explanations for phenomena without necessarily picking one out as the truth. The Jesuit Giambattista Riccioli (1598–1671) was one of the greatest observational astronomers of the seventeenth century, particularly noted for his mapping of the Moon, but he never accepted the Copernican theory, despite his awareness of the strength of the mathematical arguments for it.

By the late seventeenth century, the Jesuit order was clearly no longer a leader in science. Jesuits were excluded from the French Royal Academy of Sciences because of their Aristotelianism, which was considered dogmatic. The Jesuits in France also had to compete with the rising Oratorian order, which was also involved in education and adopted a Cartesian stance. Roman and other Italian Jesuits were affected by the general decline of Italian science in the late seventeenth and early eighteenth centuries. Scientists in the active scientific centers of France, England, and the Netherlands were increasingly likely to treat Jesuit scholars not as natural-philosophical authorities but as useful gatherers of information.

See also Acosta, José de; Clavius, Christoph; Collegio Romano; East Asian Science; Kircher, Athanasius.

References

Dear, Peter. *Discipline and Experience: The Mathematical Way in the Scientific Revolution.* Chicago: University of Chicago Press, 1995.

O'Malley, John W., S.J., and Garvin Alexander Bailey, Steven J. Harris, and J. Frank Kennedy, S.J., eds. *The Jesuits: Cultures, Sciences and the Arts, 1540–1773.* University of Toronto Press, 1999.

Jewish Culture

Although they were early modern Europe's most literate population, Jewish people made few original contributions to the scientific revolution. Reasons for this include their exclusion from most universities outside Italy, the difficulty faced by the Jewish community in supporting intellectuals with the leisure to investigate nature, and the otherworldliness of the Lurianic Kabbalah, which gained popularity among Jews in the seventeenth century. Furthermore, some rabbis denounced the seeking of natural knowledge as impious. Nor did the Christian scientific community always welcome Jewish contributions. For instance, when the great physician Amatus Lusitanus (1511–1568), of Portuguese Jewish descent, ventured some mild criticisms of Pier Andrea Mattioli's (1501–1577) *Commentary* on the ancient botanist Dioscorides, the pugnacious Mattioli responded by accusing Lusitanus of "having most perfidiously turned away from God, the Eternal." Professing Jews were also not eligible for membership in European scientific academies and societies.

The Jewish community in early modern Europe already had an established tradition of studying nature. Jewish culture, like Christian, encouraged an examination of the natural world for evidence of the divine. Alchemy flourished in the Jewish community, often being practiced by eminent rabbis and Kabbalists. Hebrew alchemical treatises were translated into European languages and eagerly read by Christian alchemists. But medicine was the main point of contact between Jewish and Christian science. Jewish physicians were common in early modern Europe, in a tradition dating back to the Middle Ages. Medicine was the one learned profession outside Jewish religious life that Gentile society allowed Jews to practice. The physician occupied a place of honor in the Jewish community, often combining medical practice with the rabbinate. Jewish physicians also served Christian patients, including kings and popes.

The question of the Jewish role in early modern medicine is complicated by the role of *conversos*. These were Jews, mostly from Spain and Portugal where many Jewish physicians had practiced under medieval Muslim rule, who had been forcibly converted to Christianity in the fifteenth century, and their descendants. Some *conversos,* the so-called crypto-Jews, practiced Judaism in secret, which made them targets for persecution by the Inquisition. Others, including Lusitanus, emigrated to the Muslim world or to the few Christian places that allowed reversion to Judaism, such as Amsterdam, so that they could live openly as Jews. Many *conversos,* both those who reconverted to Judaism and those who remained Christian, were physicians. The other significant community of Jewish physicians, which overlapped with the *conversos,* was made up of graduates of the medical schools of the Italian universities,

most importantly the University of Padua. Although Jews labored under significant disadvantages at Padua and were barred from the highest degrees, hundreds attended the medical school during the early modern period, learning not only medicine but Aristotelian natural philosophy.

Jewish and *converso* physicians published scores of medical books, generally aligned with the Galenic medical tradition. Traditional medicine, with its pagan and Muslim roots, appealed to Jews as less religiously exclusive than the new Christian medicine called for by Paracelsus, who despised Jewish physicians. However, Jewish and *converso* physicians exposed to intellectual innovation in places such as Padua or Amsterdam were influenced by new medical discoveries, such as William Harvey's circulation of the blood. Some *converso* or Jewish physicians, such as the sixteenth-century natural historian of India, García d'Orta, did publish important scientific books, but the Jewish and *converso* medical community was mainly one of practitioners rather than theorists. Most Jewish contributions to early modern medical knowledge were in the field of clinical medicine, such as Lusitanus's *Centuries*—a seven-volume work, each volume containing 100 case histories covering both medicine and surgery, including one of the first discussions of the valves of the veins.

An alternative route for Jews into the early modern scientific community was astronomy. Rabbis had long been concerned with astronomy for calendrical calculations, and new astronomical developments among Christians attracted rabbinical interest. Most familiar with Christian astronomy in the sixteenth century was the Prague rabbi David Gans (1541–1613), an acquaintance of both Tycho Brahe and Johannes Kepler and one of the few Eastern European Jews involved with the new science. Gans emerged from a rabbinical culture that was delineating a sphere for science separate from specifically Jewish knowledge. He hoped to revive astronomy as a study among Eastern European Jews and wrote a textbook on the subject. However, the disasters that befell Bohemia (the present-day Czech Republic) and Poland in the first half of the seventeenth century, such as the Thirty Years War (1618–1648) and the Chmielnicki massacres of Jews in the Ukraine in 1648, ended Jewish astronomical study in Eastern Europe. The next Jewish astronomical student of note was an Italian, Joseph Solomon Delmedigo (1591–1655), a physician, alchemist, and student of Galileo Galilei. Unlike Gans, Delmedigo accepted the Copernican system and wrote about it in Hebrew, combining advocacy of increased study in mathematics and astronomy in the Jewish community with belief in the Kabbalah.

Tobias Cohn (1652–1729) was one of the first Jews to attend a German university. He was driven out by anti-Semites and eventually took his medical degree at Padua. He claimed that it was the assertions of Christian students that Jews were ignorant in the sciences that prompted him to remedy this condition. Living in Istanbul as physician to the Turkish sultan, Cohn wrote a Hebrew textbook of medicine, first published in 1707, with a lengthy introduction covering theology and natural philosophy. This frequently reprinted work was the most extensive discussion of seventeenth-century science aimed at Jewish readers. Cohn's attitude toward the scientific revolution was mixed. He described Copernicus as a child of the devil, but he delineated Harvey's circulation of the blood for the benefit of his physician readers. He also endorsed chemical medicine in the Paracelsian tradition, one of the few Jewish physicians to do so.

See also Kabbalah; Orta, García d'; Spinoza, Baruch.
References
Friedenwald, Harry. *The Jews and Medicine: Essays.* 2 vols. Baltimore: The Johns Hopkins University Press, 1944.
Patai, Raphael. *The Jewish Alchemists: A History and Source Book.* Princeton: Princeton University Press, 1994.
Ruderman, David B. *Jewish Thought and Scientific Discovery in Early Modern Europe.* New Haven, CT: Yale University Press, 1995.

Journal des scavans

Denis de Sallo (1626–1699), a French lawyer, received a patent in 1664 to publish a weekly journal for the learned. The *Journal des scavans,* the first periodical devoted to intellectual subjects, published its first issue on January 5, 1665, beating the English *Philosophical Transactions* by a few months. The principal business of the journal was to publish reviews of new books in all fields of knowledge, but it also described scientific experiments and new observations. It was from the beginning intended to be a European rather than a parochially French publication; contributions were solicited from Henry Oldenburg in England and correspondents in the Dutch Republic, and the *Journal* circulated Europewide.

After a rocky start caused by the journal's stance on religious issues, it settled down as a report of new books, written in readable French prose and expressing opinions on leading scientific, medical, and other issues of the day. The editor also adapted and translated material from other journals, notably *Philosophical Transactions,* and held a weekly open house to show new discoveries and inventions. Some inventions, such as perpetual motion machines, were reported on rather uncritically. The *Journal des scavans* was originally independent of the Royal Academy of Sciences, but it did publish announcements of the academy's deliberations. By the early eighteenth century, the academy established informal control over the journal, with the academy's secretary, Bernard Le Bouvier de Fontenelle, serving as the journal's reviewer of scientific books.

See also Periodicals; Royal Academy of Sciences.
Reference
Brown, Harcourt. *Scientific Organization in Seventeenth-Century France (1620–1680).* Baltimore: Johns Hopkins University Press, 1934.

K

Kabbalah

Kabbalah, a form of Jewish mysticism and magic, was widely accepted in the early modern Jewish community, and many prominent rabbis were Kabbalists. It was also increasingly influential among Christians during the Renaissance and in the seventeenth century. Originally an adaptation of Neoplatonism to Judaism, Kabbalah was based on the idea that the universe proceeds from God not by creation from nothing but by divine emanation. The divine can thus be approached through the successive emanations separating God from the universe. Some Kabbalists treated the biblical Yahweh, not as the ultimate God, but as the first emanation of a more transcendent ultimate God. Kabbalists opposed any system that separated matter from spirit absolutely. Among Jews such as Abraham Yagel (1553–c. 1623), an Italian physician and admirer of Galileo, Kabbalah could legitimize study of the natural world as an emanation of God and a repository of divine signs.

Kabbalah began to be studied by Christians in late-fifteenth-century Italy. Like Hermeticism, Kabbalah was considered by many Christians to be a remnant of the ancient holy wisdom, an attitude also held by some Jews. Indeed, Kabbalah had even more authority than Hermeticism because it was written in Hebrew, which many considered to be the original language and therefore holy and which Christians were beginning to study more carefully beginning in the early sixteenth century. Kabbalah's Jewish origin gave it prestige among those Christians who saw the ancient Jews as the source of all wisdom, the Greeks and Romans being merely disciples. Although the *Zohar,* the fundamental Kabbalistic text, dates from the Middle Ages, it was thought to be much older, even derived from Moses himself. Magicians such as Giordano Bruno and Johannes Kepler incorporated Kabbalistic ideas into their syncretic intellectual systems, Bruno going so far as to publish a Kabbalistic work, *Cabala del cavallo pegaseo* (1585). Such was the prestige of Kabbalah that it was appropriated in many contexts. The Cambridge Platonist Henry More (1614–1687), for example, published *Conjectura Cabbalistica* (1653), a Cartesian interpretation of the first three chapters of the Book of Genesis, despite his complete ignorance of Kabbalah at the time.

Lurianic Kabbalah was formulated in the Palestinian town of Safed by late-sixteenth-century rabbis led by Isaac ben Solomon Luria (1534–1572). It focused on the restoration of those aspects of the divine that were thought to have become entangled in the material world through piety and magical activities rather than the study of nature. Although

Lurianic Kabbalah tended to discourage, or at least not encourage, the study of nature among Jews, its effect on Christians was different.

Lurianic Kabbalah was presented to the Christian world in the most influential seventeenth-century work of Christian Kabbalism, the three-part collection of Latin translations of Kabbalistic texts, with commentaries, by the German alchemist Christian Knorr von Rosenroth (1636–1689), *Kabbalah Denudata* (1677, 1678, 1684). Von Rosenroth, a friend and associate of Francis Mercury van Helmont and Gottfried Wilhelm Leibniz, was an excellent Hebraist whose translations were remarkably accurate. He hoped that *Kabbalah Denudata,* dedicated to "lovers of Hebrew, Chemistry, and Philosophy," would both encourage conversion of Jews to Christianity and promote harmony among Christians. *Kabbalah Denudata*'s Christianized Lurianic Kabbalah contributed to optimism about human action, including scientific investigation, as a means to recover the divine. This late-seventeenth-century Kabbalistic revival influenced such important Christian natural philosophers and opponents of purely mechanistic science as Leibniz and Anne Conway, who found that the Kabbalah supported their views of matter as ultimately spiritual.

See also Helmont, Francis Mercury van; Jewish Culture; Magic.

References
Coudert, Allison P. *The Impact of the Kabbalah in the Seventeenth Century: The Life and Thought of Francis Mercury Van Helmont (1614–1698).* Leiden, the Netherlands: Brill, 1999.
de Leon-Jones, Karen Silvia. *Giordano Bruno and the Kabbalah: Prophets, Magicians, and Rabbis.* New Haven, CT: Yale University Press, 1997.
Ruderman, David B. *Kabbalah, Magic, and Science: The Cultural Universe of a Sixteenth-Century Jewish Physician.* Cambridge: Harvard University Press, 1988.

Kepler, Johannes (1571–1630)

The greatest astronomer of his time, Johannes Kepler created the most mathematically powerful and physically accurate system of planetary astronomy to date, the necessary precondition for the achievement of Isaac Newton. The last major astronomer to be a practicing astrologer, Kepler was deeply influenced by Neoplatonic and magical traditions of cosmic harmony as well as by quantitative astronomy.

Born to a poor soldier and his lower-class wife, Kepler attended the University of Tübingen, where he studied under the astronomer Michael Maestlin (1550–1631), one of the few Copernicans in German academia, who seems to have converted Kepler to Copernicanism. A somewhat unorthodox Lutheran who originally hoped to enter the ministry, Kepler would spend much of his career working for Catholic princes, notably the Emperor Rudolf II, after whom he named his *Rudolphine Tables,* and the commander of the imperial forces during the Thirty Years War, Albrecht von Wallenstein (1583–1634).

In 1594, Kepler was recommended by the Tübingen faculty to a position as instructor and "mathematician," or calendar maker, to the town of Graz. These not very demanding duties freed him to publish his first book, *The Cosmographical Mystery* (Mysterium Cosmographicum) in 1597, the same year he married Barbara Mueller, a wealthy widow. *The Cosmographical Mystery* defended Copernicanism as an accurate picture of the heavens and set forth Kepler's famous theory of the relation of the spacing of the planets to the geometrical ratios of the regular solids. (Regular solids are those whose faces and angles are equal and parallel: the tetrahedron, the cube, the octahedron, the dodecahedron, and the icosahedron.) As a Protestant, Kepler was forced to leave Graz in 1598. Fortunately, his work had attracted the attention of Tycho Brahe, then the mathematician and astronomer to Rudolf II, and Brahe invited Kepler to Prague. Kepler worked closely with Brahe, inherited his astronomical data, defended him against the rival astronomer, Ursus, and on his death succeeded him as imperial mathematician in 1601.

The first astronomical problem that .

The complex frontispiece to Johannes Kepler's Rudolphine Tables *includes famous astronomers, astronomical instruments, and a map of Hven, the island where Tycho Brahe's observatory had been located. Two men in Arab dress represent the Arab contribution to astronomy. The imperial eagle showers gold from above, although Kepler himself, in a panel to the left of Hven, gets little of it. (Kepler,* Gesamelte Werke, *vol. 10)*

Kepler attacked using Brahe's data was that of the orbit of Mars, the most difficult planetary orbit to predict. In his approach to this problem and others, Kepler viewed science as an attempt to understand the mind of God. The infinite and unordered universe of Giordano Bruno filled him with horror; in Kepler's view, the universe, as God's creation, has to be rationally structured and ordered, rather than created by chance. This rational structure and order is mathematical and geometrical. Kepler's universe is also imbued with religious meaning. For example, it is fundamentally triadic, reflecting the Christian Trinity. The relation between the Father, Son, and Holy Spirit is analogous to the relation of the Sun, the outer heaven of the stars, and the space between them.

Despite his belief in cosmic harmony, Kepler detested the imprecision of magical writers. He was obsessed with numerical precision; even a slight deviation of the position of Mars from that theoretically predicted led him to change his entire theory. He was obsessed with showing that his astronomical theories did not merely accurately predict planetary motions, but that they were physically true. He abandoned Nicolaus Copernicus's use of the center of the Earth's orbit, the "mean Sun," as the center of the solar system, in favor of the actual Sun, from which a force that governs the motions of the planets emanated. Influenced by William Gilbert's *On the Magnet* (1600), Kepler identified this force as magnetic.

Kepler, unlike Ptolemy (A.D. 90–168), Copernicus, Tycho, or his correspondent Galileo Galilei, broke the tyranny of the idea that heavenly motion is perfect and therefore circular. He reduced planetary motion to three laws, although he did not state them as laws per se. The first two laws appeared in *The New Astronomy* in 1609, the third in the 1619 *Harmonies of the World,* a work explicating the universe as a structure organized by geometrical and musical harmonies. The three laws are as follows: (1) Planets move around the Sun in ellipses, with the Sun at one focus; (2) A line drawn between a planet and the Sun will always sweep the same area in the same amount of time. Therefore, planets accelerate as they approach the Sun and decelerate as they move away from it; (3) The squares of the times the planets take to go around the Sun are proportional to the cubes of their average distances from the Sun.

During his years in Prague, Kepler showed outstanding productivity and versatility. His 1604 *Supplement to Witelo* discussed optical problems, giving a modern explanation of vision as working through the focusing of light on the retina. (Witelo [c. 1230–after 1275] was a medieval writer on optics.) Kepler rejected the atomistic idea that the eye perceives coherent images given off by the perceived object in favor of analysis based on cones of rays of light emanating from points on the perceived object. The base of these cones is the pupil. Kepler also wrote a treatise on astrology and a book on the new star of 1604, referred to now as Kepler's star, setting forth the theory that new stars are caused by the burning of celestial waste. *The Dream* (Somnium), an account of a voyage to the Moon and one of the first works using a fictional narrative to set forth scientific principles, was begun at this time, although it was not published until 1634, after Kepler's death. His works supporting Galileo in the controversy provoked by the Italian's *Starry Messenger* (1610) earned one of Galileo's rare expressions of gratitude. Galileo's telescopic discoveries prompted Kepler's 1611 *Dioptrics,* the first discussion of the optics of the telescope, to which he suggested improvements that later became standard.

The Prague years ended in 1612 with the death of Rudolf II, who had abdicated the previous year. Kepler kept his post as imperial mathematician at a reduced salary, but left Prague for Linz, to again work as a teacher and district mathematician, supplementing his income by making calendars for the popular market. In Linz, Kepler demonstrated the error of the traditional chronology of the birth of Jesus, claiming that the actual year was 4 B.C., and he wrote a mathematical work about determining the volume of wine casks. A great deal of his time and energy was taken up with a successful defense of his mother from charges of witchcraft. He also engaged in a controversy with the English magician Robert Fludd, whose mystical approach he rejected.

Kepler's presence in increasingly intolerant Catholic Austria was controversial, and he was forced to leave Linz in 1628. He was briefly in the employ of the Bohemian warlord Wallenstein, a devout believer in astrology who employed Kepler to draw up horoscopes. Kepler died as he was returning to Linz to collect money owed him—he died poor with much money still owed by the emperor. The great work of his later years was the *Rudolphine Tables* (1627), combining Kepler's orbital theory with Tycho's data to produce the best stellar tables of the seventeenth century. Kepler's theories of planetary motion, not immediately accepted, grew increasingly influential in the course of the century, until Newton incorporated them as consequences of the laws of universal gravitation.

See also Astronomy; Brahe, Tycho; Planetary Spheres and Orbits; Rudolf II.
References
Caspar, Max. *Kepler.* Translated from the German by C. Doris Hellman. London: Abelard-Schuman, 1959.
Koyre, Alexander. *The Astronomical Revolution: Copernicus-Kepler-Borelli.* Translated from the French by R. E. W. Maddison. Ithaca: Cornell University Press, 1973.
Stephenson, Bruce. *Kepler's Physical Astronomy.* Princeton: Princeton University Press, 1987.

Kirch née Winkelmann, Maria (1670–1720)

The story of the German woman astronomer Maria Kirch shows the limitations that institutional scientific authority imposed on women. Like other German women astronomers such as Maria Cunitz, Maria Winkelmann received her first training at her father's house. In 1692, she married the astronomer

Gottfried Kirch (1639–1710), a student of Johannes Hevelius, in the hopes of continuing her studies. She and Gottfried worked together in astronomical observations and calendar making, before and after Gottfried was invited to Berlin in 1700 as the first astronomer of the newly founded Berlin Academy of Sciences.

In 1702, Maria discovered a comet, but since the report of the discovery bore Gottfried's name as the academy astronomer, it was assumed that he had discovered it. He later gave Maria the credit. When Gottfried Kirch died in 1710, Maria petitioned the Berlin Academy to appoint her an assistant astronomer for calendar making. Although she was clearly qualified and enjoyed the support of the academy's founder, Gottfried Wilhelm Leibniz, the academy, embarrassed to have a woman in an official role, denied her the position. She continued to publish in astronomy, working in the private Berlin observatory of Baron Bernhard Friedrich von Krosigk from 1712 to 1714, when von Krosigk died. She then moved to Danzig, where the Hevelius family invited her to use Johannes's old observatory. In 1716, when her son Christoph was appointed academy astronomer, she moved back to Berlin and continued to work on calendars to her death, although as a woman she continued to face much hostility from members of the academy.

See also Cunitz, Maria; Women.
Reference
Schiebinger, Londa. *The Mind Has No Sex? Women in the Origins of Modern Science.* Cambridge: Harvard University Press, 1989.

Kircher, Athanasius (1601–1680)

Athanasius Kircher was the most distinguished and versatile Jesuit natural philosopher of the seventeenth century. From a German academic family, Kircher studied at a number of German Jesuit institutions. He entered the Jesuit order in 1616 and was ordained a priest in 1628. In 1631, then teaching at the Jesuit college in Würzburg,

Kircher was driven from Germany by the Thirty Years War. He moved to Avignon, where he came into contact with the circle of natural philosophers around Nicolas-Claude Fabri de Peiresc. He was invited to Vienna in 1633 to be imperial mathematician to the Holy Roman Emperor Ferdinand II (r. 1619–1637), but he was diverted to Rome by Cardinal Barberini, who had been tipped off by Peiresc. Kircher became professor of mathematics and oriental languages at the Jesuit Collegio Romano. In addition to his Roman patrons, he continued to attract the patronage of the Hapsburg imperial family, dedicating many of his works to them.

Kircher was a polymath, a Renaissance intellect in the grand style. Humanism, magnetism, optics, Egyptology, astronomy, mathematics, chemistry, and sinology do not exhaust the range of his interests. At Rome, he was at the center of the worldwide Jesuit scientific network, in a unique position to gather observations of astronomical and geophysical phenomena. Nor were Kircher's scientific contacts limited to Jesuits or even Catholics; he also had a long association with the Lutheran astronomer Johannes Hevelius. Kircher built an enormous natural history collection at Rome that became one of the glories of the city. He published over 40 books, many of them massive illustrated tomes of over 1,000 pages. Among them, *The Subterranean World* (1665) theorizes that earthquakes and volcanic eruptions are caused by a network of underground fires. On a trip to Sicily in 1637 and 1638, Kircher had witnessed an eruption of Mount Etna and climbed to the edge of Vesuvius, inspiring his study of nature. *The Subterranean World* also contains discussion of fossils, which Kircher believed are formed within the Earth.

Kircher's *Magnetism, or the Magnetic Art* (1665), contains a scheme for measuring a magnet's strength by using a balance. Kircher also promoted the idea of collecting information on the magnetic declination to solve the longitude problem. Kircher's view of the natural world was allegorical. For example, *Magnetism* concludes with a discussion of God as "nature's magnet." His most famous work was his study of hieroglyphics, which he treated as complex symbols revealing ancient holy truths known to the Egyptians. Kircher was among the first to treat the Coptic language as relevant to Egyptology in his *Introduction to Coptic, or Egyptian* (1636). His Egyptian studies were given to the world in the massive *Oedipus Aegyptiacus* (1652–1654), which made Kircher a European celebrity. Despite his skepticism regarding alchemy and natural magic and his interest in experiment and new instruments such as the microscope, Kircher gained a reputation among late-seventeenth-century scientists, particularly Protestant ones, as highly credulous.

See also Collegio Romano; Jesuits; Museums and Collections.

Reference

Findlen, Paula. *Possessing Nature: Museums, Collecting, and Scientific Culture in Early Modern Italy.* Berkeley and Los Angeles: University of California Press, 1994.

L

Laboratories

The word "laboratory" began to take on its modern meaning in the late sixteenth century, originally referring to places for chemical or alchemical procedures as differentiated from kitchens and artisans' workshops. The word "laboratory" identifies the space as one where work takes place, associating laboratories with a more practical than textual approach to science. Laboratories included heat sources, such as alchemical furnaces, and the equipment the operator needed to carry on the work at hand. The association of laboratories with chemical or alchemical work as opposed to general experimental science continued until the late seventeenth century. Originally, laboratories were attached to dwellings and were frequently the possession of apothecaries, used for the refinement of medicines. More elaborate laboratories dedicated to more esoteric chemical procedures were attached to royal or princely courts, such as that of Emperor Rudolf II. Landgrave Moritz of the German state of Hesse Kassel (1572–1632), the alchemist prince, appointed a Paracelsian professor of chemistry at the University of Marburg in 1609 who included laboratory work in his students' instruction, the first professor to do so. Another aristocrat with an elaborate laboratory was Tycho Brahe, whose observatory and scientific complex at Uraniborg included a well-furnished underground laboratory with 16 furnaces, where Tycho and his sister Sophie carried out alchemical work.

One question regarding laboratories was whether they should be private areas where an adept could contemplate the divine mysteries, as in the alchemical model, or whether they should be more public spaces. Andreas Libavius, in his description of an ideal chemical laboratory published in 1606, attacked Tycho's Uraniborg laboratory as promoting a selfish withdrawal from society. He emphasized the openness of ideal laboratory space and its integration with the domestic and civic life of the chemist.

Laboratories became associated with the scientific societies of the late seventeenth century. The French Royal Academy of Sciences had a well-equipped permanent laboratory established in the King's Library; as always, the less well-funded English Royal Society had to make do with what was available. The leading chemist of the society, Robert Boyle, maintained a laboratory attached to the house he shared with his sister. Boyle viewed the laboratory as a sacred space and was annoyed by a constant stream of visitors expecting to be entertained with dramatic experiments. Robert Hooke maintained a small private laboratory attached to

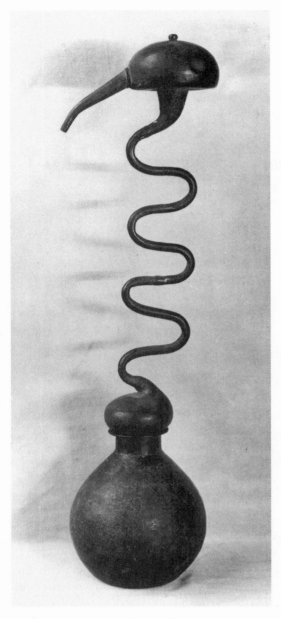

A sixteenth-century distilling apparatus owned by Giambattista della Porta. (Library of the University of Pennsylvania)

date the experiments by their witness. Such viewers were expected to be male and upper-class, or else their witnessing would have little weight. By contrast, the artisanal-class lab technicians employed by Boyle and other wealthy natural philosophers were expected to be as invisible as possible, except when receiving blame for an experiment gone wrong.

> *See also* Alchemy; Chemistry; Experiments.
> *References*
> Hannaway, Owen. "Laboratory Design and the Aim of Science." *Isis* 77 (1986): 585–610.
> Shackelford, Joel. "Tycho Brahe, Laboratory Design, and the Aim of Science: Reading Plans in Context." *Isis* 84 (1993): 211–230.
> Shapin, Steven. "The House of Experiment in Seventeenth-Century England." *Isis* 79 (1988): 373–404.

Laws of Nature

Although the concept of a law-governed nature had deep roots in the Western tradition, it was fully developed during the scientific revolution. The idea of laws of nature, which is often treated as a distinguishing feature of the Western scientific tradition, goes back to the ancient Greek and Roman Stoic philosophers. The concept was not employed very often in medieval natural philosophy, the most prominent example being the thirteenth-century friar Roger Bacon's (c. 1220–1292) reference to the law of refraction in optics. The fifteenth-century astronomer Regiomontanus (1436–1476) pioneered the terminology of law in mathematics and astronomy. He influenced subsequent astronomers, including Copernicus, who made greater use of it in his own work. Not all scientists made use of the concept, though; Galileo Galilei did not refer to any of his scientific discoveries as "laws," including the ones we now describe as laws. Nor did his contemporary Johannes Kepler, despite the subsequent fame of Kepler's three laws of planetary motion.

In seventeenth-century science, laws of nature assumed more prominence. The first

his rooms at Gresham College. Hooke's private laboratory served as a place where he could practice and rehearse his experiments before displaying them in a more public space in his capacity as the Royal Society's curator of experiments. The public experimental spaces attached to the society demanded that the viewers of successful experiments vali-

person to define the discovery of laws as the principal goal of scientific inquiry was Francis Bacon, a lawyer himself. In his view, laws are immutable, arrived at through a lengthy process of empirical inquiry. Given Bacon's notorious indifference to mathematics, it is not surprising that he did not think laws of nature would necessarily be expressed mathematically, in contrast to others who discussed laws of nature. The pioneer in actually basing science on law was René Descartes, whose three laws of motion assert that simple and undivided things are only changed by forces outside themselves, that things in motion tend to continue moving in straight lines, and that the total amount of motion in a collision between bodies remains the same, although it can be redistributed between the bodies. Descartes saw these laws, basic to his natural philosophy, as imposed on the universe by God and arrived at through logical deduction rather than empirical investigation; they are the basic and obvious premises on which scientific inquiry starts rather than the result of inquiry. The assertion that laws of nature ultimately derive from God, who imposed them on the world, became central to the definition of the concept. Robert Boyle, who did not make much use of the concept of law in his own experimental work, argued that since inanimate things, lacking intelligence and will, cannot obey laws the way humans do, the operation of laws requires constant divine action.

The most prominent advocate of the law of nature concept in the later seventeenth century was Isaac Newton. Like Boyle, Newton distinguished between levels of law. The most important were the three laws of motion, which are the only laws capitalized in Newton's *Mathematical Principles of Natural Philosophy* (1687). These laws are the intellectual basis of Newton's system, much as Descartes's laws were the basis of his system. Other laws, most famously the law of gravity, are also laws of nature in Newton's view, but they are not as general as the three laws of motion. All these laws are derived empirical-

ly and expressed mathematically, and all have been imposed by God. The power of Newton's example led to the popularity of the idea of science as a search for laws of nature. Newton's great rival Gottfried Wilhelm Leibniz, also a supporter of the law of nature concept, was leery of the excessive intellectual dependence on God that Newton's system seemed to require, favoring instead a concept wherein the laws of nature are inherent in the created universe, rather than continuously imposed by God.

See also God; Natural Theology.
References

Oakley, Francis. *Omnipotence, Covenant and Order: An Excursion in the History of Ideas from Abelard to Leibniz.* Ithaca: Cornell University Press, 1984.

Weiner, Friedel, ed. *Laws of Nature: Essays on the Philosophical, Scientific, and Historical Dimensions.* Berlin: Walter de Gruyter, 1995.

Leeuwenhoek, Antoni van (1632–1723)

The greatest microscopist of the scientific revolution, Antoni van Leeuwenhoek was also one of the least educated major European scientists. Of lower-middle-class Dutch origin, he never attended a university and was unfamiliar with both of the international languages of science at the time, Latin and French. Leeuwenhoek supported himself as a draper and municipal officer in his hometown of Delft in the Netherlands. The first evidence of his interest in microscopes, possibly inspired by Robert Hooke's *Micrographia* (1665), is in 1673, when he and his microscopes were mentioned in a letter from the Dutch physician and anatomist Reinier de Graaf (1641–1673) to Henry Oldenburg.

With 50 years of observation, Leeuwenhoek's microscopic skill and incredible diligence and patience enabled him to make a presence for himself on the international scientific scene, even without a formal education. His fundamental contributions include the discovery of red blood cells, of the circulation of blood through the capillaries, of the existence of protozoa, and of the nature of the male sperm cells. Leeuwenhoek's house at Delft, full of microscopes set up with objects for visitors to examine, was a major tourist attraction, the crowds of visitors often making it difficult for him to carry on his research. Visitors included royalty as well as eminent natural philosophers such as Christiaan Huygens. Leeuwenhoek received recognition from the natural-philosophical community by being unanimously elected a fellow of the Royal Society in 1680, a distinction that meant much to him although he never visited London.

Leeuwenhoek made over 500 microscopes, favoring the single lens design. Some were of exceptionally high quality, capable of magnifications of 270. Leeuwenhoek's manual dexterity, patience in observation, and mastery of microscopic technique in the preparation and observation of specimens were unrivaled at the time, and combined with the high quality of his microscopes, these skills made some of his observations difficult to duplicate. He was the first to take a quantitative approach to microscopy and to devise a crude scale of microscopic measurement, comparing microscopic phenomena to grains of sand or strands of hair.

While other leading microscopists, such as Marcello Malpighi and Jan Swammerdam, used the microscope to investigate things that interested them, Leeuwenhoek saw microscopy as an end in itself, and he put an enormous number of things under the lens. All the products of nature he could find ended up under his microscopes: He asked his neighbors for items to examine, like hair clippings or blood samples; he obtained from local businesses items such as infested grain and pieces of butchered animals; and he went to the dock to sample the wide range of goods available there, such as whale skins. Nor was Leeuwenhoek averse to putting portions of his own body under the lens, publishing the results of microscopic examination of tartar from his teeth in 1683. Sometimes his catholicity of interest and his fondness for

sharing his knowledge with members of the elite led to bizarre situations. On one occasion Anthonie Heinsius (1641–1720), the grand pensionary of Holland, one of the highest offices in the Dutch state, received a letter from Leeuwenhoek, detailing the results of the microscopic analysis of a substance found between Leeuwenhoek's toes after he had left his stockings on for two weeks.

Perhaps because of his lack of language skills and education, Leeuwenhoek published no original treatises but made his results public as letters to *Philosophical Transactions,* published in English translation, and to the Dutch journal *De Boeksall van Europe.* His discovery of the protozoa, then known as *animalcules,* was made known in a letter to *Philosophical Transactions* published in 1675, followed by his description of the sperm cells in 1677. Many of these letters were eventually gathered into published collections in Latin and Dutch. The letters were accompanied by sketches that Leeuwenhoek employed local draftsmen to make.

Leeuwenhoek's natural philosophy was a crude Cartesianism, and his analysis of microscopic phenomena was mechanistic. Like his Dutch contemporary Swammerdam, he opposed the idea of spontaneous generation, holding to a "spermist" doctrine of preformationism that held that the infant is already present in the male sperm, the female serving as an incubator. This differed from the dominant ovist school of preformationism, which held that the infant is originally present in the female egg.

Although the microscope was becoming a less fashionable instrument, Leeuwenhoek continued to make public the results of his microscopic examinations until his death in 1723.

See also Microscopes.

References
Dobell, Clifford. *Antoni van Leeuwenhoek and His "Little Animals."* New York: Russell and Russell, 1958.
Ruestow, Edward G. *The Microscope in the Dutch Republic: The Shaping of Discovery.* Cambridge and New York: Cambridge University Press, 1996.

Schieerbeek, A., and Maria Rosenboom. *Measuring the Invisible World: The Life and Works of Antoni Van Leeuwenhoek, FRS.* London: Abelard-Schuman, 1959.

Leibniz, Gottfried Wilhelm (1646–1716)

A German philosopher and scientist, Gottfried Wilhelm Liebniz has often been described as the last man knowledgeable in all branches of knowledge, including science, philosophy, language, history, religion, law, librarianship, and poetry. His father, a university professor, educated Leibniz by giving him the run of his library, and Leibniz taught himself Latin at the age of seven or eight. He attended the Universities of Leipzig, Jena, and Altdorf, taking his doctorate from Altdorf. After an early visit to Paris, where he studied with Christiaan Huygens, he spent most of his life as a diplomat and counselor in the employ of minor German courts. Leibniz urged the establishment of the Berlin Academy of Science, the Prussian scientific academy based on the model of the French Royal Academy of Sciences. Founded in 1700, the Berlin Academy would become a leading scientific body in the eighteenth century.

Leibniz's career as a philosopher differed from those of other early modern intellectuals. In his lifetime, he published only one book under his own name, but he made extensive contributions to periodicals, one of the first major figures to exploit the growing periodical press. Not bound to one particular style of philosophizing, Leibniz could write in a Scholastic way for Scholastic periodicals and in a Cartesian way for Cartesian periodicals. He also corresponded extensively. Most of his voluminous writings were in manuscript form and written in Latin or French, the international languages of his time.

Leibniz also differed from other innovative seventeenth-century philosophers in his reluctance to view the history of philosophy as discontinuous. René Descartes, Francis Bacon, Thomas Hobbes, and many others promoted the idea of a sharp break between ancient and medieval philosophy on one hand, and their own superior modern philosophy on the other. Leibniz asserted that his philosophy drew on what was good from the ancient and Scholastic traditions, criticizing Descartes and other moderns for unacknowledged borrowings. For example, he pointed out that Descartes's proof of the existence of God was derived from medieval Scholastic arguments. Part of a German tradition of "eclectic" philosophy, adding good things from different philosophic traditions to an Aristotelian base, Leibniz was among the last of the Renaissance humanist philosophers.

Leibniz was familiar with the writings of Descartes and other mechanical philosophers and early in his career made important contributions to mechanics. Extending Christiaan Huygens's work, he distinguished the concept of force from that of quantity of motion, and he is considered a founder of dynamics, a word he invented. He championed the notion of the conservation of force, suggesting that when an inelastic body comes to rest after an impact, its motion is transferred to its individual parts.

Leibniz was optimistic about scientific progress, which he hoped would make the world better, but he found both atomistic and Cartesian physics too materialistic and attempted to combine the mechanical philosophy with Aristotelianism. He modified the mechanical philosophy by turning it in a spiritual direction, reducing the world to "primary substances." In his view, material entities, influenced by outside sources, cannot be substances. Leibniz was influenced by the Kabbalah (Christian Knorr von Rosenroth [1636–1689], a friend of Leibniz's, was the first to translate Kabbalistic writings into Latin) and by the work of Anne Conway. He theorized that the ultimate entities of reality are monads, "simple substances" that combine to produce the world. Monads are essentially atoms—not material atoms but spiritual. Leibniz described the monads and therefore all nature as alive and arranged in a harmony directed by God, arguing that this harmony of monads proves the existence of God.

Leibniz's work provides a parallel to that of Isaac Newton, whose natural philosophy also blended elements deriving from magic with mathematics and theology (although Newton was not as influenced by Scholasticism as was Leibniz). Both invented the calculus around the same time, leading to a vicious priority dispute, provoked by Newton, in which the English mostly took his side for nationalistic reasons. For decades, the English refused to use Leibniz's superior notation system, with disastrous results for their mathematics. Leibniz also disagreed with other aspects of Newtonianism, such as the use of gravity, which he held to be a revival of occultism, and Newton's use of space as an absolute. Leibnizian physics defined motion and therefore space as relational.

Leibniz placed great stress on the perfection of the world as created by God. This did not require that any world God created would be perfect, as Leibniz believed in standards of good and evil independent of God's will. Since God, in this view, is both good and omnipotent, the world he chose to create from all possible creatable worlds had to be the best of all possible worlds. In his *Candide* (1759), Voltaire (1694–1778) caricatured this position as mindless optimism, distorting Leibniz's position, which was working from a different definition of goodness than the ordinary one. Leibniz defined goodness as the combination of the maximum diversity of phenomena with the maximum simplicity of laws, rather than as the provision of a pleasant life. Given Leibniz's definition of goodness, human happiness was only one among many considerations for God in creation—one that could be sacrificed for more important things.

Leibniz did not found a school of "Leibnizians," and he thought the dogmatism of the Cartesians repulsive. He was not very influential in England, where praise for Leibniz would conflict with the Newton cult, or in France, where his rationalism was inconsistent with the sensation-based epistemologies that had become dominant there. Leibniz's attempt to compromise between Aristotelianism and the mechanical philosophy was seen in France as intellectually dishonest. He was always influential in Germany, though, particularly in the German universities. Religious authorities disdained Leibniz as a determinist, but versions of his philosophy were widely circulated through the eighteenth century.

See also Academies and Scientific Societies; Leibniz-Clarke Correspondence; Mathematics; Mechanics; Newton, Isaac; Priority.

References
Jolley, Nicholas, ed. *The Cambridge Companion to Leibniz.* Cambridge: Cambridge University Press, 1995.
Westfall, Richard S. *The Construction of Modern Science: Mechanisms and Mechanics.* New York: Wiley, 1971.

Leibniz-Clarke Correspondence

In 1715 and 1716, a dispute was carried on in letters between Samuel Clarke (1675–1729), a friend of Isaac Newton's generally thought to be acting for him, and Gottfried Wilhelm Leibniz over a number of topics, most importantly the religious implications of Newtonian physics. Leibniz held that the Newtonian universe was imperfect because it occasionally requires God to intervene to prevent it from running down. He claimed that this image of an imperfect universe, which required by implication a fallible Creator, was causing religious decline in England. Clarke, a Newtonian philosopher and Church of England divine, replied that Leibniz's alternative theory of a universe capable of running perfectly without periodic divine intervention eliminates the need for providence, leading to materialism and atheism. The Newtonian system, by contrast, supports natural theology by making providence essential to the universe. Leibniz also attacked Newtonian physical ideas, including absolute space and time, the Newtonian theory of gravitation, which he charged introduced an occult force, and atomism.

The opposition between Leibniz and English Newtonians such as Clarke was exacerbated by the bad blood between Newton and Leibniz, caused by Newton's accusation that Leibniz had plagiarized the calculus. Politics also complicated things. The German prince whom Leibniz worked for, the Elector of Hanover George I (1660–1727), inherited the British throne in 1714, and his daughter-in-law Princess Caroline (1683–1737) moderated the Leibniz-Clarke correspondence. Thus, the English Newtonians were afraid that Leibniz would come over to London as court philosopher. This was not really a possibility because of bad relations between King George I and Leibniz. The widely circulated controversy was cut short by Leibniz's death in 1716. First printed in 1717, the correspondence was frequently reprinted in the eighteenth century.

See also God; Newtonianism.
Reference
Vailati, Ezio. *Leibniz and Clarke: A Study of Their Correspondence.* New York: Oxford University Press, 1997.

Libavius, Andreas (c. 1560–1616)

One of the leading opponents of Paracelsianism, the German Lutheran schoolmaster Andreas Libavius also published what is often considered the first chemical textbook, *Alchemia* (1597). He was the son of a poor weaver, but he managed to be educated at the University of Wittenberg, the University of Jena, from which he received a Ph.D. in 1581, and the University of Basel, from which he received an M.D. Although he had earned a medical degree, after leaving the university he supported himself as a schoolmaster rather than as a physician.

Libavius was a trained humanist, teaching history and poetry at Jena from 1588 to 1591. His eclectic philosophy took elements from Ramism and from the humanistic Aristotelianism in the Lutheran tradition that

This illustration in Andreas Libavius's work Of the Philosopher's Stone *shows a number of allegorical representations of chemical processes. The phoenix at the top represents the stone itself.*

traced back to Philip Melanchthon (1497–1560). He despised the Paracelsians, both for what he saw as the arrogance of their claims to remake nature and for attacking the humanistic learning he himself taught. As a loyal Lutheran, by the end of his life Libavius also looked askance on the alliance of Paracelsianism and Hermeticism with German Calvinism that would eventually issue in the Rosicrucian manifestos. Libavius was a prolific, vigorous, and wordy controversialist who attacked Paracelsianism ferociously. "Paracelsianism will be philosophical when all the whores are chaste virgins and all sophistries are indubitable truths," he wrote in *Letters on Chemical Things* (1595). Ironically, as an alchemist and a believer in the medical applications of chemistry, Libavius himself was sometimes accused of Paracelsianism.

Libavius, the discoverer of stannic chloride, was in many ways a traditional alchemist, believing in the possibility of the transmutation of metals. Where he broke from the alchemical tradition was in his belief in open communication, as opposed to the alchemical practices of secrecy and the use of allegorical language. He saw chemical medicine as a supplement to the traditional *materia medica* rather than as a replacement for or successor to the older tradition. *Alchemia* was an attempt to reduce chemical knowledge, primarily viewed in a medical context, to an organized form that could be taught in a humanistic institution. It was the first attempt to present chemistry systematically, as a body of knowledge rather than as a set of procedures for accomplishing certain ends, and it included the first detailed description of a model chemical laboratory. *Alchemia* was often followed, adapted, and abridged in seventeenth-century chemical textbooks.

> *See also* Alchemy; Chemistry; Education; Hermeticism; Laboratories; Paracelsianism.
> *References*
> Debus, Allen G. *The Chemical Philosophy: Paracelsian Science and Medicine in the Sixteenth and Seventeenth Centuries.* 2 vols. New York: Science History Publications, 1977.
> Hannaway, Owen. *The Chemists and the Word: The Didactic Origins of Chemistry.* Baltimore: Johns Hopkins University Press, 1975.

Libraries

Although some early modern scientists boasted that they attained knowledge through direct observation in the field or the laboratory rather than by reading books, libraries were still an integral part of the scientific revolution. The fifteenth-century humanists and their patrons were avid collectors of classical manuscripts, including many of mathematical, scientific, or medical interest. The Vatican Library, founded in this period, contained hundreds of scientific manuscripts, but smaller and more specialized collections were also built by working

scholars such as the German astronomer Regiomontanus (1436–1476).

As the mass of printed knowledge swelled, a well-stocked library became a necessity for some kinds of investigation. General-interest libraries, public as well as private, were of limited use to natural philosophers, as natural philosophy and science occupied only a modest share of their collections, which were dominated by theology and humanistic studies. College and university libraries were usually strongest in these areas as well, and many professors who wished to research in natural philosophy and science were forced to put together their own libraries. Indeed, university libraries were often simply uninterested in scientific books. Even the library of Cambridge University, Isaac Newton's university, did not own a copy of his *Mathematical Principles of Natural Philosophy* (1687) until 1715, when Bishop John Moore (1646–1714) willed his library to the university. Seventeenth-century Catholic university libraries and other institutional libraries suffered from the additional handicap of the church's ban on Protestant, Copernican, and Cartesian works, although sometimes the books could be acquired for the purpose of refutation.

The private libraries put together by working natural philosophers ranged from a few hundred to a few thousand books, with the tendency for libraries to grow larger in the later phases of the scientific revolution. Although few restricted their book collecting entirely to natural philosophy, a large private library could function as a research center. One notable sixteenth-century example was John Dee's library at Mortlake. Widely used by his pupils and friends, Dee's library contained over 3,000 printed volumes—a very large number for the sixteenth century—as well as hundreds of manuscripts. Books and manuscripts were often only one part of a scientific library like Dee's; along with these there might be maps, collections of curiosities, and instruments. Unfortunately, much of Dee's library was destroyed when his house was attacked in his absence by local residents who suspected his magical interests. In the seventeenth century, the Cambridge mathematician and bibliophile Isaac Barrow (1630–1677) opened his extensive library to use by other scholars, lending books to Newton among others.

Institutional libraries containing resources for natural philosophy were sometimes associated with scientific societies. In 1667, Henry Howard donated to the Royal Society the library of his grandfather, the great Renaissance collector Thomas Howard, 14th earl of Arundel (c. 1585–1646). The contents of the donation provoked debate between Robert Hooke, who believed that only books on "Arts and Naturall History" should be retained, and those fellows who wished to keep other books for their value and rarity. The gift was given under tight conditions, and the nonscientific books were retained into the nineteenth century. Hooke himself was a great scientific book collector, acquiring over 3,000 volumes with an unusual degree of specialization in science, medicine, and technology. Another illustrious natural philosopher, Gottfried Wilhelm Leibniz, actually was a working librarian, serving the Elector of Hanover George I (1660–1727). Leibniz also built a very large personal library of over 6,000 volumes.

See also Printing.
References
Hunter, Andrew, ed. *Thornton and Tully's Scientific Books, Libraries, and Collectors: A Study of Bibliography and the Book Trade in Relation to the History of Science.* 4th ed. Brookfield, VT: Ashgate, 2000.
Peck, Linda Levy. "Uncovering the Arundel Library at the Royal Society: Changing Meanings of Science and the Norfolk Donation." *Notes and Records of the Royal Society of London* 52 (1998): 3–24.

Literature

Science and literature were overlapping realms during the scientific revolution, and the "two cultures" idea of a radical separation had not yet emerged. Although only Blaise Pascal was able to attain greatness in both science and literature, many scientists had literary interests. Galileo Galilei, a distinguished writer of Italian prose, belonged to the Florentine literary academy Della Crusca and wrote criticism of the Italian poets Torquato Tasso (1544–1595) and Ludovico Ariosto (1474–1533) as well as some of his own poetry. Science was communicated in a far greater range of literary forms than today, including the vast mass of alchemical poetry and Galileo's dialogues. Science, whether traditional or innovative, also provided a ready stock of metaphors, ideas, and images for poets and other writers. One example was the Galenic medical theory of the four humors. The supposed influence of the humors on temperament made them convenient tools for creating comic character types, eventually giving the word "humor" its most common meaning in English. The four elements and the celestial spheres of traditional astronomy, with the music of their turnings— the "music of the spheres"—were also stock literary devices.

Changes in contemporary science inspired varying degrees of interest among literary writers. At one extreme, William Shakespeare (1564–1616) took almost no interest in science. What little scientific imagery occurs in his writings is traditional, like the heavenly spheres. Shakespeare's contemporary John Donne (1572–1631), by contrast, took much more interest in the new science, frequently incorporating recently discovered astronomical phenomena such as new stars into his poetry. Donne also movingly expressed the uncertainty that many educated Europeans felt with the decline of the Aristotelian universe in his *An Anatomy of the World: The First Anniversarie* (1611):

And new Philosophy calls all in doubt
The Element of Fire is quite put out
The Sun is lost, and th'earth, and no
    mans wit
Can well direct him where to look for it.
(ll. 205–208)

New scientific instruments, most importantly the telescope and microscope, also attracted the interest of literary writers. Both were natural candidates for use as poetic images. The English poet John Milton (1608–1674), who had visited Galileo in Tuscany, speculated in his poem *Paradise Regained* (1667) that when Satan took Jesus to a mountaintop to show him the kingdoms of the Earth, he had employed a telescope. (Likewise, in Milton's *Paradise Lost* [1667] an angel and Adam speculate on Earth-centered versus Sun-centered astronomy without coming to a conclusion.)

The seventeenth century also saw increasing use of science in fictional narratives. Arguably the first science fiction produced in Europe was Johannes Kepler's *Somnium,* published posthumously in 1634 but written much earlier. *Somnium,* which contains autobiographical elements, is the narrative of a dream voyage to the Moon, described in a way that draws on Galileo's recent telescopic discoveries. This was followed by a spate of lunar voyage narratives by writers such as the English bishop Francis Godwin (1562–1633) and the French nobleman Cyrano de Bergerac (1619–1655). These works depict the Moon as an Earth-like body floating in space rather than resorting to the traditional picture of a perfect and unchanging body embedded in a crystalline sphere. Francis Bacon drew on the resources of fiction to set forth a scientific program in his utopia, *New Atlantis* (1627), and Margaret Cavendish incorporated consideration of natural philosophical theories in her fictional utopia, *The Description of a New World, Called the Blazing World* (1666).

Although poets sometimes wrote panegyrics of scientists and science, investigators into nature had long been the targets of literary satire and ridicule. Alchemists and astrologers were attacked as unscrupulous and greedy frauds, and physicians had been ridiculed at least since the Greeks and Romans. The most famous attack on physicians' claims to knowledge during the scientific revolution was *The Hypochondriac* (1673), by the late-seventeenth-century French dramatist Molière (1622–1673). The play features a medical student's claim that opium causes sleep due to its "dormitive virtue" and a doctor who denies the circulation of the blood because it is not found in the ancient medical works. This was ridicule of conservative and Aristotelian science rather than of contemporary science, but Molière and other satirists also frequently attacked the modern, mechanical scientist. Molière's *The Learned Ladies* (1672) satirized the Cartesianism of women virtuosi (a frequent target of male ridicule), and Thomas Shadwell's (c. 1642–1692) *The Virtuoso* (1676) displayed to London theatergoers the character of Sir Nicholas Gimcrack, the amateur investigator who claimed to have turned a spaniel into a bulldog and a bulldog into a spaniel by transposing their blood. *The Virtuoso* and other late-seventeenth-century English satires were squarely aimed at contemporary scientific beliefs and practices, such as transfusion, identified with the "new philosophers" of the Royal Society.

The greatest satirist of science was the Irish clergyman Jonathan Swift (1667–1745). The third section of his *Gulliver's Travels* (1726), the "Voyage to Laputa," contains a detailed satire of the science of the Royal Society, indicating Swift's familiarity with the society's *Philosophical Transactions* and the published writings of scientists such as Robert Boyle. Here Swift ridicules scientists as impractical—the Laputans are highly skilled theoretical mathematicians, but they cannot build a house—and as disturbed with ridiculous fears—the Laputans fear that the Earth might fall into the Sun or be destroyed by the return of a comet, a direct reference to some Newtonian theories about the role of comets in Earth's history. They also perform ridiculous experiments (some of them taken directly from the records of the Royal Society) such as extracting sunbeams from cucumbers.

See also Rhetoric; Utopias.
References
Nicolson, Marjorie Hope. *The Breaking of the Circle: Studies in the Effect of the "New Science" upon Seventeenth-Century Poetry.* Evanston, IL: Northwestern University Press, 1950.
————. *Science and Imagination.* Ithaca: Great Seal Books, 1956.
Patrides, C.A., ed. *The Complete English Poems of John Donne.* London and Melbourne: Bent, 1985.

Locke, John (1632–1704)

Although John Locke is now principally remembered as a political philosopher and founder of the liberal tradition, he also made important contributions to psychology and attempted to devise a new philosophical style compatible with the new science of the seventeenth century. From a Puritan background, Locke was educated at Oxford in the 1650s, a time of great scientific activity. He was excited by the philosophy of René Descartes, kept a weather diary, studied medicine, and assisted Robert Boyle with experiments during Boyle's time at Oxford. Locke never took an M.D., but he practiced medicine and was a collaborator with Thomas Sydenham.

Although Locke became a member of the Royal Society in 1688, his own interests, as they eventually developed, did not lie in scientific practice but in philosophy, politics, and religion. Locke viewed the task of the philosopher to be, not setting up a comprehensive system in the manner of Plato, Aristotle, or Descartes, but clearing away errors and preconceptions in order for the scientist, such as "the great Huygenius [Christiaan Huygens] and the incomparable Mr. Newton," to make progress in knowledge. His *Essay Concerning Human Understanding* (1690), drawing on Baconian and Gassendist empiricist ideas, analyzes the human mind in terms of the reception and combination of sense-impressions—Locke strongly opposed belief in innate ideas. His argument that knowledge enters our minds through our senses and that these sensations are combined in our minds into complex ideas would become the basis of eighteenth-century psychology. Basic to his analysis is the distinction between primary qualities existing in a thing in itself, such as mass, and secondary qualities created by our perceptions, such as color. Locke did not invent this distinction, which was fundamental to the mechanical philosophy, but his discussion of it is classic.

Although he was an acquaintance and correspondent of Newton, with whom he served on a committee on problems of the English coinage, Locke was skeptical of grand theories in the sciences, believing that they often exceed the boundaries of what is knowable. He thought of mathematics as one of the few spheres of human thought where certainty is possible, but he was not skilled at it himself and was unable to grasp Newton's argument in *Mathematical Principles of Natural Philosophy* (1687). Locke and Newton together would be exalted as founders of the European Enlightenment in the eighteenth century.

Reference
Cranston, Maurice. *John Locke: A Biography.* New York: Macmillan, 1957.

Logarithms

The logarithm, a numerical device to promote ease of calculation by substituting addition and subtraction of exponents for the multiplication and division of numbers, was first invented by the Scottish mathematician John Napier (1550–1617). Napier, who was also the inventor of the decimal point, coined the word "logarithm" from the Greek *logos arithmos,* which can be translated as "number of the word." A zealous Protestant and millenarian, Napier wanted to calculate the date of the Second Coming and the millennium by manipulating the numbers of the prophetic books of the Bible. However, the principal use of his logarithms was to simplify trigonometric calculations for navigators and astronomers. His original table of logarithms,

Description of the Marvelous Canon of Logarithms (1614), contained not the logarithms of numbers, but the logarithms of sines of angles. This work was written in Latin, but shortly after it was published the English East India Company commissioned an English version for the use of navigators. It was followed by *The Construction of the Marvelous Canon of Logarithms* (1619), published posthumously, in which Napier explained how he had come up with the table.

The Swiss instrument maker Joost Bürgi (1552–1632) came up with the idea of logarithms independently, but he did not publish until 1620. Johannes Kepler's astronomical *Rudolphine Tables* (1627) incorporated logarithms, and by vastly simplifying astronomical calculations, they contributed to the advance of astronomical precision. The British Isles remained the center of logarithm use. Henry Briggs (1561–1630), professor of geometry first at Gresham College and then at Oxford, published a table of logarithms to the base 10 in 1624. He invented a number of calculational techniques to obtain logarithms, but for decades, logarithmic tables would remain plagued by calculational errors and misprints. Logarithmic tables also helped spread the use of decimals.

> **See also** Decimals; Gresham College; Mathematics; Slide Rules.
> **Reference**
> Grattan-Guiness, Ivor. *The Norton History of the Mathematical Sciences: The Rainbow of Mathematics.* New York: W. W. Norton, 1998.

A view of the exterior of the London College of Physicians built after the Great Fire of London (1666), with a cutaway to show the anatomical theater. Formerly thought to be designed by Christopher Wren, this building is now credited to Robert Hooke. (J. Britton and A. Pugin, Illustrations of the Public Buildings of London, vol. 2, 1828)

London College of Physicians

The premier organization of English university-trained physicians, the London College of Physicians was chartered in 1518 by Henry VIII (r. 1509–1547), thanks to the efforts of the humanistically trained royal physician Thomas Linacre (c. 1460–1524). Alarmed by the low status of London physicians and the anarchy of the London medical market, Linacre founded the College of Physicians on the model of Italian colleges of physicians and London guilds. Henry's charter, confirmed by an Act of Parliament in 1523, gave the college the right to license all medical practitioners within a seven-mile radius of London.

Originally restricted to English physicians, the college was opened to Scottish physicians in 1606, shortly after James VI of Scotland inherited the English throne as James I (r. 1603–1625). In 1621, membership was restricted to those with medical degrees from the English universities, but since it was common for English physicians with foreign degrees to be "adopted" by the English universities, many physicians with Continental degrees continued to be members. Fellows had to pass an oral Latin examination in classic medical texts and were required to be elected by the current fellows. The college did not aim to incorporate all London physicians, still less all medical practitioners, although less prominent physicians and other practitioners could receive licenses from the college. The college also advised the government on medical issues such as preventing plague.

Although regulating the medical marketplace was always the college's main function, in the seventeenth century it also became an active sponsor of scientific research. Fellows were required to attend the Lumleian Lectures, founded in 1582. New medical ideas could be introduced at these lectures, the most famous example being the Lumleian Lectures of William Harvey. During the English Civil War of the 1640s, when the college lost much of its power, a group of physicians began meeting at the college to investigate rickets, which was considered a new disease as no treatment was prescribed in the classical medical literature. The product of these studies was Francis Glisson's (1597–1677) *Treatise on Rickets* (1650), which combined empirical research with the traditional framework of Galenic academic medicine. The college also established a laboratory, a library, a medical museum, and regular dissections. Despite the loss of its power to regulate medical practice, some viewed the college as a realization of Bacon's utopian research institute, "Solomon's House."

After the Restoration of the English monarchy in 1660, some fellows of the college feared an institutional rivalry with the newly established Royal Society, along with a short-lived Helmontian group, the Society of Chemical Physicians. A more immediate threat was the destruction of the college library, museum, and laboratory in the London fire of 1666. After this loss, the college did not reestablish itself as a center for medical research. It did survive, and in 1682 it gained the name Royal College of Physicians (under which it persists to the present). But it was not able to control London medicine. By the 1680s, the college had become more intellectually diverse, overlapping with the Royal Society in membership, and it included physicians representing several different medical approaches.

See also Harvey, William; Physicians; Royal Society.
References
Clark, Sir George. *A History of the Royal College of Physicians of London.* 3 vols. Oxford: Clarendon Press for the Royal College of Physicians, 1966.
Cook, Harold. *The Decline of the Old Medical Regime in Stuart London.* Ithaca: Cornell University Press, 1986.

Longitude
See Navigation.

M

Magic

The scientific revolution saw a growing separation of the spheres of science and magic. By the early modern period, European magic was the product of millennia of blending, including Greek, Mesopotamian, Chinese, Jewish, Egyptian, Arabic, and indigenous European elements. Magic existed in highly elitist forms that relied on the knowledge of learned languages such as Greek or Hebrew, as well as in the common charms of the village witch or cunning man. Magic in this period was based on ideas of correspondence, or connection, between different things in different realms. The most famous form of correspondence was the microcosm-macrocosm analogy, whereby the human body and the universe were regarded as analogous; to take a crude example, the heart was the equivalent of the Sun. Magic did not draw rigid distinctions between that which is alive and that which is dead, and magicians viewed various natural items as endowed with "occult" or hidden powers unexplainable in mechanical terms. Astrologers viewed the influences of the planets in this way as well. Magicians were often anti-Aristotelian and opposed to the traditional natural philosophy of the universities.

As practiced, European magic was divided into theoretical or "high" magic, and applied or practical magic. High magic generally involved mystical contemplation and harmony with the divine. Hermeticism, Kabbalah, and some forms of supernatural alchemy emphasized magical rituals as a form of spiritual improvement. High magic required the magician to be a certain kind of person, pious and virtuous and also ritually pure, avoiding certain foods and maintaining sexual moderation. High magic was elitist; its practitioners often claimed that their knowledge was not for the "vulgar."

Practical magic had much closer affinities to scientific practice than did high magic. Various magical subdisciplines such as alchemy and astrology were aimed at practical results, rather than mystical contemplation, and the personal qualities of the practical magical practitioner were less important. Many high magicians, such as John Dee, also practiced practical magic, but Europe was full of practical magicians who didn't bother about theory. Practical magicians were less elitist than high magicians, and some wrote manuals in vernacular languages aimed at a wide audience, such as William Lilly's (1602–1681) *Christian Astrology* (1647).

Most early modern criticism of learned magic was not based on the idea that magic is invalid or not in accord with scientific ideas, but on the belief that magic is satanic. This may explain the appearance of the Faust legend, a German story about a magician who sold his soul to the Devil to gain knowledge, which originated in the sixteenth century and was transmitted to a number of European cultures. In England, for example, Christopher Marlowe (1564–1593) wrote the play *Doctor Faustus* (1604). To the degree that "Faustian" magic, based on calling up demons or dead spirits by the aid of Satan, actually existed, it was the province of charlatans. But learned magicians were also accused of working with the Devil, as when John Dee's house was ransacked by local mobs. Many prominent magicians got into trouble with church authorities, including Giordano Bruno, who was actually burned at the stake, and Tommaso Campanella, who spent many years in prison.

Magicians replied to this criticism by arguing that magic was a higher or more perfect form of Christianity, emphasizing the process of personal purification that serious magic required. This argument was much more common than the notion that magic was an alternative to Christianity, although Bruno may have argued this. All magicians insisted on the compatibility of magic and belief in and worship of God. Some believed that magical knowledge could reconcile differences between Catholics and Protestants. The Prague court of the Holy Roman Emperor Rudolf II, a Catholic fascinated by magic, welcomed distinguished magicians of all confessions, including Dee and the German Lutheran astronomer and astrologer Johannes Kepler, as well as prominent Kabbalists from the Prague Jewish community. Other magicians took a more committed position; for instance, the Rosicrucians, who got started in Germany in the early seventeenth century, sided with radical Protestantism.

The distinction between magic and natural philosophy was unclear in the sixteenth century. Many notable figures, such as Paracelsus, combined them. Copernicus quoted the writings of Hermes Trismegistus, and some have argued that magical interests played a role in inspiring his heliocentrism.

Many of the earliest Copernicans in the sixteenth century had magical interests. Bruno made heliocentrism and the infinity of the universe central to his magical and religious system.

Several seventeenth-century figures also combined magic and natural philosophy, notably Kepler and Isaac Newton. However, by this time there was also a growing hostility toward magic among natural philosophers. Despite some interest in alchemy, Francis Bacon distrusted magic as insufficiently public and as based on an arrogant attitude toward nature. Kepler viewed the universe in a magical way, but he debated the English magician Robert Fludd, attacking him for lack of empiricism. Marin Mersenne in particular despised magic and magicians for what he viewed as their heresies, and he also engaged in controversy with Fludd. The mechanical philosophy that Mersenne promoted was put forward in large part as something that would combat the magical view by reducing ordinary interaction to matter and motion, with no occult influences, and reserving action outside this scheme to God, and possibly Satan, but not to occult forces or magicians. Magic continued to play a role in the development of natural philosophy—both Newton and Robert Boyle engaged in alchemical work, for instance. But their interest was less public than that of earlier natural philosophers and was not based on a consistent magical worldview. Whatever Newton's own beliefs, the spread of Newtonian and Cartesian natural philosophy left little room for magic in the European scientific community of the eighteenth century.

See also Alchemy; Astrology; Hermeticism; Kabbalah; Natural Magic; Witchcraft and Demonology.

References

Kearney, Hugh. *Science and Change, 1500–1700.* New York: McGraw-Hill, 1971.

Thomas, Keith. *Religion and the Decline of Magic.* New York: Charles Scribner's Sons, 1971.

Yates, Frances Amelia. *Giordano Bruno and the Hermetic Tradition.* London: Routledge and Kegan Paul, 1964.

Magnetism

See Gilbert, William.

Malebranche, Nicholas (1638–1715)

A priest of the Congregation of the Oratory, Nicholas Malebranche was the most influential philosopher in late-seventeenth-century France. He hoped to be the new Thomas Aquinas to René Descartes's new Aristotle, and to produce a thoroughly Christian Cartesianism by combining Descartes's philosophy with that of St. Augustine (354–430). From a wealthy French family of royal servants, Malebranche received an M.A. from the University of Paris in 1656 and then studied theology at the Sorbonne for three years. He entered the Oratorians, the most Cartesian of Catholic orders, in 1660 and was ordained in 1664. Most of his career was devoted to research and writing, supported by his inherited wealth. Malebranche's massive *Search after Truth* was first published in 1668, and it went through six editions in his lifetime. Among many other topics, it set forth a classic exposition of the egg of a living being, describing each egg as containing a series of fully formed embryos, one inside the other—the so-called "Russian doll" theory.

Malebranche's version of Cartesianism emphasizes the radical dependence of the universe on God. His "occasionalist" view of causation asserts that all effects are directly created by God. What humans view as causes are in reality occasions for God's actions. For example, Malebranche solved the problem of mind-body interaction by asserting that each instance of interaction is a divine miracle. Malebranche was always more concerned with metaphysics and theology than with science, but he briefly served as a professor of mathematics and played an important role in the dissemination of the Leibnizian calculus in France. He joined the Royal Academy of Sciences in 1699, when Cartesians were admitted. He also wrote an optical work, *Reflections on Light, Colors, and Fire* (1699).

Malebranche's became the dominant interpretation of Cartesian metaphysics in the eighteenth century.

See also Cartesianism; Causation.
Reference
Hobart, Michael E. *Science and Religion in the Thought of Nicholas Malebranche.* Chapel Hill: University of North Carolina Press, 1982.

Malpighi, Marcello (1628–1694)

The greatest anatomist of the late seventeenth century, Marcello Malpighi was the first to introduce microscopy into anatomy, thereby changing ideas about many aspects of the human body. He received an M.D. and a Ph.D. from the University of Bologna in 1653. From 1656 to 1659, he was a professor of medicine at the University of Pisa, where he met and was influenced by Giovanni Alfonso Borelli (1608–1679), a member of the Accademia del Cimento who impressed on him the importance of the experimental method in medicine and the mechanical nature of bodies. In 1659, Malpighi settled in Bologna as a professor of theoretical medicine at the University of Bologna. He combined a professorship of practical medicine at Bologna with private practice from 1666 until 1691, when he moved to Rome to accept an appointment as personal physician to the new pope, Innocent XII (pope, 1691–1700).

Malpighi's first publication was *On the Lungs* (1661), which for the first time elucidated the internal structure of the lungs and described the circulation of the blood through the capillaries. It was followed by works on the tongue and the brain, as well as a fundamental study of the anatomy of the silkworm, published in 1669. Malpighi believed it possible to better understand human anatomy through the study of other animals and even plants. His techniques included dissection, microscopic examination, and the injection of colored fluids that enabled him to trace the connections of vessels and channels in the body. Malpighi's studies of the development of the chicken embryo laid the foundation for the science of embryology, and he also made important studies of the internal structure of plants. In anatomy, his discoveries include the Malpighian bodies in the kidney and the Malpighian layer of the skin.

Applying the mechanical philosophy to anatomy, Malpighi conceived of the body as a series of glands. His innovations faced some opposition in Bologna, both from traditionalists and from those who opposed theoretical approaches to medicine. However, he was an internationally respected scientist, elected a fellow of the Royal Society in 1668. The English connection was especially important as all of Malpighi's works—beginning with *On the Formation of the Chick in the Egg* in 1673 and including his collected works in 1686 and his posthumously published manuscripts in 1697—were published in London under the society's auspices.

See also Circulation of the Blood; Microscopes.
References
Adelmann, Howard B. *Marcello Malpighi and the Evolution of Embryology.* 5 vols. Ithaca: Cornell University Press, 1966.
Fournier, Marian. *The Fabric of Life: Microscopy in the Seventeenth Century.* Baltimore: Johns Hopkins University Press, 1996.

Mathematics

The scientific revolution saw the creation of the modern field of mathematics, advances in many of its branches, and the systematic application of mathematics to many areas of natural knowledge. At the beginning of the period, the term "mathematics" in its most common use referred not to the modern discipline but to mathematical astronomy; when Johannes Kepler was appointed imperial mathematician in 1601 it really meant that he was imperial astronomer and astrologer. What we now call the mathematical disciplines were practiced in a number of different arenas in the sixteenth century. Humanists, such as Federigo Commandino (1509–1575),

occupied themselves with the recovery of ancient Greek mathematical texts. Commandino himself translated many classical texts into elegant humanist Latin, including Euclid's (fl. c. 300 B.C.) *Elements,* various works of Archimedes (c. 287–212 B.C.), and Apollonius's (c. 262–c. 190 B.C.) *Conics* as well as many others. Since many of these texts were fragmentary, had been corrupted by generations of copyists, or had to be reconstructed through references in other texts, editing and translation required original mathematical work. Even as late as the early eighteenth century, Edmond Halley's principal mathematical work as Savilian Professor of Geometry at Oxford was an edition and reconstruction of the Greek text of Apollonius. This kind of mathematical work, also found in original treatises such as Commandino's *On the Center of Gravity of Solids* (1565), did not go beyond the application of Greek techniques and continued to be founded on Greek assumptions, most importantly the primacy of geometry. In this it was actually retrogressive from the algebraic techniques derived from Arabic mathematics.

Mathematics was also studied in universities, but it had a low position in the disciplinary hierarchy. Mathematics and mathematical astronomy were seen as sciences that deal with the appearances of things, whereas natural philosophy was thought to deal with their actual nature, reflecting the rather low status given to mathematics by Aristotle. This is one reason why Galileo Galilei left the University of Padua for the court of the duke of Tuscany, where disciplinary boundaries were fluid and where Galileo as mathematician could also speak on natural philosophy. Little formal instruction in mathematics was offered in the curriculum of European universities, although many universities had an active mathematical culture flourishing outside the formal institutions. Nonuniversity institutions of education, such as France's College Royale or England's Gresham College, taught mathematics, although they tended to focus more on applied math than theoretical.

A countervailing tradition from antiquity, ultimately stemming from Pythagoras, saw mathematics as the hidden structure of the universe, known only to the initiate. This interpretation had little institutional support. The most notable application of mathematical Pythagoreanism in the scientific revolution was Kepler's attempt to correlate the planetary orbits with the five regular solids.

Outside the university and learned traditions, a variety of applied approaches to mathematics existed. Mathematics was necessary for business and accounting, and one of the principal Renaissance works on mathematics, the *Summa* (1494) of Fra Luca Pacioli (c. 1445–c. 1514), was designed as a textbook for accountants. This type of mathematics was most developed in Italy, which had the most sophisticated commercial culture. Mathematical techniques were applied by highly competitive specialists, and mathematics became a highly competitive culture. Thus, the possessor of a new technique would do better to keep it secret than to share it with competitors. The controversy between Niccolò Tartaglia (c. 1499–1557) and Girolamo Cardano, which divided the Italian mathematical world in the mid-sixteenth century, originated because Tartaglia believed that Cardano had passed on Tartaglia's technique for solving cubic equations without his permission. Some of the most vicious disputes in the early modern intellectual world were in mathematics, including that between Pierre de Fermat and René Descartes over mathematical optics in the late 1630s and the extremely nasty priority dispute between Isaac Newton and Gottfried Wilhelm Leibniz over the discovery of the calculus, which separated British and Continental mathematics, to the disadvantage of the former, for a century. This competitive culture was sometimes encouraged by the institutions of higher learning; the distinguished French mathematician Gilles Personne de Roberval (1602–1675), a champion of the geometry of infinitesimals and the founder of kinematic geometry, was required to defend his chair in

mathematics at the Royal College in Paris every three years in open mathematical competition. He thus retained his chair from 1634, when he first won it, until his death. Similar requirements were not imposed on professors in other fields.

Applied mathematics, drawing from Arabic and medieval sources as well as the Greeks, was also used in the more practical intellectual cultures of surveying, navigation, astronomy, and engineering. Trigonometry was an especially important discipline, due to its applicability to surveying, navigation, and cartography. This tradition was particularly strong in the Netherlands and England, where Robert Recorde's (c. 1510–1558) mathematical textbooks introduced the equals sign (=), so used because parallel lines symbolized equality. By contrast, England was very weak in more theoretical mathematics.

Although Renaissance Europe used the superior Hindu-Arabic system of numerals still in use today, it was handicapped by the lack of a standard notational system, particularly given the high possibility that printers unfamiliar with mathematics or its terminology could make typographical errors or fail to present equations and formulae in a clear way. Much of the modern notational system was put into place or standardized during the scientific revolution, including the plus (+) and minus (−) signs popularized by the German mathematician Michael Stifel (c. 1487–1567). The clarity of mathematical expression was also improved by the decimal, invented by Simon Stevin, and calculation was made easier by the logarithms first devised by John Napier (1550–1617), who published his first logarithmic tables in 1614.

Mathematics occupied a prominent role in the curriculum of the Jesuit colleges by the early seventeenth century. Jesuit educators employed the textbooks of the Jesuit Christoph Clavius, a promoter of mathematical education. Clavius and his mathematical allies within the Jesuit order fought a fierce battle against Jesuit natural philosophers to establish the Aristotelian legitimacy of mathematical sciences. Interest in mathematical education spread through the learned world via Jesuit schools, among others. The first permanent chairs in England devoted to pure mathematics, as opposed to mathematical astronomy, were the Savilian Chair of Geometry established at Oxford in 1619 and the Lucasian Chair of Mathematics at Cambridge in 1664.

The major intellectual revolutions within mathematics in this period were the creation of modern algebra by Francois Viète, the creation of analytic geometry by Descartes and Fermat, and the discovery of the calculus by Leibniz and Newton. Viète's symbolic algebra vastly increased the power and generalizability of algebraic methods and tended to shift the balance of power in mathematics from geometry to algebra. After Viète, an important step was the recognition of negative and imaginary numbers as real solutions of algebraic problems, notably by Albert Girard (1595–1632), who demonstrated that the roots of an equation are equal to the number of the highest power of the unknown. Analytic geometry, the mapping of algebraic equations by geometrical curves on a rectangular grid, also tended to promote numerical over spatial mathematics, as now mathematicians used equations to approach geometrical problems.

An important step toward the calculus, treating quantities as generated by continuous motion rather than as static, was made by a disciple of Galileo, the Jesuati priest Francesco Bonaventura Cavalieri (1598–1647), a professor of mathematics at Bologna. Building on some of Kepler's techniques for measuring volumes, Cavalieri used a method of indivisibles, or infinitesimals, to solve problems about the areas of geometrical figures. The final step to the infinitesimal calculus was made independently by Newton and Leibniz, Newton's accusations of plagiarism notwithstanding. (In a remarkable coincidence, on the other side of the world the Japanese mathematician Takakazu Seki [1642–1708] was working out a calculus similar to Leibniz's. There is no evidence that Seki, working from the Chinese mathematical tradition, was influenced by any Western mathematical knowledge circulating in China and Japan at the time.) Newton's calculus relied on fluxions, and Leibniz's on differential increments. Leibniz, always concerned with clarity, produced a system with superior notation that would prove more fruitful in the eighteenth century. The mid-seventeenth century also saw the development of probability theory, originating in the work of Blaise Pascal and Fermat.

Outside pure mathematics, the physical sciences were being mathematicized in the seventeenth century, the major figures in this process being Galileo and Newton. Galileo spoke of the language of nature as expressed in geometrical figures, without the mysticism of Kepler. Unaware of contemporary developments in mathematics, Galileo continued to see mathematics primarily in terms of Euclidean geometry. Unlike Galileo, Newton was a great and up-to-date pure mathematician as well as a physicist, and his great work on physics, *Mathematical Principles of Natural Philosophy* (1687), made enormous mathematical demands.

Each succeeding dominant physical system—Aristotelianism, Cartesianism, and Newtonianism—was more quantitative than the last, and a knowledge of mathematics was a requirement for anyone to work constructively in physics. Quantitative approaches were also being applied in other fields, even in the study of society with its development of political arithmetic.

See also Cycloid; Descartes, René; Fermat, Pierre de; Leibniz, Gottfried Wilhelm; Logarithms; Newton, Isaac; Pascal, Blaise; Political Arithmetic; Probability; Wallis, John.

References

Dear, Peter. *Discipline and Experience: The Mathematical Way in the Scientific Revolution.* Chicago: University of Chicago Press, 1995.

Grattan-Guiness, Ivor. *The Norton History of the Mathematical Sciences: The Rainbow of Mathematics.* Chicago: University of Chicago Press, 1998.

Mahoney, Michael Sean. *The Mathematical Career of Pierre de Fermat, 1601–1665.* 2d ed. Princeton: Princeton University Press, 1994.

Matter

Early modern scientists inherited conflicting theories about the nature of matter from ancient and medieval science. The scientific revolution did not resolve this confusion, but it did lead to more complex and empirically based ideas. The originally dominant theory, Aristotelianism, divided matter into the four actually existing elements of earth, air, fire, and water, which were thought to compose things by their mixture. Aristotelian thought also posited the theoretical "prime matter," that is, matter existing in a state prior to the imposition of a specific form. Although Aristotelians viewed matter as a continuum with no irreducible atoms, some, particularly when discussing mixtures, talked of *minima,* the smallest possible pieces of matter capable of showing the characteristics of a substance. Thus, a *minima* of flesh would be the smallest possible piece that can still be called flesh. This piece could be divided, but then it would cease being flesh and become something else. Aristotelians also distinguished between imperfect terrestrial matter and perfect celestial matter.

A rival inheritance from the classical world was atomism, the belief in atoms—hard, indivisible, and undifferentiated tiny bodies that were thought to form substances by agglomerating. Atomism, which generally had a hard time explaining the diversity of material things, worked better for physics than for chemistry or the study of living things. Some early modern thinkers, such as the French chemist and textbook writer Etienne de Clave, upheld both atomistic physics and minimalist chemistry.

The dominant early modern theories of matter used for chemistry were derived from Arabic and medieval alchemy. Alchemical theory itself was vastly diverse. Ever since the Middle Ages, some alchemists had adopted theories of *minima,* later influencing the alchemy and chemistry of Robert Boyle. One of the most influential alchemical matter thories in early modern Europe was that of the "Paracelsian triad" of salt, sulfur, and mercury—a modification of an Arabic alchemical theory describing metals as composed of sulfur and mercury. For alchemists in the tradition of Paracelsus, salt, sulfur, and mercury were not the same things we now refer to by those names, but physical principles whose actions and intermingling produce all matter. Mercury represents the principle of volatility; sulfur, that of combustibility; salt, that of stability. The substances we call salt, sulfur, and mercury contain the Paracelsian principles in their most concentrated form. The fact that there were three principles was seen as a material embodiment of the fundamental doctrine of the Christian Trinity.

Johannes Baptista van Helmont rejected this arrangement, declaring instead that all substances are reducible to water and shaped into different forms by principles called "seeds." Another alchemist, the German Johann Becher (1635–1682), believed that minerals are formed from water and earth, earth being subdivided into "fatty," "mercurial," and "stony" earths, corresponding to sulfur, mercury, and salt. Others combined the Paracelsian three with the Aristotelian four to produce a five-element system of mercury, salt, sulfur, earth, and water, in which the Paracelsian elements are the active principles shaping the passive earth and water.

René Descartes set forth a radically new approach to matter that drew on ancient atomism. In Descartes's mechanical philosophy, matter is dead, completely incapable of acting on its own and transmitting only that motion that had initially been given to the universe by God at the Creation and conserved ever since. Unlike atomists, such as his contemporary Pierre Gassendi, Descartes also denied the existence of a void—Cartesian matter fills the universe, leaving no empty space. This required him to postulate the existence of a subtle and weightless matter to fill the empty spaces left by cruder and more particulate matter and to explain why material substances occupying the same volume do not always weigh the same.

This subtle matter worked adequately for physics—Cartesians explained physical forces such as electricity and magnetism simply by introducing new forms of subtle matter—but not very well for chemistry, in which Descartes had no interest. Subsequent Cartesian chemists explained the chemical properties of substances by describing the shapes of the particles that composed them. Acids, for example, were thought to be composed of pointed parts that dissolve other substances by penetrating them. The belief in angular particles became part of eighteenth-century chemical orthodoxy.

One thing that Cartesians and atomists did agree on was that Aristotle's distinction between celestial and terrestrial matter was useless. Developments in astronomy, such as the analysis of comets and new stars, rendered this separation obsolete. Then, in late-seventeenth-century England, belief in the void as demonstrated by Boyle's experiments with the air pump made atomism, in classical or Gassendist form, more appealing than strict Cartesianism.

The leading matter-theorist of late-seventeenth-century England, Boyle was reluctant to accept the idea of purely dead matter, leaving some room for self-acting matter. His theory, however, remained basically mechanical. He was suspicious of the whole idea of a small number of basic elements composing all matter, pointing out that many substances cannot be reduced to basic elements, while others are reducible to a large number of different substances, depending on the processes used. Rather than relying on atoms, Cartesian matter, or elements, Boyle described the material world in terms of corpuscles that are created from the coalescence of tiny particles of matter. Boyle's alchemically influenced corpuscularianism was subsequently refined by Isaac Newton, who claimed that attractive and repulsive forces exist between corpuscles. He proposed that these forces, rather than the physical entanglements of corpuscles themselves, structure material things.

See also Alchemy; Atomism; Cartesianism; Mechanical Philosophy.

References
Brock, William H. *The Norton History of Chemistry.* New York: W. W. Norton, 1992.
Emerton, Norma E. *The Scientific Reinterpretation of Form.* Ithaca: Cornell University Press, 1984.
Toulmin, Stephen, and June Goodfield. *The Architecture of Matter.* New York: Harper and Row, 1962.

Mechanical Philosophy

The mechanical philosophy of the seventeenth century viewed matter and motion as sufficient to explain all natural occurrences. This made it distinct from Aristotelianism, which treated qualities as existing independently from substances, and also from magical philosophies, which relied on attractive and repulsive occult forces acting at a distance, as astrologers believed the forces of the stars affected things on Earth. In the view of strict mechanical philosophers, matter itself is "dead," or inert, acting only as motion is impressed upon it. Bits of matter interact only by direct contact or impact. Mechanical philosophy became the dominant, although never the exclusive, approach to natural philosophy in the seventeenth century.

Although the first mechanical philosopher to appear in the historical record was the Dutchman Isaac Beeckman, the most influential mechanical philosophers of the first half of the seventeenth century were two Frenchmen—Beeckman's pupil René Descartes and Pierre Gassendi. Their mechanical philosophies took different forms and sprang from different sources. Descartes, emphasizing his own originality, saw the universe as a plenum, full of different kinds of matter. Like Aristotle's, Descartes's natural philosophy denies the existence of a void. In contrast, Gassendi was a humanist who proclaimed himself not an innovator but a reviver of the natural philosophy of the ancient Greek Epicurus (341–270 B.C.), an atomist. Gassendi saw the universe as composed of atoms—the smallest bits of matter of which all else is composed—circulating in a void. Arguments between Cartesian and Gassendist mechanical philosophers often turned on the existence of the void. Gassendism, mostly expressed in difficult Latin rather than Descartes's elegant French, would be eclipsed in France by Cartesianism but would have a great influence in England.

Although many natural phenomena could be explained by mechanical interaction, some presented difficult problems. Magnetism was a favorite phenomenon of natural magicians, who saw in it indisputable evidence of occult forces acting at a distance, and it was therefore incumbent on mechanical philosophers to find a mechanical explanation for it. Descartes came up with a complicated explanation of magnetism based on a "subtle" matter, not immediately apparent to the senses. In this view, screw-shaped magnetic particles are emitted by the Sun and channeled through the Earth. The opposite polarities are accounted for by the supposition that some of the particles have a left-hand thread and others a right-hand thread. Descartes's point in devising such explanations was not to prove that they are necessarily true, but to demonstrate that the most puzzling phenomena can be explained mechanically.

Gassendi and other atomists also explained many physical phenomena in terms of the interactions of particles of various shapes. Among the phenomena that mechanical philosophers explained by matter and motion were living phenomena. Descartes believed that the human body is an automaton informed by a nonmaterial soul, speculating that the link between soul and body is located in the pineal gland. Animals, lacking souls, can thus be treated purely as machines. This mechanistic biology was taken further by Giovanni Alfonso Borelli (1608–1679) and other "iatromechanists," or mechanical physicians, of the late seventeenth century.

The relation of mechanical philosophy to religion was complicated. In order to be acceptable in early modern Europe, mechanical philosophy had to be reconcilable with

The reflex action of a child pulling his hand from a fire was one aspect of human behavior easy to explain in mechanical terms, as in this illustration to René Descartes's Treatise on Man *(1664). (National Library of Medicine)*

Christianity. The alliance between Christian theology and Aristotelianism was long-standing, nearly as influential in Protestant as in Catholic Europe. The person responsible for the theological development of the mechanical philosophy was a disciple and correspondent of Descartes and Gassendi, Marin Mersenne.

Mersenne, a priest from a particularly ascetic order of Franciscan friars, was concerned about magic and popular superstition. He saw the magical tradition, and particularly the belief that nature is somehow alive, as the root of all heresy. Mersenne was particularly horrified by the heresy of Giordano Bruno, and unlike other mechanical philosophers, he did not see Aristotelianism as the main enemy. Mersenne claimed that nature is not alive and that it acts in a mechanical way at all times, except when God or other supernatural entities such as devils or the souls of humans act directly on it. God is not the soul of the world, but its master or governor.

The mechanical philosophy enabled a clear distinction between nature in its ordinary course and miraculous actions. Catholics were particularly concerned to emphasize the continuing possibility of miracles. The dominant belief among Protestant theologians was that miracles had ceased to occur after the time of the Apostles. Catholics, on the other hand, believed that miracles were still ongoing, and that the persistence of miracles in the Catholic tradition was strong evidence that Catholicism was true. Thus, finding a rigorous definition of miracles was an important part of Catholicism's defense against Protestantism, and much of the ecclesiastical bureaucracy for evaluating miracles was put into place at this time. Mechanical philosophy, with its deterministic approach to nature in its ordinary course, allowed for miracles in which divine action wrenches nature from its ordinary mechanical course. Some Catholic natural philosophers claimed this was one of mechanical philosophy's advantages.

However, there was a terrible trap for Catholics in the mechanical philosophy. Mechanical matter theory in the Catholic world faced the problem of transubstantiation, the doctrine that the priest performing the Mass actually changes the bread and wine into the flesh and blood of Christ. For centuries, Catholic theologians had defined transubstantiation theologically in Aristotelian terms of substance and qualities, or "accidents." The elements of the Mass retain the "accidents," or appearance, of the bread and wine, while their substance is changed into flesh and blood. Mechanical physics, denying the distinction between substance and accident, was incompatible with this explanation. Descartes tried to solve this problem by arguing that the inside is transformed while the edges retain the properties of the original substances, but this was rather unsatisfactory. His works were put on the Index of Forbidden Books in 1663 for this reason, and openly supporting mechanical philosophy in the Catholic world remained risky.

In the Protestant world, religious problems with the mechanical philosophy were different. Many Protestant intellectuals viewed mechanical philosophy as useful in combating superstition, but problematic in that a consistent materialism could be seen as disputing divine action in the world, leading to what the seventeenth century defined as "atheism." This was less lack of belief in God's existence than lack of belief in God's activity. Those who believed that God exists but is indifferent to human beings, never bothering to interfere in the mechanical workings of the cosmos, were called atheists. Descartes himself got into trouble in the Netherlands with conservative Calvinist-Aristotelian university authorities because of this position. The Dutch Jew Baruch Spinoza, whose natural philosophy was mechanical, was even more radical and frightening to Christians, with his combination of a strict mechanical determinism and a belief in God's immanence in nature.

Mechanical philosophy in both its Cartesian and Gassendist forms was introduced to England in the 1650s. (Thomas Hobbes was also an original mechanical philosopher, but his system had less influence.) The English reception of mechanical philosophy was eclectic, in that ideas were drawn from both the Cartesian and atomist traditions, and it was also nonexclusive, in that mechanical philosophy was combined with ideas taken from other traditions. Strict mechanical philosophy, with its insistence on mechanical, and only mechanical, explanations, was not congenial to the antidogmatic late-seventeenth-century English scientists.

Because of their interest in medicine as well as in the vitalistic tradition of William Harvey and in alchemy, English mechanical philosophers were willing to consider the possibility of an active, vital matter, as long as it acted mechanically. The chemistry of Robert Boyle, for instance, combined mechanical and alchemical ideas. Physicists and cosmological theorists such as Christopher Wren and Robert Hooke combined mechanical theories with the "magnetical philosophy" stemming from the work of William Gilbert. A strict mechanical philosophy, however, was held to lead to materialism and atheism by prominent English philosophical thinkers such as Henry More (1614–1687). The culmination of these tendencies was the natural philosophy of Isaac Newton, which combined mechanical explanations with universal gravitation, which many mechanical philosophers, particularly the Cartesians dominant on the European Continent, saw as an intellectually illegitimate revival of occult and magical forces.

See also Beeckman, Isaac; Boyle, Robert; Descartes, René; Gassendi, Pierre; Hobbes, Thomas.

References

Gaukroger, Stephen. *Descartes: An Intellectual Biography.* Oxford: Oxford University Press, 1995.

Osler, Margaret J. *Divine Will and the Mechanical Philosophy: Gassendi and Descartes on Contingency and Necessity in the Created World.* Cambridge: Cambridge University Press, 1994.

Westfall, Richard. *The Construction of Modern Science: Mechanisms and Mechanics.* New York: Wiley, 1977.

Mechanics

The science of mechanics was one of the most intellectually active disciplines during the scientific revolution. During the early phases of mechanical science, mechanicians built on the achievements of medieval scholastics, as well as reviving material from antiquity such as the work of Archimedes (c. 287–212 B.C.) and the treatise on *Mechanics* falsely ascribed to Aristotle. Mechanics as a science had been developed extensively in the Middle Ages, although on a more theoretical than experimental basis. Medieval mechanics was the principal influence on Leonardo da Vinci (1452–1519), whose ingenious mechanical ideas exerted little influence only because they were not published or circulated. Mechanics did suffer from the lack of a standardized vocabulary, and many mechanical terms were used to denote different things, even by the same author.

The mechanics of bodies that are not in motion, called statics, was studied by the Dutchman Simon Stevin, who described the conditions that exist with equilibrium on an inclined plane. In hydrostatics, a matter of great concern for the Dutch, Stevin provided the first demonstration of the hydrostatic

paradox—the fact that the pressure exerted by a column of water depends on its height and not on the total weight of the water. But the greatest mechanician of the early modern period was Galileo Galilei, whose greatest achievement in mechanics was working out the properties of uniformly accelerated motion, such as that of a falling body. Galileo extended the study of falling bodies to pendulums, which were particularly important because of their relation to clocks, and to projectiles, important because of the military use of cannon. He suggested the idea of a pendulum clock and demonstrated the parabolic motion of projectiles. Although Galileo flaunted his departures from Aristotle, he kept the traditional Aristotelian distinction between natural motion, or motion according to a body's natural inclination, and violent motion, or motion against a body's natural inclination, compelled by some outside force. His most important departure from Aristotelianism was the quantitative character of his mechanics.

The mechanical philosophers René Descartes and Pierre Gassendi abolished the distinction between natural and violent motion, espousing a rectilinear inertia as opposed to Galileo's circular inertia. In this view, any circular motion a body makes is not a natural motion, as it was for Aristotle and Galileo; rather, circular motion requires a constraint on the body, whose inertial tendency is to go in a straight line. Another of Descartes's mechanical concepts is the conservation of quantity of motion—the product of a body's size and velocity.

In some ways, Descartes's mechanics represents a return to the qualitative Aristotelian tradition as opposed to the quantitative Galilean one; he distrusted Galileo's mathematical mechanics as overly abstract and more concerned with mathematical formulae than with the real world. The mechanical philosophies of Descartes and Gassendi also restricted the interactions between bodies to those caused by direct contact, which made impact a major concern for mechanicians in the later seventeenth century. Descartes's own theory of impact was clearly inadequate; among other things, he claimed that the impact of a smaller body can have no effect on a larger body at rest.

Meanwhile, the Galilean tradition in Italy continued to produce new mechanical concepts. Galileo's disciple Evangelista Torricelli (1608–1647) demonstrated "Torricelli's principle": that connected heavy bodies can be treated as one body, concentrated at the common center of gravity. Torricelli attempted to expand Galileo's mechanics from kinematics—the study of moving bodies—to dynamics—the study of moving bodies and the forces that put them in motion. Torricelli's dynamics, however, were expressed in an idiom foreign to mechanical philosophy, and despite their sophistication, they did not have much influence outside Italy. Also building on Galileo's work, Torricelli founded hydrodynamics, providing a formula for a liquid's rate of flow from a puncture in the side of a vessel: its velocity is proportionate to the speed acquired by a body falling from the top of the vessel to the point of the puncture. Also outside the Cartesian tradition, which was better suited to solids than to fluids, Blaise Pascal further developed hydraulics, examining the conditions under which a fluid body remains in equilibrium. There was a steady demand for hydraulic engineers in Italy, and that country would lead Europe in the field of applied mechanics during the seventeenth century.

Cartesians dominated the mechanics of solid bodies in the third quarter of the seventeenth century, although much of their efforts went into disproving Descartes's mechanical theories. Christiaan Huygens, no slavish Cartesian, developed a Cartesian mathematical mechanics but disproved Descartes's theories of impact early in his career. He created an alternative theory in the Cartesian idiom, generalizing Torricelli's principle to examine the impact of two bodies in terms of their common center of gravity. This also led him to deny Descartes's theory of the conservation of the quantity of motion, substituting for quantity of motion the product of a body's magnitude and the square of its velocity as the quantity conserved. Huygens also developed Descartes's idea of the constrained nature of circular motion into the mathematical theory of centrifugal force. Another mechanical innovator within the Cartesian tradition was Gottfried Wilhelm Leibniz. In mechanics a disciple of Huygens, Leibniz coined the word "dynamics" and reinterpreted Huygens's kinematics in terms of the concept of a body's force, although his concept of force is more similar to our concept of kinetic energy.

The greatest mechanical innovator of the late seventeenth century was Isaac Newton, who was not a Cartesian, however much he drew on Cartesian mechanics. Newton's laws of motion offered a consistent framework for mechanics superior to that of the Cartesians, and the calculus developed by Newton, and independently by Leibniz, offered a powerful mathematical tool to mechanicians. The calculus was superior to the classical geometry that had dominated mechanical mathematics up to that point, and it made possible the study of a new range of mechanical problems involving bodies moving at nonuniformly varying speeds. Newton's concept of force as something that acts on a body rather than as a property of a body paved the way for a fully dynamic approach. Newton also fully integrated the terrestrial and celestial realms into the same mechanical system.

See also Galilei, Galileo; Newton, Isaac.
References
Dugas, Rene. *A History of Mechanics.* Translated by J. R. Maddox. New York: Dover Publications, 1988.
Westfall, Richard. *The Construction of Modern Science: Mechanisms and Mechanics.* New York: Wiley, 1977.

Medicine

Although the practice of medicine did not change greatly during the scientific revolution, medical theory did. Theory gradually moved from a foundation in the texts of ancient and Islamic physicians, most importantly Galen (129–c. 199) and Avicenna (980–1037), to a foundation in the empirical study of the body. Of all the various practitioners of medicine in the early modern period, including surgeons, midwives, apothecaries, and locally recognized "experts" who practiced among their neighbors, physicians benefited most from this medical theory, at least at first. In fact, physicians were distinguished from other medical practitioners by the training in medical theory and practice they received at the medical schools attached to early modern universities. Surgeons, apothecaries, midwives, and other people who dealt with specific conditions had amassed a body of practical knowledge, but this did not rise to the dignity of a learned science. That is why they wrote in vernacular languages rather than Latin. Also, classical texts dealing with surgical and obstetrical subjects were not translated or reprinted as frequently as those in purely medical subjects.

The traditional Galenic medicine practiced by early modern physicians was oriented toward preserving health more than preventing and curing specific diseases. In the sixteenth century, traditional medical learning came under attack from several directions. The appearance of new diseases revealed traditional medicine to be at best incomplete. The most notable example is syphilis, thought at the time to have come from the newly encountered Americas. (The name "syphilis" derives from a description by the Italian physician Girolamo Fracastoro [c. 1478–1553] in the poem "Syphilis, or the French Disease" [1530]. The poem's character, Syphilis, is a shepherd cursed by the god Apollo with a loathsome disease. Fracastoro speculated that the disease was carried by seeds or spores— one of the earliest approaches to the germ theory.) New diseases contributed to a growing tendency to think of diseases as specific entities, rather than, as Galen would have had it, as manifestations of imbalances within the body. Neither had Galen employed the concept of contagion, which Renaissance physicians took from the Arabs and used to explain the spread of disease. Traditional pharmacy also clearly needed to be supplemented with knowledge of the medical properties of plants from the New World, or for that matter, the properties of the many Northern European plants unknown to the Mediterranean-based ancients.

Within Renaissance Europe itself, the "medical humanists" wished to drive out the Arabic and medieval Latin contributions, which they saw as accretions on classical medicine, and to recover the original medicine of Galen and other ancient Greek authorities in improved texts. One of the great triumphs of this movement was the publication of Galen's complete works by the famous Aldine Press in Venice in 1525, followed by a complete Greek edition of the works of Hippocrates (c. 460–377 B.C.) the following year. New Latin translations were quickly made from these texts, circulating to physicians throughout Europe.

The humanist revival of ancient philosophy made physicians more conscious of the need to fit medicine into an overall philosophical framework. For example, the University of Paris medical professor Jean-Francois Fernel (1497–1558) attempted to Platonize Galenism, explaining some bodily functions as the work of spirits emanating from the celestial regions—a striking departure from Galenic materialism, although Fernel continued to call himself a Galenist. Others invoked Aristotle against Galen—for example, supporting Aristotle's belief in the primacy of the heart against Galen's theory of a heart-brain-liver triad. More radically, Paracelsus and his followers wished to abandon the entire traditional medical corpus, Greek and Arab alike, and create a new and exclusively Christian medicine. Occultists, many of whom were Paracelsians themselves, insisted on the importance of astrological or other magical forces for medicine. Proprietary cures were sold by unlicensed and often uneducated "empirics," or "charlatans" (the word "charlatan" was originally an Italian

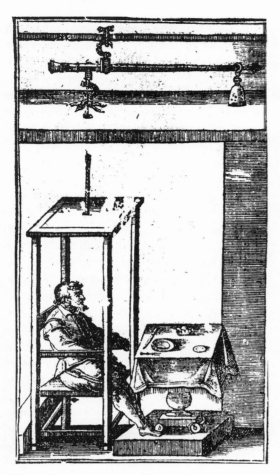

This weighing chair was one of Santorio Santorio's attempts to apply technology to medicine. (National Library of Medicine)

came in a flood in the seventeenth century, most notably William Harvey's discovery of the circulation of the blood through the body. This new knowledge, increased even more with the use of microscopes and animal experiments, made Galenism seem out of date, although as yet it had little effect on medical practice or strictly medical concerns.

Another application of science to medicine in this period was the use of more sophisticated measuring instruments for medical diagnosis. This movement was led by a Padua physician and acquaintance of Galileo, Santorio Santorio (1561–1636), who like Vesalius was a professor at the greatest center of medical learning, the University of Padua. A Galenist in doctrine, Santorio applied the newly invented thermometer to medicine and also devised a number of other instruments, including a *pulsilogium*, or timepiece for counting pulses. For 30 years, Santorio weighed himself, his food, and his excrement to understand the relation between nutrition, elimination, and health. Also involved in the movement to reduce Galenic diagnosis and treatment into a methodical system, he advocated a more mathematical and quantitative approach to medicine. At the time, however, Santorio was a voice crying in the wilderness, only to be taken up decades later by physicians interested in mechanical philosophy.

The great struggle in the medical community in the late sixteenth and seventeenth century was between the traditional Galenic medical professors and the new Paracelsian or chemical physicians. (In addition to the university professors, another loyal constituency for traditional medicine was Europe's many Jewish physicians, who disliked the aggressive and exclusive Christianity of Paracelsian medicine.) Paracelsians, many of whom lacked medical degrees, originally worked principally outside the university establishment at both ends of the social spectrum, as court physicians or in private practice. They found a place in the medical marketplace with a defiantly populist, anti-intellectual appeal. By the early seventeenth

term meaning "medicine-seller"). Physicians failed to drive these theoretically ignorant competitors from the medical marketplace, but none of these new approaches had much effect on the central arena, the university medical curriculum. The curriculum remained under the control of traditional physicians, although there was a humanist-inspired tendency to emphasize Galen and Hippocrates and to move away from the Arabs.

The most radical departures from received medical doctrine in the sixteenth and early seventeenth centuries were in anatomy, a low-status field associated with surgeons. The innovations that began with Andreas Vesalius and his *Of the Fabric of the Human Body* (1543)

century, some physicians, most notably the Wittenberg professor Daniel Sennert (1572–1637), were combining Galenic and Paracelsian medicine. Johannes Baptista van Helmont, on the other hand, reformulated "chemical medicine" in a very anti-Galenic way and was also influential.

The struggle between Galenic medicine and the Paracelsian "iatrochemists" was complicated in the later seventeenth century by a new movement to apply the mechanical philosophy to medicine, called "iatromechanism." Descartes, for one, wrote medical tracts and hoped that his reformation of natural philosophy would culminate in a reformed and effective medicine. Giovanni Alfonso Borelli (1608–1679) applied mechanical reasoning to the bodies of animals in *Of the Motion of Animals* (1680). Similar ideas were applied to the human body in a medical context by the physician Giorgio Baglivi (1668–1707), who claimed that the human body operates by number, weight, and measure. Newtonian physicians in Scotland and the Dutch Republic supported a mathematical approach to the body, viewing it as a network of canals, with movements ruled by mathematical laws. The greatest of the mechanical physicians was Hermann Boerhaave (1668–1738), who dominated early-eighteenth-century medicine and applied hydraulic models to the functioning of the body.

The University of Leiden, where Boerhaave taught, had replaced lectures on classical medical texts with lectures on specific medical subjects, accompanied by dissections and visits to patients' bedsides. Leiden supplanted a declining Padua as the intellectual center of the European medical world, and medical education generally was moving away from reliance on classical texts. Even in universities where the classical texts were still officially part of the curriculum, lectures on these texts often dealt more with recent developments than with the classics themselves. Some intellectually radical physicians, notably Thomas Sydenham, even attacked the entire theoretical orientation of medical education, teaching that medicine can only be learned at the bedside. Hippocrates the observer, rather than Galen the theoretician, became the most admired ancient physician.

No longer a discipline that looked backward to the ancients, medicine entered the eighteenth century with a belief in the progress of medical knowledge. However, the effect of all this new knowledge on the treatment of patients was not that great, nor was medicine notably more effective. For example, bloodletting, originally a Galenic measure to restore the balance of humors by removing excess blood, continued to be practiced well into the nineteenth century, long after the collapse of Galenism had deprived it of a theoretical rationale.

See also Anatomy; Apothecaries and Pharmacology; Dissection and Vivisection; Galenism; Harvey, William; London College of Physicians; Midwives; Paracelsianism; Physicians; Physiology; Surgeons and Surgery; Sydenham, Thomas; University of Leiden; University of Padua; Vesalius, Andreas.

References
Conrad, Lawrence I., Michael Neve, Vivian Nutton, Roy Porter, and Andrew Wear. *The Western Medical Tradition: 800 B.C. to A.D. 1800.* Cambridge: Cambridge University Press, 1995.
Porter, Roy. *The Greatest Benefit to Mankind: A Medical History of Humanity.* New York: W. W. Norton, 1997.
Wear, Andrew, R. K. French, and I. M. Lonie, eds. *The Medical Renaissance of the Sixteenth Century.* Cambridge: Cambridge University Press, 1985.

Medieval Science

The idea of a scientific revolution implies a sharp break with the previously existing body of science. Although the leaders of the early modern scientific revolution did not use the phrase "scientific revolution," they thought of what they were doing and presented it either as something totally new, as did Francis Bacon and René Descartes, or as the revival of one or another of the ancient Greek intellectual traditions, as did Galileo

Galilei, who claimed to be the successor of Archimedes, and Pierre Gassendi, the "new Epicurus." Even Aristotelians such as William Harvey associated themselves with the ancient Aristotle rather than with the body of Aristotelian natural philosophy built up by medieval Scholastics. There were exceptions to this; Johannes Kepler put forth one of his optical works as a commentary on the thirteenth-century optical theorist Witelo (c. 1230–after 1275), and certainly the late medieval Scholastic natural philosophy to which most early modern scientists were exposed during their university educations left traces in their mature thought. However, the idea of a sharp break with medieval science persisted through the eighteenth and nineteenth centuries, when medieval science was caricatured as dogmatic Aristotelianism, and medieval medicine as dogmatic Galenism. Denigration of medieval science was particularly congenial to Protestants and anticlericals, who blamed its alleged sterility on the dead hand of the Catholic Church. Liberals framed the Renaissance, the Reformation, and the scientific revolution as one vast movement of emancipation from the Catholic Church.

This picture was seriously questioned in the early twentieth century by a devout and conservative French Catholic who was also a distinguished physicist, Pierre Duhem (1861–1916). In various writings on the history of science based on extensive primary research into what were then obscure manuscripts, Duhem claimed that the true scientific revolution took place not in the early modern period but in the late thirteenth century, with the condemnation of some Aristotelian ideas by the bishop of Paris in 1277. The Aristotelian denial of the possibility of multiple worlds, for example, was thought to be a heretical limitation on God's power. This did not mean that anyone was required to believe that there were multiple worlds, however; they were just forbidden to deny that God could have made them. The effect of this decree, according to Duhem,

was to inspire a group of avant-garde scholars at the University of Paris, notably Jean Buridan (1300–1358) and Nicole Oresme (c. 1320–1382), to come up with a new, non-Aristotelian physics. (Duhem was a patriotic Frenchman, and his study mostly ignores the innovative physicists at Merton College in Oxford who were the Parisians' contemporaries.)

The most notable innovation of their science was the idea of "impetus," a quality of a moving body that keeps it in motion. This differed from the Aristotelian theory that a body's motion is maintained by the medium in which it moves. Impetus theory, claimed Duhem, led directly to the mechanics of Galileo Galilei. This was not a sharp break but the continuous development of a fourteenth-century tradition. The humanist Aristotelianism of the Renaissance was thus actually a step backward, and Galileo's work was a triumphant revival of the new physics of the Parisians.

Duhem's research fundamentally changed the question of the relationship of medieval to early modern science, and no serious historian of science today views the science of the Latin Middle Ages as dogmatic or sterile. However, Duhem's claims about the connection of medieval physics to the scientific revolution did not stand up under examination. The brilliant work of the late medieval Paris and Oxford physicists, which continued to be expressed in the form of commentaries on Aristotle's works, modified the Aristotelian system rather than overthrowing it. Many of their claims remained speculative. When d'Oresme, for example, discussed the rotation of the Earth, his influential arguments were directed at demonstrating that such rotation was possible, or at least impossible to disprove. Copernicus, on the other hand, wanted to demonstrate that the Earth actually does rotate. Galileo's key contribution to the overthrow of Aristotelian physics—his insistence on mathematization—was not based on the work of the late medieval physicists. Although he was acquainted with their ideas and some of his early work makes use of impetus theory, he later rejected it as a theory of motion, and the concept of inertia as developed during the scientific revolution is quite different from the late medieval concept of impetus.

An alternative way of directly connecting medieval with early modern science, mostly appealing to Anglophone scholars, is to trace the origins of the experimental method to Oxford scholars of the thirteenth century, notably Robert Grosseteste (1175–1253) and Roger Bacon (c. 1214–c. 1294). This assertion is problematic because there is little evidence of direct connections between early modern and medieval experimentalism. With the rise of the study of medieval science as a discipline, scholars have also rejected the idea that it is only of interest as a foundation for early modern science.

What medieval science did clearly accomplish was the revival of interest in scientific questions in European intellectual life. Medieval natural philosophers, mostly connected with universities, translated and circulated a great deal of Greek and Arabic scientific and mathematical material, also making innovations of their own, particularly in optics. Medieval scientists also established the disciplinary framework within which early modern scientists worked. When the scientific revolution did arrive, it arrived in a society where scientific issues already had a recognized intellectual and institutional place.

See also Arabic Science; Aristotelianism.
Reference
Cohen, H. Floris. *The Scientific Revolution: A Historiographical Enquiry.* Chicago: University of Chicago Press, 1994.

Merian, Anna Maria Sibylla (1647–1717)

Anna Maria Sibylla Merian was the most highly regarded illustrator of natural history in the later seventeenth century and a fine natural historian herself. From a family of publishers and artists in the German city of Frankfurt,

Merian acquired her artistic skills in her family environment. In 1665, she married another artist, Johann Graff (1636–1701), subsequently moving to his home town of Nuremburg. There, from 1675 to 1680, she published her first book, a three-part copperplate collection of flowers and wreaths intended to serve as models for artists and needleworkers. Far more innovative and scientific was her *Caterpillars* (1679), a collection of illustrations of the life cycle of insects, accompanied by Merian's own observations.

After the publication of a second volume in 1683, Merian abruptly abandoned her husband (they later divorced) to join a sectarian community in Friesland in the Dutch Republic. There she continued her natural-historical studies, breeding and observing insects until, disillusioned with the community, she left for Amsterdam in 1691. She made friends among the natural historians, collectors, and artists resident in that center of world trade, and she enjoyed seeing collections of insects from foreign lands. She was, however, disappointed in the little research available on the metamorphoses of these insects.

At the age of 52, Merian left Amsterdam with her 21-year-old daughter, Dorothea, for the Dutch sugar colony of Surinam in South America. For two years, the women ranged the plantations of Surinam, breeding and observing insects and talking about them with the Native Americans and resident Africans (Merian owned a few slaves). On her return to Europe in 1701, she hoped to sell the specimens she had collected. The self-published *Metamorphoses of the Insects of Surinam* (1705), Merian's masterpiece, contained many interesting observations on society as well as on plants and insects—many described for the first time in a European text. She also produced many of the illustrations for Georg Eberhard Rumph's *The Ambonese Curiosity Cabinet* (1704). In Merian's last years, she and Dorothea brought out an expanded version of *Caterpillars* with a Dutch translation of the German text.

See also Illustration; Natural History; Women.
Reference
Davis, Natalie Zemon. *Women on the Margins: Three Seventeenth-Century Lives.* Cambridge: Harvard University Press, 1995.

Mersenne, Marin (1588–1648)

Marin Mersenne, one of the main circulators of scientific knowledge and promoters of the mechanical philosophy in the first half of the seventeenth century, came from a poor French family but was educated by the Jesuits at La Flèche, the same school René Descartes attended. He studied theology at the University of Paris, joined the Franciscan order of Minims in 1611, and was ordained a priest in 1612.

Mersenne moved the mechanical philosophy in a religious direction, believing that it could solve two religious problems. One was magic and popular superstition. Mersenne saw magic, and particularly the belief that nature is somehow alive, as the root of all heresy, and he was particularly horrified by the heresy of Giordano Bruno. He was also hostile to contemporary magicians such as Tommaso Campanella and Robert Fludd. Mersenne claimed that nature is not alive but that it acts in a mechanical way at all times, except when God directly acts on it. God is not the soul of the world, as magicians and pantheists had it, but stands outside it as a master or governor.

The second problem was skepticism, an important issue in the French Catholic Church at the time. Mersenne defended the possibility of real knowledge about the world in *The Truth of the Sciences* (1625). He believed that in order to defeat magic and skepticism, scientific knowledge had to be widely distributed, and along with his contemporary Descartes, Mersenne was a pioneer in the use of the French language for scientific purposes. Unlike Descartes and other mechanical philosophers, Mersenne did not see Aristotelianism as the main enemy. Neither was he particularly hostile toward Protestants, and he corresponded with Protestant natural philosophers.

Mersenne's most important contribution to the development of science was his role as an "intelligencer," or circulator of scientific ideas. He engaged in a vast correspondence with the leading mathematicians and natural philosophers of Europe, including Galileo Galilei, Descartes, Pierre Gassendi, Isaac Beeckman, Johannes Baptista van Helmont, Pierre de Fermat, Thomas Hobbes, Sir Kenelm Digby, and Blaise Pascal. Mersenne was the main contact Descartes had with French intellectual life during the time he spent in the Dutch Republic. Mersenne did not merely passively circulate information but took an active role in promoting debate. For example, he invited Gassendi to formulate and publish his objections to Descartes's *Meditations.* He also encouraged precise and rigorous measurements and descriptions of natural phenomena and experiments.

In addition to his correspondence, Mersenne's other contribution to the scientific community was his weekly meetings to discuss natural philosophy, which he hosted in his cell at the Minim monastery at the Place Royale in Paris beginning around 1635. It was at one of these meetings that Pascal met Descartes for the first time. Not opposed to Galileanism or Copernicanism, Mersenne also played an important role in circulating Galileo's works and ideas in France, as in his *The Mechanics of Galileo,* published in 1634, after Galileo's condemnation.

Mersenne's name today is often associated with Mersenne primes, which are prime numbers of the form $2^n - 1$, which he discussed in his *Synopsis of Mathematics* (1624). However, such primes had been discussed before, and Mersenne was not a significant mathematician. His most original contributions to scientific knowledge were in the field of acoustics and harmony, where he established the mathematical relations between the notes emitted by a vibrating string and its length, tension, and thickness—one of the earliest examples of the successful use of mathematical laws in the scientific revolution.

See also Correspondence; Descartes, René; Music.
Reference
Brown, Harcourt. *Scientific Organization in Seventeenth-Century France (1620–1680).* Baltimore: Johns Hopkins University Press, 1934.

Meteorology
See Weather.

Microscopes

The exact origins of the microscope are unclear, but it is known to have been invented shortly after the telescope, the earliest examples being associated with the Dutch engineer Cornelius Drebbel (1572–1633). Drebbel's early designs employed two convex lenses on the model of Johannes Kepler's telescopes. By the late seventeenth century, these were replaced with the more efficient single-lens design, which avoided the refraction of the light rays that casts colored rings around the objects under investigation. Several natural philosophers of the early to mid-seventeenth century, including Andreas Libavius, Athanasius Kircher, and Galileo Galilei, used microscopes in their investigations, and a broadsheet with a microscopic study of the bee was published in 1625 by Francesco Stelluti (1577–1652), a member of the Accademia dei Lincei. However, systematic and widely distributed microscopic research was a phenomenon of the second half of the century, closely associated with the Royal Society of London. Marcello Malpighi's *On the Lungs* (1661) and Robert Hooke's lavishly illustrated *Micrographia* (1665), which included a number of technical improvements to the microscope, inaugurated this new era.

Of the five major microscopists of the late seventeenth century—Hooke, Malpighi, Nehemiah Grew, Jan Swammerdam, and Antoni van Leeuwenhoek—all but Swammerdam were fellows of the Royal Society. Hooke and Grew were employees of the soci-

An early-eighteenth-century microscope designed by eminent microscope maker John Marshall. (Science Museum, London)

ety, and Malpighi and Leeuwenhoek used the society to publish their results. By contrast, the other internationally prominent scientific body, the French Royal Academy of Sciences, displayed little collective interest in microscopical research, although the leading light of its early years, Christiaan Huygens, did make some significant but unpublished examinations of microorganisms. The Baconian aspects of microscopic research, particularly the amassing of data as practiced by Leeuwenhoek, made this study more congenial to the English society than to the more theoretical French. The mechanical philosophy, with its emphasis on the construction of the material world out of very small parts, whether

Cartesian corpuscles or atoms, also encouraged microscopic examination. Microscopists, especially Swammerdam, pointed out the religious meaning in microscopy, praising the wondrous and minute handiwork of God in crafting the smallest beings.

The areas of science most affected by microscopy in the seventeenth century were natural history and anatomy. The most important anatomist to use the microscope was Malpighi, who published a number of important microscopic studies on the structure of the human body, on the anatomy of the silkworm, and on the development of the chick embryo. By 1685, the Dutch physician Goverd Bidloo (1649–1713), drawing principally upon Malpighi, published *Anatomy of the Human Body,* the first atlas of the human body to incorporate microscopic data. Swammerdam, although interested in anatomy, principally examined insects, while Grew studied plants. Hooke and Leeuwenhoek studied a range of phenomena.

Many seventeenth-century microscopists, notably Swammerdam and Leeuwenhoek, were inspired by their opposition to the doctrine of spontaneous generation, which held that small living things, such as worms or insects, emerge spontaneously out of nonliving matter. For this reason, much microscopic research went into establishing the reproductive cycle of very small animals. However, not all microscopists opposed the idea of spontaneous generation; Hooke endorsed it along with sexual reproduction in *Micrographia,* and Kircher's disciple Filippo Buonanni (1638–1725), author of *Micrographia Curiosa* (1691), continued to defend it.

Microscopic anatomy did not catch on in European medical schools until later. Because of its difficulty and this lack of an institutional base, microscopy was beginning to decline by the late seventeenth century. Although Leeuwenhoek soldiered on to his death in 1723, the great seventeenth-century microscopists had no counterparts in the early eighteenth century. However, good microscopes were widely available commercially and continued to be used for diversion and entertainment.

See also Grew, Nehemiah; Hooke, Robert; Leeuwenhoek, Antoni van; Malpighi, Marcello; Swammerdam, Jan.
References
Fournier, Marian. *The Fabric of Life: Microscopy in the Seventeenth Century.* Baltimore: Johns Hopkins University Press, 1996.
Ruestow, Edward G. *The Microscope in the Dutch Republic: The Shaping of Discovery.* Cambridge and New York: Cambridge University Press, 1996.

Midwives

Most obstetrical care in early modern Europe was provided by female midwives, a situation that did not change during the scientific revolution. Ordinarily, male medical professionals were called on only in the case of abnormal births that presented a risk to the mother or child. What did change during this time is that the body of knowledge among midwives concerning normal births moved from a predominantly oral and female craft tradition, not unlike the male craft traditions of apothecaries and surgeons, to an increasingly written body of knowledge available to both sexes. A number of midwifery manuals were published, both by midwives and by male medical professionals.

The seventeenth century saw the emergence of the "man-midwife"—the male who claimed that his medical training, unavailable to women, gave him superior expertise in the birthing process, although whether this was actually true is questionable. Man-midwives were also more likely than women to use instruments, such as the obstetrical forceps developed in the sixteenth century. But man-midwives were still rare, their practice largely restricted to a social elite. For example, Louis XIV's (r. 1643–1715) use of a man-midwife to deliver one of his illegitimate children set an example for the French nobility.

Although arrangements varied greatly throughout Europe, midwifery practice was regulated principally by the state and the

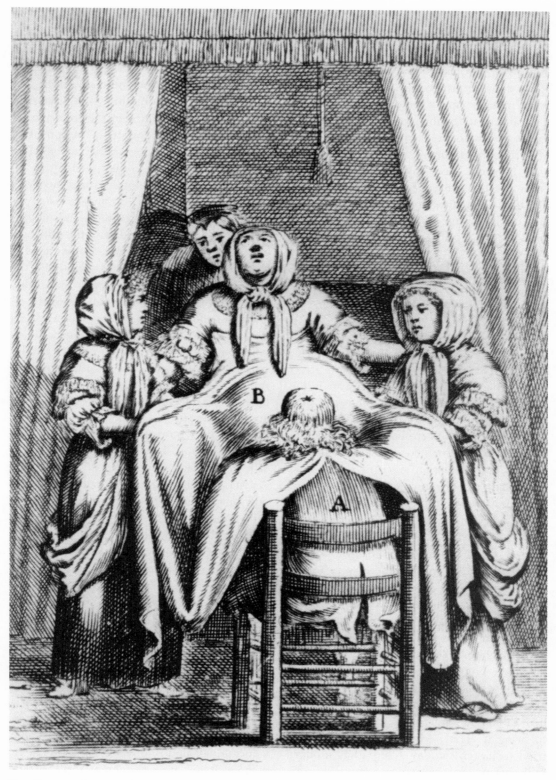

This illustration from an early-eighteenth-century edition of Louise Bourgeois's classic midwifery manual shows a man attending at the birth. (National Library of Medicine)

church, and only secondarily by male medical professionals. Midwives who organized into groups for the advancement of their profession or for the protection of their interests, however, were usually blocked by the established male groups of physicians. Some medical training, in the form of public lectures or courses open to or even required for midwives, emerged during the scientific revolution. In some cases midwives, particularly those in urban centers, were related or married to male medical professionals and shared medical and anatomical knowledge with them.

See also Medicine; Women.
Reference
Marland, Hilary, ed. *The Art of Midwifery: Early Modern Midwives in Europe.* New York: Routledge, 1993.

Millenarianism

In Christian theology, millenarianism is the belief in 1,000 years, a millennium, of perfect life on Earth before the Last Judgment. The founders of Protestantism in the sixteenth century frowned on millenarianism, but its "premillenarian" variant, which held that Christ would return at the beginning of the 1,000 years rather than at the end, became a common belief among Protestants in the seventeenth century. By contrast, millenarianism played only a minor role in early modern Catholicism, and there were only a few Catholic millenarian natural philosophers, such as Tommaso Campanella.

The society that would exist during the millennium, it was claimed, would be perfect, with total human command over nature paralleling the command over nature that Adam had enjoyed in the Garden of Eden before the Fall. Passive millenarians anticipated an imminent millennium but did not exert themselves to bring it about, while active millenarians tried to hasten the day. Many early modern millenarians, particularly in central Europe and the British Isles, were involved in scientific and technological projects, believing that

these projects would be part of the improvement of life and control over nature in the millennium. Millenarianism contributed to an optimism about the capacity of the human mind to understand and dominate nature, and millenarianism was sometimes seen as an antidote to skepticism since in the millennium, certainty would undoubtedly be possible. Millenarianism was one source of belief in intellectual progress, often leading to an emphasis on application rather than theory.

Despite its biblicism, millenarianism was often blended with magical ideas. Astrologers attempted to predict the date of the millennium based on the schedule of conjunctions of the planets, and alchemists identified the philosopher's stone with the returned Christ. One of the earliest connections between science and the Second Coming was made by Paracelsus. He identified the reborn Elijah, who some believed would be Christ's herald, with the mastery of nature—calling him Elias Artista, or Elijah the Artisan. The German chemist Johann Rudolf Glauber (1604–1670) actually identified a salt whose marvelous life-sustaining powers he had observed in an Austrian spring; later he manufactured the salt in his lab, identifying it as the Elias Artista. His discovery of the salt was supposedly one of the events that would lead to the millennium, and "Elias Artista" was an anagram for "Et Artis Salia," or Salts of Art.

Glauber was far from the only Protestant to link science and the millennium. Seventeenth-century Protestant millenarians were very internationally minded, and much information was disseminated between millenarians in various countries, such as the members the Hartlib circle. Millenarianism had a particularly strong influence on natural philosophy in England during the period of Puritan rule in the 1650s. Puritan ideologists in and out of the Hartlib circle combined millenarian theology with Baconianism and the "pansophic" educational philosophy of the Czech reformer John Amos Comenius (1592–1670) to support universal reform through improved knowledge of nature.

Many important English scientists in the later seventeenth century were millenarians as well, although after the Restoration of Charles II (r. 1660–1685) in 1660 it was necessary to be discreet about it. Late-seventeenth-century English millenarian natural philosophers tended to be passive rather than active millenarians, abandoning the practical emphasis of the Hartlib circle. One example was Isaac Newton, who devoted the final decades of his life to studies of the apocalyptic books of the Bible to predict the approximate time of the millennium. Unlike many other millenarians, he dated it far in the future. He believed that the millennium might even bring fundamental changes in the laws of nature, thus making his own laws of nature provisional. In the eighteenth century, open expression of belief in an imminent millennium was widely ridiculed, even when held by an eminent natural philosopher, Newton's disciple William Whiston (1667–1752). The millenarian ideal was eventually secularized into the idea that scientific research promotes economic development.

See also Puritanism and Science.
References
Osler, Margaret J., ed. *Rethinking the Scientific Revolution.* Cambridge: Cambridge University Press, 2000.
Webster, Charles. *The Great Instauration: Science, Medicine, and Reform, 1626–1660.* New York: Holmes and Meier, 1975.

Mining

The sixteenth century was a golden age of mining literature, most of it emanating from Germany, where mining had boomed in the period from 1460 to 1530. The demand for metal led to an increase in technically demanding and capital-intensive deep mining. This led to a boom in literature designed not only to explain mining processes but also to assert the dignity of mining scholarship against various charges leveled against it. For example, mining was sometimes seen as an impious assault on Mother Earth. It also had

a servile taint, being practiced in both the ancient and the early modern world by slaves or forced laborers, like the Indians drafted to work and die in the great Spanish silver mine of Potosí in Peru. (Mining in the Americas was technically backward compared to mining in central Europe, although the important mercury amalgamation process for refining silver was developed there in 1554 by Bartolomé de Medina [d. c. 1580], a merchant from Seville, and further advanced by the priest Alvaro Alonso Barba [c. 1569–1662], another Peruvian Spaniard.) Mining was also associated with the evils of civilization, as gold was associated with greed and iron with violence. Georgius Agricola, the foremost mining writer in the period, wanted to establish that mining and metallurgical knowledge were morally and intellectually worthy of a free scholar, pointing out that mining required knowledge of mathematics, natural philosophy, medicine, and law.

The mining literature was influenced by alchemy and often framed its discussions in alchemical terms. Yet it differed from much of the mainstream alchemical literature on metals in that mining writers asserted that their knowledge should be widely distributed rather than restricted to a circle of adepts. For example, one of the few major non-German mining writers, the Italian Vannoccio Biringuccio (1480–1537), author of *Pyrotechnics* (1540), wrote specifically to spread central European mining knowledge to the backward mining community of Italy, with the hope of enhancing Italian prosperity.

One way that alchemy affected thought about mining was through the widespread acceptance of the ancient idea that metals are like plants, that they vegetate and renew themselves in the womb of the Earth, possibly influenced by the stars (metals were correlated with planets in magical and astrological thinking). This suggested that a closed and exhausted mine could be reopened for harvesting after a period of time. Although this belief was incompatible with the shift in seventeenth-century natural philosophy away from organic and toward mechanical approaches to nature, it persisted among miners into the eighteenth century. Mining and metals generally were closely associated with magic and superstition, particularly given the dangers miners faced in dark and lonely surroundings. Another "magical" approach that was widely accepted by miners was the use of a dowsing rod to locate ores.

Mining presented a number of challenging practical and technological problems, which influenced the development of science and engineering. Draining mines involved a significant application of energy for the pumps. During the late seventeenth century, several early steam engines were developed in England to solve the problem of pumping out coal mines without requiring hundreds of horses. Sixteenth-century mining engineers who were working on drainage problems observed that a column of water cannot be raised much higher than 30 feet, eventually inspiring a research program that culminated in the development of the barometer. Mining often made miners sick, and both Agricola and Paracelsus, physicians involved in the German mining industry, wrote on the diseases of miners in some of the earliest literature on occupational diseases.

See also Agricola, Georgius; Technology and
    Engineering.
References
Long, Pamela O. "The Openness of Knowledge:
    An Ideal and Its Context in Sixteenth-Century
    Writings on Mining and Metallurgy." *Technology
    and Culture* 32 (1991): 318–355.
Merchant, Carolyn. *The Death of Nature: Women,
    Ecology, and the Scientific Revolution.* San
    Francisco: Harper & Row, 1980.

Monsters

The scientific revolution was haunted by monsters. Conjoined twins, headless babies, and other humans and animals born with deformations or other irregularities were seen as monsters—portents, curiosities, mysteries, and entertainment. Stuffed and otherwise preserved monsters appeared in cabinets of curiosities, and living monsters, such as the always popular dwarfs, entertained nobles at courts or commoners at fairs. Monsters were pictured and discussed in learned treatises and popular broadsides, and writers on natural magic, such as Giambattista della Porta, gave instructions on how to create monsters. These varied between techniques to produce actual monsters, such as the breeding of two-legged dogs by repeated mutilation, and illusionary techniques, such as the distorting mirrors of Athanasius Kircher's museum in Rome, which could replace the head of a viewer with an animal's.

The belief that monsters were prodigies bearing providential or allegorical messages, such as a two-headed baby presaging division in a kingdom, waned along with providential prodigy-belief in general. Monsters could still be born to women as punishment for sin or as divine warnings of future events, but they could also be explained naturally. The traditional Aristotelian theory of the generation of monsters was that they are caused by an excess or defect in the seed. Another theory was that a particularly vivid image or longing in a pregnant woman's consciousness could cause a monstrous birth, as a woman who craved a lobster gave birth to a monster resembling one. Sixteenth-century medical men, such as the French surgeon Ambroise Paré (1510–1590), wrote treatises discussing human and divine causes of monstrosity, and organized natural philosophy continued this interest during the seventeenth century. Reports of monstrous births and monstrous individuals, both human and animal, appeared frequently in scientific publications such as *Philosophical Transactions*. English, French, and German scientists performed and published dissections of monsters, tending by the end of the century to explain the purpose of their research as casting light on normal development rather than as exploring monsters out of curiosity about the monsters themselves. The desire to examine or dissect a monster could involve scientists in conflict with the monster's family or owners, and sci-

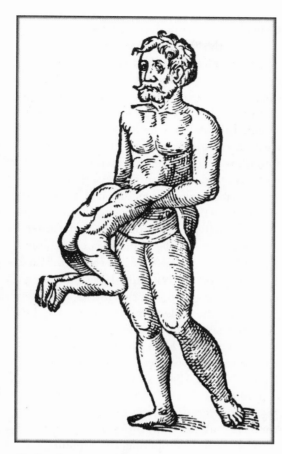

The French surgeon and medical writer Ambroise Paré thought an excess of seed caused this monster with a second body growing out of his stomach, illustrated in his work Of Monsters and Prodigies.

entists grew increasingly suspicious of fake or fraudulent monsters. The connection between science and popular monster-culture went both ways, and by the late seventeenth century, promoters of monster exhibits were using scientific language in their promotions and presenting their monsters as educational.

See also Prodigies.

References

Hanafi, Zakiya. *The Monster in the Machine: Magic, Medicine, and the Marvelous in the Time of the Scientific Revolution.* Durham, NC, and London: Duke University Press, 2000.

Wilson, Dudley. *Signs and Portents: Monstrous Births from the Middle Ages to the Enlightenment.* London and New York: Routledge, 1993.

Museums and Collections

During the scientific revolution the European museum evolved from a number of different collecting traditions, including the "cabinet of curiosities," a tradition strongest in the German princely courts; the educational collections made by natural historians, particularly in Italy; the institutional collections of universities and colleges of physicians; and the collections of the relics of saints and devotional objects found throughout Catholic Europe. All of these collections faced many vicissitudes, and the art of collection was much more highly developed than the art of curating.

Cabinets, designed to impress the viewer and display the wealth and status of the owner, contained a variety of rare, exotic, and valuable objects. These included manufactured items, including things from far away places that had been brought back to Europe in the age of European expansion, such as stone knives and statues of gods from Mexico or booty from the wars against the Turks. Cabinet collections also included old coins and artifacts from classical antiquity, works of art, and various devices showing human ingenuity, such as scientific or craft instruments or artifacts, like a cherrystone inscribed with dozens of faces. Natural objects were also displayed, preferably striking or beautiful things, such as a "unicorn's horn," which actually came from a narwhal. Objects were arranged to create elegant and dramatic contrasts with symbolic meaning rather than by their type or origin. The greatest cabinets, belonging to princes and kings such as Emperor Rudolf II, claimed to reproduce the world in miniature, a microcosm of the universal macrocosm. As time went on and cabinets became more elaborate, a greater variety of objects were regarded as suitable for display.

Another type of collection had an explicitly educational purpose and concentrated on natural objects—animals, plants, feathers, shells, and so on. These collections were arranged in a more functional way and placed more emphasis on the typical than on the exotic quality of the items displayed. Italy was the center for this type of collecting in the sixteenth century, with notable collectors such as Ulisse Aldrovandi of Bologna. These collections often became attached to universities both in and outside Italy—the University of Leiden, for instance, had a notable collection. Like the cabinets of curiosities, some of these collections became widely publicized, sometimes through the publication of catalogues, and they were popular destinations for visitors.

Museums were associated with notable natural philosophers. Athanasius Kircher's Museum Kircherianum in Rome was a vast collection of a huge range of objects, many originating in the worldwide network of Kircher's fellow Jesuits. The museum reflected Kircher's interests, being particularly strong in material from Egypt. Kircher's museum became a must-see for visitors to Rome, the most distinguished of whom might be escorted by Kircher himself. Like many collections gathered by an individual, it suffered from neglect after his death. Although scientific academies might seem to offer a solution to this problem, their institutional health was often less robust than one might imagine, and thus their record as curators is distinctly mixed. The short-lived Italian societies—the Accademia dei Lincei and the Accademia del Cimento—did not exhibit collections, nor did the French Royal Academy of Sciences. The English Royal Society had a rich if heterogeneous collection of natural wonders that it originally purchased from a French immigrant, Robert Hubert. Nehemiah Grew published a catalog of the expanded collection in 1681, attempting to judge the objects by their intellectual interest rather than by their rarity and exoticism. However, the society lacked the institutional structure to curate the collection properly, and it fell into neglect.

The major collections of late-seventeenth and early-eighteenth-century England were put together by individuals. The Tradescant Collection of natural and artificial rarities, known as the "Ark," passed from the gardener John Tradescant (1570 or 1575–1638) to his son John Tradescant (1608–1662) and then to the astrologer and alchemist Elias Ashmole (1617–1692). In 1686, Ashmole founded the Ashmolean Museum (one of the earliest recorded uses of "museum" as an English word in the modern sense) at Oxford to house the collection. The Ashmolean featured a laboratory and lecture hall in addition to the collection, and it became the center of scientific activity in Oxford. The first and second keepers of the Ashmolean Museum, Robert Plot (1640–1696) and Edward Lhwyd (1660–1709), built large collections of fossils and geological specimens, but after their time the collection was neglected, and even partially sold off. The greatest English collector was Sir Hans Sloane (1660–1753), a physician and president of the Royal Society from 1727 to 1741, succeeding Isaac Newton. Sloane collected entire collections, buying them from the estates of deceased collectors. He built up a repository of over 100,000 separate items, comprising the founding collection of the British Museum.

By the late seventeenth century, a new ethic of collecting had emerged that frowned on the search for wonder as a criterion for selecting and displaying objects in a collection. Rather than collecting across a wide range of objects, collectors were encouraged to specialize in a particular category and emphasize the value of their collections for advancing knowledge rather than exciting wonder. John Woodward (c. 1665–1728), an English physician and rival of Sloane, built a vast and well-documented collection of fossils on this principle. He was the first to insist that the provenance of the specimens be recorded.

See also Aldrovandi, Ulisse; Kircher, Athanasius; Natural History; Royal Society.

References
Findlen, Paula. *Possessing Nature: Museums, Collecting, and Scientific Culture in Early Modern Italy.* Berkeley and Los Angeles: University of California Press, 1994.

Impey, Oliver, and Arthur MacGregor, eds. *The Origins of Museums: The Cabinet of Curiousities in Sixteenth- and Seventeenth-Century Europe.* Oxford: Clarendon, 1985.

Music

Early modern Europe inherited from the classical and medieval worlds an idea of music as a mathematical and philosophical discipline. In the ancient Pythagorean and Neoplatonic traditions, the laws of harmony were considered to govern the universe as well as the narrower realm of musical sounds. Humanists revived this ancient musical theory as they revived everything else they could from the ancient world (although they were unable to recover ancient musical practice). Music was also a topic of interest for early modern natural philosophers because they operated in a social context where knowledge of music and the ability to make it was considered a necessary attainment in the social elite. Skilled musicians among early modern natural philosophers included Galileo Galilei, Thomas Hobbes, and Christiaan Huygens. Writers on musical theory included just about every leading figure of seventeenth-century physical science including Giambattista della Porta, Simon Stevin, Johannes Kepler, Francis Bacon, René Descartes, Pierre Gassendi, Athanasius Kircher, Isaac Beeckman, Marin Mersenne, Huygens, Robert Hooke, John Wallis, Isaac Newton, and Gottfried Wilhelm Leibniz. Musical theory was closely related to mathematics and other forms of practical learning. Music was taught as part of the curriculum at both the universities and the new educational institutions of the period—one of the seven professors at Gresham College, for example, was a professor of music.

Practical music teachers were closer to artisans than to upper-class and clerical natural philosophers, and so the treatises on music produced by early modern natural philosophers were not practical works aimed at teaching musical skills but theoretical or "speculative" works on the mathematical and physical principles governing musical sounds and their effects. Music composition had advanced beyond that of the ancients, and the new musical practices introduced in the Middle Ages had to be systematized, as the Italian theorist Gioseffo Zarlino (1517–1590) did, for example, when he systematized counterpoint.

The Pythagorean alliance of music with mathematics and cosmology remained powerful in the seventeenth century. Drawing on the tradition of the harmony of the celestial spheres, Johannes Kepler organized the planets in their journey around the Sun in terms of musical harmonies, even giving the musical notation associated with different planets in his *Haromonies of the World* (1619). He derived the harmonies of the planets by comparing the notes that correspond with their maximum and minimum velocities around the Sun. (The planets do not actually produce sounds, Kepler explained, because of the airlessness of space.) Kepler's intellectual opponent, Robert Fludd, also viewed the universe as fundamentally musical. The massive musical treatise of Athanasius Kircher, *Musurgia* (1650), drew heavily on Fludd and also asserted the relevance of music for cosmic structure.

Kepler's main work on music itself dealt with the problem of the consonances, or harmonious pairs of notes. This was a particularly important problem because the four consonances known to the ancients since Pythagoras—identical notes or unisons, the octave, the fourth, and the fifth—had recently been joined by four others—the major and minor thirds and sixths. Zarlino had explained these intervals arithmetically, and Kepler did so geometrically, by inscribing polygons in a circle.

Despite the brilliance of Kepler's work, further developments in musical science abandoned pure mathematics and took a more physical and experimental approach, eventually explaining consonance by the ratios of the vibrational frequencies of the

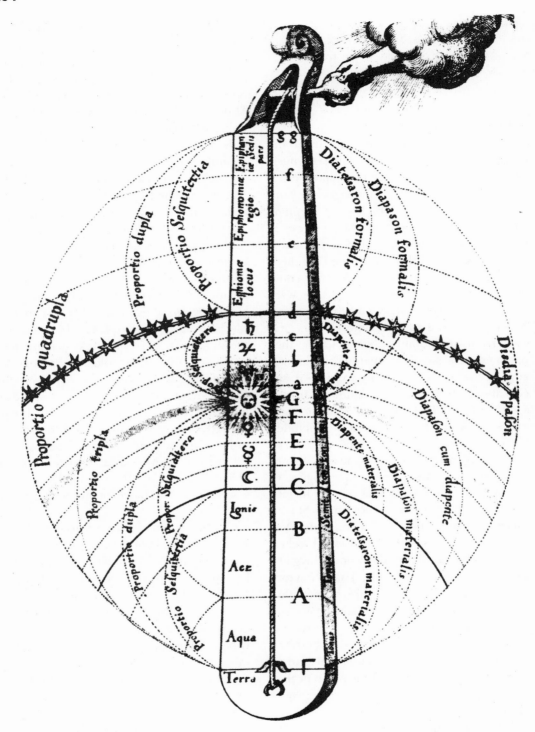

Robert Fludd's cosmology viewed the universe in terms of the harmonies created from a single-stringed instrument, or monochord, as seen in this diagram from his Utriusque Costal.

instruments that produce the notes. This tradition can be traced to late-sixteenth-century Italians, notably the mathematician Giovanni Battista Benedetti (1530–1590) and Vincenzo Galilei (c. 1520–1591). The only writer on musical science in the early modern period to be a significant creative musician, Galilei was part of an active group of Florentine musicians contributing to the transition from Renaissance to Baroque music. Benedetti and Galilei began to assert the importance of the physical construction of the material music generator, the musical instrument, in creating specific musical tones, thereby rooting musical theory in physical science and, in Galilei's case, in experiment, rather than in pure mathematics. This work was carried on by Vincenzo's son Galileo, who proved in *Discourses on Two New Sciences* (1638) that the frequency of a vibrating spring's motion is the cause of the pitch of the sound it produces.

The most significant acoustical theorist among the early modern natural philosophers writing on music was Marin Mersenne, author of *Universal Harmony* (1636–1637). A pioneer in the science of music, Mersenne based music theory not on pure mathematics but on the physics of the music-producing instrument, notably the vibrating string. His was the first scientific treatise on music to include exhaustive discussion of actual musical instruments, about which Mersenne learned by talking to the people who built them. Through experiment, he established the mathematical relationships between the tones of a vibrating string and its length, tension, and thickness. Mersenne was the first to identify the overtones—the tones that differ from the main tone produced by a vibrating instrument. His work integrated musical theory and experimental philosophy, and it provided one of the earliest examples of the successful mathematization of a science.

Mersenne's friend Descartes chose music as the subject of his first book, *Musical Compendium*. It was written in 1618 as a gift to his friend and fellow mechanical philosopher, Beeckman, but was not published until after Descartes's death. Beeckman and Descartes arrived at contrasting mechanical explanations of the propagation of sound. Beeckman explained sound by the production of sound corpuscles; Descartes, by a vibration in the material plenum.

Music remained an object of keen scientific interest in the late seventeenth century, in both the English Royal Society and the French Royal Academy of Sciences. In addition to Mersenne's work, the English were influenced by the acoustical experiments suggested by Francis Bacon. Experimental musical performances took place at some of the early meetings of the Royal Society, whose first president, Sir Robert Moray (1608–1673), was a musician, and papers on musical subjects appeared in *Philosophical Transactions*. Hooke, and later Newton, led the field among the English who wrote on music. Newton determined the wave nature of sound and the speed of its propagation in *Mathematical Principles of Natural Philosophy* (1687). The leading musical and acoustic investigators in the Royal Academy were Huygens and, later, Joseph Sauveur (1653–1716), inventor of the term "acoustics."

See also Fludd, Robert; Kepler, Johannes; Mersenne, Marin.

References

Cohen, H. Floris. *Quantifying Music: The Science of Music at the First Stage of the Scientific Revolution.* Dordrecht, the Netherlands: Reidel, 1984.

Gouk, Penelope. *Music, Science, and Natural Magic in Seventeenth-Century England.* New Haven, CT: Yale University Press, 1999.

N

National Differences in Science

Although nearly all early modern scientists of significance operated as part of an international community of inquirers, they were also members of nations and national scientific communities. ("Nation" here does not necessarily mean a sovereign political entity; there was an Italian scientific community although there was no central government covering the entire peninsula.) There were many reasons for national scientific differences. The relations of church, state, and university varied throughout Europe. The extremes here were Spain, where the ferociously repressive Catholic Church and powerful universities kept conservative Aristotelianism in intellectual authority well into the eighteenth century, and the Dutch Republic, whose weak state Calvinist church, although not completely without influence, was unable to control scientific life. The scientific culture of England, where there were only two universities, was less influenced by university life than was Germany, with its many universities. Where each state stood vis-à-vis the process of colonization, state building, and trade also affected science. Because of the Spanish Empire in the New World, Spaniards were better situated to carry out science in the Americas than were other Europeans until the seventeenth century. Amsterdam, in the Dutch Republic,

was the commercial capital of Europe and a center for the collection of natural objects from the East, and thus was a center for natural history. France, marked by a high degree of political and intellectual centralization in Paris, had a more authoritarian style of scientific organization than more decentralized societies such as Britain or the Dutch Republic.

Italy, in the sixteenth century a leader in science as in many other cultural areas, began to fall back in the seventeenth century, particularly after the trial of Galileo in 1633. However, repression by the church did not end Italian science, which was also supported by Italy's multiplicity of princely courts. In the seventeenth century, the courts displaced the once-great but now declining Italian universities as centers for science. The career of Galileo is exemplary in this respect; he left the University of Padua to climb a two-rung ladder of Italian courts, moving from the court of the duke of Tuscany to that of the pope. Italian science tended to the dramatic and spectacular, which courtly patrons valued. Courts were relatively uninterested in theoretical explanations, only reinforcing the church's efforts to stem the tide of new philosophy. For example, post-Galilean Italy produced the finest telescopes and a number of significant astronomical observers, but the ban on Copernicanism meant that their observations were not placed in a theoretical framework. Italian courtly science also valued a certain elegance and the natural philosopher's exhibition of ingenuity in argument.

Germany, loosely united along with parts of eastern Europe into the Holy Roman Empire, had a relatively high degree of technological development in areas such as mining; a remarkably large number of German natural philosophers, from Paracelsus and Agricola on, were associated with mining. This led to a German lead in chemistry through the mid-seventeenth century. Germans tended to distinguish less between science and magic than did other European peoples in the course of the scientific revolution, and alchemy remained a vital force in Germany through the eighteenth century. Both German magicians and German scientists were great believers in doing things; after all, that quintessential German professor, Faustus, whose legend originated in the sixteenth century, sold his soul not only for knowledge but for power. This focus on the practical led Germans such as Otto von Guericke (1602–1686) to produce and refine a number of scientific instruments. The domination of intellectual life in both Protestant and Catholic Germany by the German universities, with their Aristotelian curricula, made German natural philosophers, most notably Gottfried Wilhelm Leibniz, much less hostile to Aristotelianism than were natural philosophers elsewhere.

The Dutch Republic was one of Europe's most decentralized polities, and its science remained small in scale because the republic lacked a large, culturally central scientific organization along the lines of the Royal Society or Royal Academy of Sciences. Dutch scientists tended to be more interested in investigating the details of nature than in creating natural-philosophical theory, and the greatest Dutch theorist, Christiaan Huygens, spent much of his working career in Paris. The characteristically Dutch branch of science was microscopy, drawing on the excellent Dutch lens-grinders, and the greatest Dutch scientist to spend his career in the Netherlands was Antoni van Leeuwenhoek, whose microscopic studies took place in his own home and who turned to England and its Royal Society to publish his discoveries. The Dutch, who possessed the least aristocratic culture in Europe, were unafraid to investigate those things defined as vermin, and they were Europe's leading students of insects.

The dominant countries in natural philosophy by the end of the scientific revolution, England and France, displayed strikingly different scientific styles. In the France of Louis XIV (r. 1643–1715), science was subordinated to the glory of the "Sun King" (although the King himself had little interest in the subject). Early in the seventeenth century, French science had been decentralized. In addition to Parisian science, there was an active scientific circle in Provence around Nicolas-Claude Fabri de Peiresc. French science at this time was also informal, as in the conferences of Théophraste Renaudot. By the end of the century, however, the Parisian and relatively structured Royal Academy of Sciences had established unquestionable intellectual dominance, while lecturers and demonstrators such as the Cartesian Jacques Rohault (1618–1672) spread the new philosophy, usually in a Cartesian form, to the male and female French social elite. In this period, France pioneered the idea of the professional scientist working outside of the university environment. French science eventually also became dominated by Cartesian theory, particularly after the reorganization of the Royal Academy in 1699, and it emphasized logic over empirical observation. French science sought the causes of things, and one of the main French objections to Newtonianism was that Newton (one of the few major scientists of his time who never visited Paris) gave no cause for gravity.

England had been a scientifically backward country in the sixteenth century, but by the late seventeenth century, it had more than made up for lost time. Although patronized by monarchs, English science was also open to an elite of independent gentlemen. Although it lacked a university, London did have the Royal Society, and the city became the center of English science. It did not enjoy the dominance of Paris, however, since the English university towns maintained an active scientific culture. Royal Society members Robert Hooke and Isaac Newton could be categorized as professional scientists, but the society was also open to the gentlemanly amateur in a way that the Royal Academy of Sciences was not. Although the ideology of the society was collaborationist, the gentlemen who comprised its membership were free to follow their own interests, rather than being organized into the kind of massive, centrally directed, collaborative projects more typical of the rival French organization. English scientists, drawing on the Baconian tradition, prided themselves on the study of fact and emphasized the experimental approach to knowledge as opposed to theoretical learning, which they claimed leads only to fruitless disputes. Of course, the English were not the only people to perform experiments, but they gave experimentation a more central role in the quest for knowledge. English scientists abandoned the quest for certain knowledge that had preoccupied Aristotle, Descartes, and even Bacon in favor of finding explanations for phenomena that exhibited the highest degree of probability.

See also Academies and Scientific Societies; Courts; Politics and Science.

References

Porter, Roy, and Mikulas Teich, eds. *The Scientific Revolution in National Context*. Cambridge: Cambridge University Press, 1992.

Shapin, Steven. *A Social History of Truth: Civility and Science in Seventeenth-Century England*. Chicago: University of Chicago Press, 1994.

Natural History

In the early modern period, natural history covered territories now occupied by botany, zoology, conchology, mineralogy, and geology as well as subjects no longer studied at all,
like the healing properties of springs. Natural history was a science of the particular, concerned with the description and classification of the phenomena of nature, and it did not have the theoretical aspirations of natural philosophy. During the scientific revolution, natural history was transformed by the new knowledge coming from areas of the world Europeans were beginning to explore. It also became less textual and less tied to the world of allegory and symbol. Although during this period the subdisciplines of natural history emerged as more important in their own right, the parent field did not disappear.

Like many disciplines, natural history was defined during the Renaissance in terms of a body of ancient texts. The field was unusual in that its fundamental text was not Greek but Latin—the voluminous *Natural History* of the ancient Roman Pliny the Elder (A.D. 23–79). Pliny was not considered an infallible authority; the physician and humanist Nicolò Leoniceno (1428–1524) published *On the Errors of Pliny* in 1492. However, his general philosophy was widely accepted as the foundation of natural history during the Renaissance.

Pliny had a very broad definition of the subject matter of natural history: anything worthy of memory. A more circumscribed subject matter had been described by other natural historical authorities, notably the Greek botanists Theophrastus (c. 372–c. 287 B.C.) and Dioscorides (A.D. c. 40–c. 90) and Aristotle in his writings on animals. These ancient texts, however, were consistently challenged with the arrival of new information, not only from those parts of the world that Europeans had only recently encountered, but even from parts of Europe that the ancients had little familiarity with, such as central Europe and Scandinavia. As no one could examine all of this data, the progression of natural history depended on communication and informal collaboration between investigators. Creators of large natural historical compendia relied on outside correspondents to provide new information on the various regions of the world.

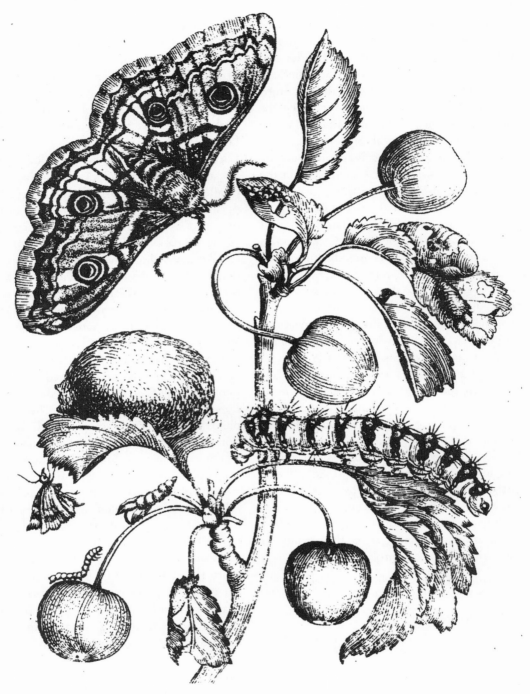

This illustration from Anna Maria Sibylla Merian's Erucarum Ortis *(1718) is characteristic of her work in that the insects and plants are presented in a unified composition rather than isolated.*

Natural history was allied to medicine through medical botany, by far the most developed and intensely studied subfield of natural history in the early modern period. Natural history, with its emphasis on particulars, offered physicians an intellectual way out of the theoretical debates that dominated university medical theory, and the scientific revolution saw many physicians who were natural historians. By the late sixteenth century, natural historians such as Ulisse Aldrovandi, lecturer in natural history at the University of Bologna, were also insisting on the importance of natural history for natural philosophers, not just for physicians. Aldrovandi's educational brief also covered the nonmedical subjects of animal life and fossils in addition to botany.

Italy led the way in the study of natural history in the sixteenth century. Italy was the center of the rapidly growing material culture of natural history, with museums of natural objects and botanical gardens. The universities with the most developed medical faculties, notably Padua and Bologna, were in Italy, and some of the most active patrons of science were the popes and the Medici dukes of Tuscany. Italy also developed a natural historical community around leading scholars, such as Aldrovandi and the botanist and physician Pier Andrea Mattioli (1501–1577).

Natural history, which often worked with beautiful and intriguing objects, was well suited to the courts of early modern Italy and the rest of Europe. As a courtly discipline, Renaissance natural history was drawn to the exotic and remarkable, such as the rhinoceros brought to Rome to be presented to Pope Leo X (pope 1513–1521). Less spectacular acquisitions filled the collections of lesser princes, as well as aristocrats and members of the upper middle class. The realm of natural history was stretched in the Renaissance to include the study of things now thought mythical, such as the unicorn, the horns of which occupied pride of place in particularly splendid natural history collections. (These were actually narwhal's horns.)

The greatest natural history writer of the mid-sixteenth century outside botany was the Swiss physician and humanist Konrad Gesner (1516–1565). Gesner's voluminous works, such as the four-volume *History of Animals* published between 1551 and 1558, were essentially illustrated compendia of textual information in the humanist tradition. What interested Gesner and other early modern natural historians about natural objects was not only what we call scientific information, but the whole elaborate body of cultural associations attached to all the known plants, animals, and stones of the Old World. Gesner's book contained not only facts, myths, and legends about the animals themselves, but a great deal of what later natural historians would consider extraneous information, such as a list of the places in the Bible where each animal is mentioned and a list of proverbs that mention the animal. Gesner also discussed the emblems, or little pictures meant to express a sentiment or quality, in which each animal occurs. (Books of emblems were very popular during the Renaissance.) Gesner's encyclopedic works were reprinted as late as 1669. His emblematic approach to natural history, treating the cultural associations of the things described as primary information, reached its full flowering in Aldrovandi's massive encyclopedias.

Not all natural histories of the sixteenth century were concerned with cultural contexts. In particular, New World natural histories tended to be less contextual than Old World books, as the animals and plants of the New World simply had fewer cultural associations for Europeans. New World natural history was also much less textually oriented than the kind of natural history Gesner practiced, because species unique to the New World were not known to ancient writers.

By the middle of the seventeenth century, the contextual, emblematic approach was obsolete, the victim of a shift in European thought away from Renaissance encyclopedism to a study of natural phenomena in and for themselves. (This shift was paralleled in the study of botany, which grew less focused on the medical uses of plants and more interested in the plants themselves.) More than ever, natural historians began to emphasize direct observation rather than the mastery of classical texts.

One aspect of the earlier natural history that did persist was its collaborative character. In particular, natural history was ideal for the Baconian fact-gathering that was charac-teristic of the early Royal Society. *Philosoph-ical Transactions* was deluged with natural-historical reports. Natural history enabled people isolated from the major scientific cen-ters, such as provincial clergymen, to partic-ipate in the scientific endeavor—a tradition that would persist for centuries. The natural history of lands outside Europe continued vigorously as well. By the late seventeenth century, these lands were studied more often by nonphysicians, such as Georg Eberhard Rumph in Indonesia or Anna Maria Sibylla Merian in South America. One natural histo-rian and collector who was also a physician was Hans Sloane (1660–1753), whose studies of the natural history of Jamaica made his sci-entific reputation. Another physician, Engel-bert Kämpfer (1651–1716), studied the nat-ural history of Japan while stationed at the trading post belonging to the Dutch East India Company.

Although natural histories of local areas were pioneered by the natural historians who wrote about lands outside Europe, a local approach was also suited for natural history within Europe. The Oxford physician Robert Plot (1640–1696) wrote two local studies, *The Natural History of Oxfordshire* (1677) and *The Natural History of Staffordshire* (1686). Plot's definition of natural history was elastic enough to include antiquities and descrip-tions of the houses of the gentry. Like many natural historians who appealed to a general audience, he concentrated on the unusual and the marvelous. He conceived his project as eventually leading to a natural history of England on a county-by-county basis, and he circulated a questionnaire to this end. Although his project came to naught, a num-ber of county natural histories were pro-duced in late-seventeenth and early-eigh-teenth-century England. Another characteris-tically English development was the merging of natural history with natural theology, as in the work of John Ray and the Reverend William Derham (1657–1735). Ray and Derham emphasized the proofs of God's

benevolent care as seen in the adaptations of living things.

See also Aldrovandi, Ulisse; Botany; Exploration, Discovery, and Colonization; Fossils; Merian, Anna Maria Sibylla; Ray, John; Zoology.
References
Findlen, Paula. *Possessing Nature: Museums, Collecting, and Scientific Culture in Early Modern Italy.* Berkeley and Los Angeles: University of California Press, 1994.
Grafton, Anthony, and Nancy Siraisi, eds. *Natural Particulars: Nature and the Disciplines in Renaissance Europe.* Cambridge: MIT Press, 1999.
Jardine, Nicholas, Emma Spary, and James A. Secord, eds. *Cultures of Natural History.* Cambridge: Cambridge University Press, 1996.

Natural Magic

During the scientific revolution, natural magic in its broadest sense meant magic that involved manipulating the physical and occult properties inherent in various substances, as opposed to magical operations involving angels or devils, or natural philosophy's em-phasis on the basic laws of the universe. For example, natural magicians were interested in the powers of magnets, powers difficult to explain by traditional natural philosophy. Natural magic overlapped with alchemy and astrology, but it also included subjects like the concoction of salves and ointments and the magical or healing properties of gems. It drew on a wide range of ancient, medieval, and Arabic sources and was also interested in find-ing the rational basis of popular superstitions.

Natural magicians distinguished themselves from natural philosophers by taking an activist position, seeking knowledge not for its own sake but for practical use. For example, the skilled natural magician was expected to be knowledgeable about optics, but not necessar-ily about optical theory of the kind studied in universities. Instead, he had practical knowl-edge of the tricks that could be played with light. Natural magic frequently aimed to startle and amaze by drawing on the powers of specif-ic substances. A classical piece of natural magic was the English magician John Dee's invention of an artificial insect to fly across the set of a play. Natural magic's entertainment value made it popular in courtly settings, but it also reached out to a wider population in books of secrets. The line between natural magic and technology could be very fine indeed.

Phenomena thought to be demonic could also be explained by natural magic, as when the eminent Neapolitan natural magician Giambattista della Porta claimed that the "flying ointment" that allegedly gave witches the ability to fly was really a hallucinogen working through the natural properties of the ingredients. Despite its distinction from witchcraft, however, natural magic attracted the interest and suspicion of the Catholic Church in the late sixteenth century, as well as that of many Protestants. Wishing to monopolize supernatural power, the church associated natural magic with witchcraft and demonic magic. The Inquisition suppressed della Porta's writings. In the long run, a more effective strategy for combating natural magic was promoting the mechanical philosophy, which by reducing the universe to matter and motion left no occult powers to serve as the intellectual basis for natural magic. But the tradition of natural magic continued at a popular level, and the late-seventeenth-century public performer of entertaining experiments for salon, lecture hall, or courtly audiences was the heir of the natural magician.

See also Books of Secrets; Magic; Porta, Giambattista della; Weapon Salve.

References

Eamon, William. *Science and the Secrets of Nature: Books of Secrets in Medieval and Early Modern Culture.* Princeton: Princeton University Press, 1994.

Thorndike, Lynn. *A History of Magic and Experimental Science.* 8 vols. New York: Macmillan, 1923–1958.

Natural Philosophy

See Aristotelianism; Cartesianism; Epicureanism; Hermeticism; Newtonianism; Stoicism.

Natural Theology

Long before the scientific revolution, Christian theologians had divided theology into two branches, "natural" and "revealed." Revealed theology was that known only because God had revealed it in the Bible or, for Catholics, through the tradition of the church, and it covered such specifics of Christian belief as the Trinity and the divinity of Christ. Natural theology was based on the evidence of the universe, and thus could be understood by anyone, Christian or not. It covered such issues as the existence and providence of God and the existence of the human soul. Natural theologians, even those who were self-professed Aristotelians, often pointed to Plato as an ancient who knew many true things about God and the soul without the benefit of the Christian revelation.

Knowledge of the natural world was obviously relevant to natural theology. A classic example is the argument from design, which seizes on the complexity and functionality of the universe to argue that it must have been designed by an intelligent mind, rather than coming together through random chance. This intelligent and benevolent mind could only be God.

The great controversies between Catholics and Protestants in the sixteenth century were about revealed, not natural, theology. Protestants tended to be suspicious of natural theology as intellectually arrogant, and they asserted the inability of the corrupted human will to understand anything about God without divine assistance. In the conflict with Protestantism, Catholics in turn emphasized the authority of the church over the achievements of human reason. But the tradition of natural theology in the Catholic Church as expressed by canonical medieval philosophers such as Thomas Aquinas (c. 1225–1274) was too powerful for Catholics to deny it totally.

Natural theology made a remarkable comeback in the seventeenth century, particularly among those who wanted to emphasize the areas of agreement, rather than disagreement, among different groups of Christians. Most natural philosophers with strong religious beliefs had a positive attitude toward natural theology and often presented their work as a contribution to it. Discerning the hand of God in nature was one of the justifications for doing science in the first place. Mechanical philosophers argued that a mechanical universe is actually better evidence of the providence of a God who created order out of matter and motion than is an Aristotelian system that treats purpose as a characteristic of the universe itself. René Descartes claimed that by reducing animals to machines, he was exalting the divine nature of the human soul.

Blaise Pascal was an exception in denying natural theology, as he thought any approach to God or divine truth outside the Christian revelation would lead to atheism. This fear was fulfilled in Baruch Spinoza, who essentially reduced all theology to natural theology. Robert Boyle occupied an intermediate and more widely shared position. A devout Christian, Boyle constantly insisted on the usefulness of science for understanding God. In his will, he established the Boyle Lectures, a chief instrument for the dissemination of natural theology in eighteenth-century England.

England was the country where the natural-theological tradition was strongest and most creative at the end of the scientific revolution. Natural theology there drew both on Newtonian physics, which presupposed active involvement by God in the operations of the universe, and on the life sciences, which emphasized God's design of all living things to be perfect for their function. One of the classics of the genre, John Ray's *The Wisdom of God Manifested in the Works of Creation* (1691), was influential in founding the discipline of "physicotheology," which applied the argument from design to the biological details of creation. Boyle, Newton, and Ray—three of the most influential scientists of late-seventeenth-century England—established a firm alliance between science and natural theology that persisted in Britain in various forms well into the nineteenth century. But Britain also saw the use of natural theology against Christianity in the works of the late-seventeenth and early-eighteenth-century Deists who followed Spinoza.

See also Book of Nature; Boyle Lectures; God; Newtonianism; Religion and Science.
Reference
Brooke, John Hedley. *Science and Religion: Some Historical Perspectives.* Cambridge: Cambridge University Press, 1991.

Nature, Images of
See Personifications and Images of Nature.

Navigation
Because of the dramatic expansion of the geographical range of European seafaring in the early modern period, problems of navigation were of great concern to individuals, businesses, and governments. Navigational problems were not merely intellectual questions but could be matters of life and death, especially to a ship running low on supplies and desperately needing to know the way to port. Governments offered lavish rewards for successful navigational methods, and navigational problems engaged the attention of many of Europe's leading scientists, including Galileo Galilei, Christiaan Huygens, Robert Hooke, John Flamsteed, and Isaac Newton. Navigational applications underlay efforts to teach mathematics at institutions such as Gresham College and were also behind the founding of the Greenwich and Paris Observatories. Newton himself urged the teaching of mathematics to young men being educated for the sea, against the advice of conservative sailors who thought it unnecessary.

The magnetic compass, which indicates the direction of a ship, was already in general use in the late medieval period and was frequently used as an example of the advances made by the moderns over the ancients. The first European people to grapple with complex navigational problems during the Renaissance were the earliest explorers, the Spanish and Portuguese. The Casa de la Contratación, founded in 1503 to regulate Spanish movement to and from the New World, trained pilots and captains in the arts of cartography and navigation, making navigation a more textual body of knowledge as opposed to something passed down orally. The Portuguese naval commander João de Castro (1500–1548) made the first study of magnetic declination, and another Portuguese, Pedro Nunes (1502–1578), demonstrated that sailing in a great circle is a shorter way of getting to one's destination than sailing in a straight line.

There are two basic steps in fixing the position of a ship at a given moment when out of sight of land—that of determining latitude (north-south) position and that of determining longitude (east-west) position. The first was far less daunting. Since the relative position of the celestial bodies changes with the latitude, the altitude of the polestar or the Sun above the horizon indicates the latitude. Portuguese sailors were using quadrants to determine the altitude of the polestar in their west African voyages of the fifteenth century. But south of the equator, where the polestar is no longer visible, it was necessary to use the Sun, whose path is more complicated. For assistance with this procedure, navigators used printed tables of the variance from the equator of the Sun's apparent path around the Earth, called the ecliptic. Also helpful was the invention of the backstaff by the English sea captain John Davis (c. 1550–1605) in the early seventeenth century. This enabled an observer facing away from the Sun to take its altitude. (The old method, which required looking directly into the Sun, took a heavy toll on sailors' eyesight.)

The longitude was a far tougher nut to crack, a classic technological problem of the early modern period. As the celestial bodies seemed to rotate around the Earth from east to west (the shift to Copernicanism made no difference to navigators, whose skills were based on the appearance of the sky, not the underlying reality), they did not offer a way to know one's position. Existing methods, based on observing the Moon or simply estimating the speed one had been traveling for a given time, were maddeningly and even dangerously imprecise. Most approaches to the longitude reduced the problem to finding the difference between the time on the ship and the time at a fixed point, usually the home port. The difference in time could be translated into spatial terms as the difference in longitude between the two points. There were all sorts of bizarre schemes for this, such as that based on Sir Kenelm Digby's powder of sympathy, a version of the weapon salve. Since the "sympathy" of the powder worked over great distances, it was suggested that ships should carry wounded dogs whose wounds would be dusted with the powder. Bandages from the wounds, which would be kept at the home port, would be dipped in water every hour. In response, the dogs on the ships would yelp, and the home-port time would be known. It is not known if this was ever tried.

The two main practicable approaches to determining the longitude were using astronomical objects to determine the correct time and creating a clock that could keep accurate time on a ship. If the home-port time of a celestial occurrence were known, all that would be necessary would be to compare the ship's own time upon observation of the same occurrence. Galileo hoped that the movements of the satellites of Jupiter he had discovered could be the astronomical clock, and he tried to sell this as a navigational method to both Spain and its mortal enemy the Dutch Republic. But the precise telescopic observations needed could not be taken from the deck of a ship. The effort to create an accurate marine clock underlay many of the clockmaking innovations of Hooke, Huygens, and others in the late seventeenth century, but these efforts were also in vain. Another approach was to map the variations in the Earth's magnetic field and observe the ship's compass. This approach was popular in England, and Edmond Halley among others worked on it, but to no real advantage. Another plan was to fill the oceans with boats that would fire cannons, thus studding the seas with known points from which sailors could figure their distance. This was obviously impracticable. The problem of the longitude would remain to be solved in the eighteenth century with the invention of an accurate shipboard clock by John Harrison (1693–1776).

See also Astronomy; Clocks and Watches; Compasses; Exploration, Discovery, and Colonization; Technology and Engineering.

References

Howson, J. B. *A History of the Practice of Navigation.* Glasgow: Brown, Son and Ferguson, 1951.

Sobel, Dava. *Longitude: The True Story of a Lone Genius Who Solved the Greatest Scientific Problem of His Time.* New York: Walker, 1995.

Neoplatonism

See Hermeticism; Platonism.

New Star of 1572

A star bright enough to be seen in the daytime appeared in the constellation of Cassiopeia in early November 1572. It steadily dimmed in brightness, disappearing from

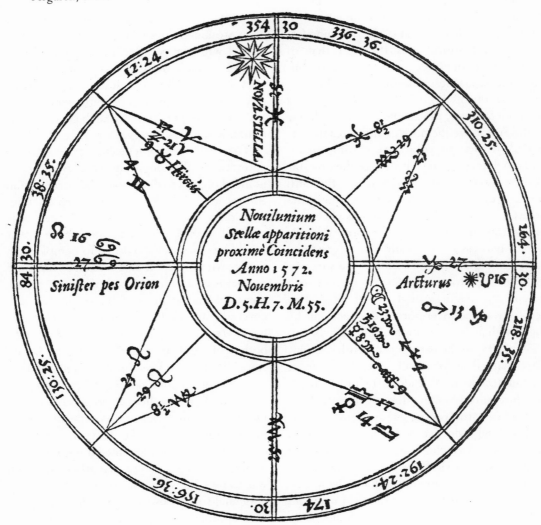

Tycho Brahe drew up this astrological chart of the position of the stars and planets at the appearance of the new star of 1572. (Tycho Brahe, De Nova Stella, *facsimile edition published by the Danish Royal Society, 1901)*

sight by March 1574, but it made a remarkable impression. Like all unusual celestial phenomena of the period, the new star generated an extensive literature interpreting it astrologically or setting forth its meaning as a portent, with the parallel frequently made to the Star of Bethlehem. Europe's astronomers and cosmologists debated the nature of the star, some claiming that it was a comet, and argued over whether it was positioned above or below the sphere of the Moon. In terms of astronomy, a superlunary new star dramatically challenged the Aristotelian distinction between the perfect and unchanging world beyond the sphere of the Moon and the imperfect and corruptible one below it.

The astronomer who published most extensively on the star was Tycho Brahe. Brahe's first publication, *On the New Star* (1573), combined a record of his observations, a demonstration of the new star's position in or near the sphere of the fixed stars, and an astrological interpretation of its meaning. Although not widely distributed, this work contributed to Tycho's reputation among astronomers, and he returned to the subject throughout his career. The English Copernican Thomas Digges's (c. 1545–1595) publication on comets combined the most accurate observations of the new star after Tycho's with a plea for the Copernican system. The leading Ptolemaic astronomer, the Jesuit Christoph Clavius, also held that the new star demonstrated the corruptibility of the heavens.

See also Brahe, Tycho; Clavius, Christoph.
References
Dreyer, J. L. E. *Tycho Brahe: A Picture of Scientific Life and Work in the Sixteenth Century.* Edinburgh: A. & C. Black, 1890.
Johnson, Francis R. *Astronomical Thought in Renaissance England: A Study of the English Scientific Writings from 1500 to 1645.* Baltimore: The Johns Hopkins University Press, 1937.

Newton, Isaac (1642–1727)

Isaac Newton, the single most important figure of the scientific revolution, set the frame of reference for physics and science in general until the emergence of relativity and quantum physics in the early twentieth century. He made fundamental contributions to mechanics, optics, and mathematics. Born on Christmas Day to a yeoman family, Newton had a mechanical genius and love of study that were manifest from an early age. Giving up hope of ever making him into a farmer, Newton's family sent him to school and then, in 1661, to Trinity College, Cambridge, where he received a B.A. in 1665. Newton departed Cambridge during the plague of 1665 and spent over a year at his mother's house in Woolthorpe—the most scientifically creative period of his life. During Newton's *annus mirabilis,* or "year of wonder," in 1665–1666, he discovered the binomial theorem, the differential and integral calculus, and the refraction of light, and he began to work out the theory of universal gravitation. It was the most remarkable epoch for a single mind in the history of human thought. The story that Newton was inspired in his work on gravity by the sight of a falling apple appears to be founded on fact, but it took many years for the theory to emerge in its final form.

Newton was open to a wide range of beliefs and systems of thought, all of which he thought of as diverse ways of getting at truth. He initially accepted the mechanical philosophy and studied both the vortices of René Descartes and the atomism of Pierre Gassendi. He was also fascinated by the mathematical developments of the time, which he had learned outside the normal Aristotelian Cambridge curriculum in private lessons from the mathematician and divine Isaac Barrow (1630–1677). After the *annus mirabilis,* Newton also became interested in alchemy, which was not based on mechanical or mathematical principles. He carried out many elaborate alchemical experiments, carefully recorded in a notebook. Newton believed that an ancient wisdom had been possessed at the beginning of human society and then lost, and much of his work was an attempt to

In the last few decades of his life, Newton was a European celebrity and the subject of works of art and literature. This ivory bas-relief dates from around 1720. (Bumdy Library of the Dibner Institute, Babson Collection)

recover this ancient wisdom. This position and Newton's concomitant interest in Hermeticism were unusual in late-seventeenth-century science. The magical and alchemical traditions, with their emphasis on secrecy, were congenial to Newton's somewhat paranoid temperament. Only with great reluctance did Newton make public many of his discoveries, including the calculus and universal gravitation, and this reluctance led to bitter disputes with Robert Hooke and Gottfried Wilhelm Leibniz.

In 1667, Newton was elected a fellow of Trinity College, and upon Barrow's resignation as Lucasian Professor of Mathematics in 1669, Newton succeeded him. Around this time, the Royal Society in London became aware of Newton's work, principally in optics. Newton had earlier built the first reflecting telescope, and the Royal Society asked to see it. Newton's election to the society in 1672 was followed by the submission of his optical papers. This led to Newton's

feud with Hooke, who held the traditional optical theory that white light is the most simple and is modified to produce colors, as opposed to Newton's new theory that white light is a mixture of colors. Hooke addressed the young and prickly Newton condescendingly, and the enraged Newton temporarily withdrew from society affairs.

Newton's *Mathematical Principles of Natural Philosophy (Principia Mathematica Philosophiae Naturalis,* often referred to as the *Principia)* was published in 1687 at the prompting of Newton's ally Edmond Halley. Here Newton set forth his theory of universal gravitation—that there is an attractive force between bodies that varies with the inverse square of the distance between them—and his three laws of motion—(1) a body at rest or in motion in a straight path will tend to stay in that state, (2) a change of motion in a body varies with the force impressed, and (3) each action has an equal and opposite reaction. This was the most important text in physics to appear for centuries. The idea that science is the precise description of what happens rather than an explanation for why it happens is an expansion of Newton's gravity theory. Newton claimed not to make hypotheses, particularly as to the causes of gravity, although he actually did. The book added to his feud with Hooke, who made unjustified accusations that Newton had plagiarized his own earlier work on gravity.

Newton saw his science and numerous other intellectual exertions as an effort to understand the mind of God, which is truth. Newton was more religious in his approach to natural philosophy than Galileo or even Descartes, claiming of the *Principia* in a letter to Richard Bentley (1662–1742): "When I wrote my treatise about our system, I had an eye on such principles as might work with considering men, for the belief of a Deity; and nothing can rejoice me more than to find it useful for that purpose" (Turnbull 1961, 233). Newton's science places an enormous emphasis on the universe as law-governed and on God as lawgiver. The whole idea of science as searching out mathematically expressible laws of nature, such as the law of gravity or the laws of motion, is in part a Newtonian legacy. Newton's God, however, did not simply set up the laws and let a clockwork universe run itself, but continues to sustain the universe as well as creating it. Newton criticized Cartesianism and other purely mechanical philosophies for exiling God from creation. As Newton saw it, given universal gravitation, divine intervention is necessary both to keep the slight irregularities in planetary orbits from accumulating until the solar system breaks down and to keep all the matter in the universe from clumping together in one place.

Newton's theology was heretical. In the 1670s, Newton became convinced that the belief in the Trinity was erroneous. Newton kept his belief secret, managing to avoid swearing oaths to orthodox belief and refusing the sacraments of the Church of England only on his deathbed. Despite his heresy and unsociableness, Newton was an active participant in English and international scientific life. The fame of the *Principia* enabled Newton to build a following of young English, Scottish, and foreign mathematicians and scientists, including the Swiss mathematician Fatio de Duiller (1664–1753), with whom Newton had the most intense emotional relationship of his rather lonely life. He was involved in religious politics as a leading opponent at Cambridge of the Catholicizing policies of King James II (r. 1685–1688). After James II was deposed in the revolution of 1688, Newton began seeking a government position that would enable him to move to London.

In 1693, his relationship with Fatio de Duillier broke off, and shortly afterward Newton had a nervous breakdown, making wild accusations against his friends Samuel Pepys (1633–1703) and John Locke. Following his recovery, he was appointed warden of the Mint in 1696 during the great English recoinage, and moved to London. A scourge of counterfeiters, many of whom he sent to

the gallows, Newton was appointed master of the Mint in 1699. It was these and other political services, rather than his science, that led to a knighthood in 1705. Unlike most prominent English natural philosophers of the late seventeenth century, Newton never left England, but in 1699 he was appointed as one of the first foreign associates of the French Royal Academy of Sciences. Following Hooke's death in 1703, Newton was elected president of the Royal Society, serving very actively to his death. Hooke's death also allowed Newton to publish his optical work in *Opticks* (1704), which also contained the influential "Queries" on matter-theory and work on the calculus.

Newton's last decades were dominated by his presidency, which steered the Royal Society in the direction of experiment, and by his study of biblical prophecies. These years were also marked by very ugly feuds with Leibniz and John Flamsteed, the royal astronomer. Newton and Halley wanted Flamsteed to publish his astronomical data straight away, whereas Flamsteed wanted to perfect his work before publishing it. Newton accused Leibniz of plagiarizing the calculus from him, an accusation as unjustified as Hooke's earlier one against Newton. Newton behaved very badly in the controversy, fixing a Royal Society investigation and attacking Leibniz through the Newtonians, as in the Leibniz-Clarke correspondence. The feud had disastrous consequences for British mathematics, delaying the acceptance of the superior Leibnizian calculus for over a century.

The aged Newton spent most of his intellectual effort on the prophecies. A millenarian and self-taught Hebrew scholar, Newton saw correct prophetic interpretation as similar to correct scientific method. First one must get the clearest idea possible of the phenomena, and then examine their relationships. For instance, he believed that just as nature is everywhere the same, so prophetic language is everywhere the same; if a symbol, say a beast, has one meaning somewhere in the Bible, it has that same meaning everywhere in the

Bible. He spent an enormous amount of time working out the chronologies of the various ancient kingdoms. But going beyond the evidence was as illegitimate in prophetic interpretation as it was in natural philosophy, and Newton resisted setting a date for the millennium. His funeral in 1727 was a state occasion, and he was buried in Westminster Abbey among the great of England.

See also Alchemy; Cambridge University; Force; Gravity; Halley, Edmond; Hooke, Robert; Mathematics; Mechanical Philosophy; Mechanics; Newtonianism; Optics; Physics; Priority; Royal Society.

References

Dobbs, B. J. T. *The Janus Faces of Genius: The Role of Alchemy in Newton's Thought.* Cambridge: Cambridge University Press, 1991.

Turnbull, H. W., ed. *The Correspondence of Isaac Newton.* Vol. 3. Cambridge: For the Royal Society at the University Press, 1961.

Westfall, Richard. *Never at Rest: A Biography of Isaac Newton.* New York: Cambridge University Press, 1980.

Newtonianism

The decades following the publication of Newton's *Mathematical Principles of Natural Philosophy* in 1687 saw the triumph of Newtonian physics and Newtonian ideology. The structure of Newtonian physics—scientific knowledge organized as a series of mathematical laws—became the model for all sciences. Even psychologists, influenced by Newton's admirer John Locke, spoke in Newtonian language of the "attraction" between ideas. The victory of Newtonianism was not simply a result of Newton's works and theories. Very few people read or were able to understand Newton's treatises themselves—a difference between Newtonianism and its rival, Cartesianism, as Descartes's works were much more accessible. Newton's use of mathematics was out of the reach of even the well-educated gentleman or salon hostess. Rather, the gap between Newton and his audience was filled by an army of Newtonian popularizers, lecturers, and textbook writers in the early eighteenth century.

In England, Newtonianism had to compete with both Cartesianism, which retained great prestige well into the 1690s, and the Baconian tradition of fact gathering and suspicion of mathematics. Some of the earliest Newtonian institutional victories came in the Scottish universities, where James Gregory (1666–1742), professor of mathematics at the University of Edinburgh, was a zealous Newtonian. By 1710, Newtonianism had displaced Cartesianism throughout the Scottish university system. In England, thanks to Newton's presidency of the Royal Society from 1703 to his death in 1727, the society and its publications, the "high ground" in British science, were dominated by Newtonian ideas and experiments. Newtonianism was also disseminated through the Boyle Lectures.

British Newtonians held a variety of religious positions, their membership ranging from a few Anglican high churchmen and other religious conservatives, like Gregory, to Edmond Halley, who had very little interest in religion at all and was denied a chair at Oxford out of suspicion of heresy. Suspicion of Newtonianism was enhanced by the fact that some Newtonians, such as the bold heretic William Whiston (1667–1752), shared Newton's own more discreet anti-Trinitarian theology. More radical Deists, believing in God but denying Christianity, also presented Newtonian physics as consistent with their own position. But Newtonianism, with its place for nonmaterial factors such as gravitation, also appealed to churchmen and devout Christians as less materialistic than Cartesianism.

The first Continental country to adopt Newtonianism was the Dutch Republic, closely linked with Britain in the late seventeenth and early eighteenth centuries. The two countries had a common leader, William of Orange, from the Revolution of 1688 to his death in 1702, and they were subsequently allies in the War of the Spanish Succession (1702–1713) against France. Newtonianism quickly displaced Cartesianism in the Dutch universities. The French Protestants, or

Huguenots, having been expelled from France, had helped make Amsterdam the center of European publishing, and they also played an important role in disseminating Newtonianism throughout Europe. Since few Continental Europeans knew English (the prestige of Newton and English science generally would contribute to changing this) and Latin readership was limited, it was vital that Newtonianism be put forward in French, the language of high culture and science in most of Europe. British, Dutch, and Huguenot lecturers and engineers were Newtonian missionaries on the Continent.

Newtonianism was accepted surprisingly early in parts of Italy, where a continuing semiunderground Galilean movement saw it as a continuation of Galilean mechanics. The Inquisition looked on Newtonianism with suspicion, not least because it came from a Protestant country, but the power of the Inquisition was waning by the beginning of the eighteenth century, even in Rome. Italian Newtonians remained Catholic, incorporating Newtonianism into their vision of a moderate, liberal, and tolerant Catholicism. In the Germanic world, Newtonianism contended against the prestige of Liebniz's philosophy. Although Newtonian mechanics were incorporated into the dominant German academic philosophical synthesis of Christian Wolff (1679–1754), Newtonianism was not quite as important in Germany, always hesitant about the mechanization of nature, as it was in other areas of Europe in the eighteenth century.

In France, Newtonianism competed with the reigning Cartesianism. Its eventual victory was delayed by French national pride and the conservatism of the Royal Academy of Sciences and the French universities, which had just replaced Aristotelianism with Cartesianism in the 1690s. In early-eighteenth-century France, Newtonianism attracted those who were unhappy with the current order of French society and the dominance of the Catholic Church. It fit the Anglophilic tendency among the philosophers of the early

French Enlightenment, who admired England as a society with greater personal freedom and a relatively open government. French Newtonians like Voltaire (1694–1778) incorporated Newtonianism into an overall critique of French society, pointing out that the high honor given a genius such as Newton in England proved the superiority of free England over intolerant France. The fact that Newton came from the lower classes and was held in reverence by the greatest was used as an argument for a more egalitarian society. Voltaire made creating a Newtonian France a major project, in collaboration with his mistress, Gabrielle-Émilie, Marquise du Chatelet (1706–1749), who unlike Voltaire actually understood Newton's mathematics. Together they published popular books in French on Newtonianism, and du Chatelet went on to translate the *Principia* into French, even though she herself preferred Leibniz to Newton.

The universal adoption of Newtonianism meant the dissolution of the link between Newtonian science and a specific religious or metaphysical position. It became clear that Newtonian physics were not dependent on either Newtonian or orthodox theology. By the late eighteenth century, Newton's laws were even translated into Hebrew for the use of religious Jews. If Scottish Presbyterians, English Freethinkers, French anticlericals, Italian liberal Catholics, and a host of others were calling themselves Newtonians, then Newtonianism didn't really mean anything in religious terms.

Whatever one's religious position, though, Newtonian providentialism continued to exert an ideological influence. Providentialism developed into the belief, which Newtonianism inherited from the Baconian tradition, that more natural knowledge would enable people to more efficiently exploit the resources of nature and thus improve society, providing more resources for all. Moreover, the quest for natural knowledge itself was thought to be providentially ordained, with the goal of exercising the human mind. Newtonianism became an ideology of intellectual progress and economic development, and Newton himself, particularly in England, became a quasi deity, praised as the greatest genius in human history, who had once and for all discovered the laws by which the universe operated. As the English poet Alexander Pope put it in an epitaph intended for Newton: "Nature and Nature's Laws, lay hid in night / God said 'Let Newton Be!' and all was light."

See also Boyle Lectures; Demonstrations and Public Lectures; Leibniz-Clarke Correspondence; National Differences in Science; Newton, Isaac; Physics.

References

Dobbs, B. J. T., and Margaret Jacob. *Newton and the Culture of Newtonianism.* Atlantic Highlands, NJ: Humanities Press, 1994.

Ferrone, Vincenzo. *The Intellectual Roots of the Italian Enlightenment: Newtonian Science, Religion, and Politics in the Early Eighteenth Century.* Translated by Sue Brotherton. Atlantic Highlands, NJ: Humanities Press, 1995.

Stewart, Larry. *The Rise of Public Science: Rhetoric, Technology, and Natural Philosophy in Newtonian Britain, 1660–1750.* Cambridge: Cambridge University Press, 1992.

O

Observatories

The observatory, a building or platform specifically devoted to astronomical observation, originated in the Muslim world and spread widely in early modern Europe. In addition to clear skies and a stable platform, observatories required both a place for fixed instruments, such as the mural quadrant, which was attached to a wall running north-south, and a supply of moveable instruments. Observatories were not necessary for astronomical observation, as many astronomers simply observed from their windows or the roofs of their houses. In the sixteenth century, the founding of an observatory required a commitment of substantial resources to an astronomical program and was undertaken only by major patrons. The astronomer-prince, William IV of Hesse-Kassel (1532–1592), for example, founded an observatory in 1564.

The greatest observatory of the sixteenth century was Tycho Brahe's Uraniborg, the "fortress of Urania," muse of astronomy. It was founded on Hven, an island granted to Tycho by the king of Denmark in 1576. Uraniborg was a massive complex, containing a chemical laboratory, facilities for manufacturing astronomical instruments, and a printing press, as well as an observatory with the most up-to-date instruments of the pretele-scopic era. Tycho did not envision Uraniborg as a continuing institution, but as a temporary expedient for collecting the precise body of celestial observations required for a correct astronomical theory. Uraniborg collapsed as an institution immediately following Tycho's departure in 1597, and in a few decades it was a ruin, although his disciple Longomontanus (1562–1647) persuaded the king of Denmark to found another observatory in Copenhagen—the "Round Tower," which was attached to a church and completed in 1656. Another short-lived observatory was founded in Rome by Pope Gregory XIII (pope, 1572–1585) as part of his reform of the calendar.

Observatories, like astronomy itself, were vastly changed by the introduction of the astronomical telescope in the early seventeenth century. This change was not immediate. Galileo himself did not use an observatory for his observations, and at first the telescope did not dominate astronomical instrumentation as it did later. The first university-funded observatory, set up on the roof of the main building of the University of Leiden in 1633 at the behest of the professor of mathematics, Jacob Golius (1596–1667), had the original purpose of housing the seven-foot quadrant of the deceased physicist and Leiden professor Willebrord Snel (1580–1626). In the

Uraniborg, Tycho Brahe's "fortress of Urania," muse of astronomy, was literally a fortress, surrounded by high walls. (Fotomax Index)

decades after its construction, Leiden's observatory was elaborated, and other university observatories soon followed suit. One of these was the observatory at the Jesuit College at Ingolstadt, founded in 1637, which also was not telescope dominated.

But eventually the telescope did become dominant in astronomy and began to shape observatories. When telescopes were lengthened in the later seventeenth century, they demanded much more elaborate arrangements. This was particularly true of the so-called aerial or tubeless telescopes, in which the telescope's lenses could be over 100 feet

apart. Large observatories housing large telescopes did not monopolize the field, however; there were also small observatories, as for the use of a single astronomer and his family and students. The most important of these was the observatory of Johannes Hevelius. Hevelius's book on astronomical instruments, *Celestial Machines* (1673), contained elaborate designs for observatories housing aerial telescopes, but observatories following these designs were never built, and the aerial telescope was a short-lived fad.

The major development of the late seventeenth century was the establishment of permanent observatories backed by the resources of national governments, the Paris Observatory in France and the Greenwich Observatory in England. The Paris Observatory, which also served as a meeting place for the Royal Academy of Sciences, was a complex not unlike Uraniborg. The Greenwich Observatory, while not as well funded, was devoted solely to astronomy. Unlike previous observatories, these were permanent institutions not dependent on the will of one man. They were also expected to justify their keep by solving practical problems, particularly in navigation. Private observatories did continue to play a role in astronomy, the most noted in the early eighteenth century being that of Danish royal astronomer Ole Rømer (1644–1710). Rømer had found both the Paris Observatory and the Round Tower unsatisfactory, and so he set up his own ground-level observatory outside Copenhagen.

See also Brahe, Tycho; Cassini, Gian Domenico; Flamsteed, John; Greenwich Observatory; Paris Observatory; Telescopes.

References
Christianson, J. R. On Tycho's Island: Tycho Brahe and His Assistants, 1570–1601. Cambridge and New York: Cambridge University Press, 2000.
Donnely, Marian Card. A Short History of Observatories. Eugene: University of Oregon Books, 1973.

Oldenburg, Henry (c. 1618–1677)

Not a scientist himself, Henry Oldenburg, corresponding secretary to the Royal Society from 1662 to 1677, played a leading role in European science. He edited *Philosophical Transactions,* which he began as a personal venture in 1665, and he also maintained an extensive network of scientific correspondence, including not only every eminent English scientist of his time but also such Continental luminaries as Johannes Hevelius, Christiaan Huygens, Gottfried Wilhelm Leibniz, Antoni van Leeuwenhoek, and Marcello Malpighi, many of whose works Oldenburg saw through to publication. He also corresponded with a plethora of minor

figures in England and Europe and organized and managed the society's elaborate archive.

German by birth, Oldenburg came to England as a diplomat in 1653, was introduced to the Hartlib circle in 1656, acquired the patronage of Robert Boyle, and was involved with the Royal Society from shortly after its inception. Although several fellows corresponded on the society's behalf in the early 1660s, by the end of the decade Oldenburg, familiar with the major continental languages, nearly monopolized Royal Society foreign correspondence. He institutionalized the society's relationships with foreign scientists by proposing many of them as corresponding fellows.

Even though he did not receive a salary until 1669, Oldenburg and the society had a mutually beneficial relationship; his position as secretary gave him authority, and his extraordinary activity gave the society much prestige and recognition in Europe. Oldenburg possessed the immense tact needed to get along with the often prickly natural philosophers of the late seventeenth century, but his belief in open communication between natural philosophers involved him in a feud with Robert Hooke, who was afraid that his ideas were being stolen.

See also Correspondence; *Philosophical Transactions;* Royal Society.

References
Hall, Marie B. *Promoting Experimental Learning: Experiment and the Royal Society, 1660–1727.* Cambridge: Cambridge University Press, 1991.
Hunter, Michael. *Establishing the New Science: The Experience of the Early Royal Society.* Woodbridge, England: Boydell, 1989.

Optics

The subject of much Arabic and medieval scientific work, the study of light was advanced in several important ways during the scientific revolution. Optics was originally a low-status discipline. In the Aristotelian intellectual system that dominated medieval and Renaissance universities, optics was a "mixed" science, combining mathematics and natural philosophy, and was thus of low standing. Outside the university and its disciplinary hierarchy, however, the science of lenses was studied by craftsmen, such as those that devised the first telescopes, and by natural magicians, notably Giambattista della Porta, who perfected the camera obscura and included the first extensive discussion of lenses in his *Natural Magic* (1558).

Several major seventeenth-century scientists studied light, notably Johannes Kepler, René Descartes, and Isaac Newton. Light was particularly worthy of study for its symbolic value, as the glory of monarchs or the quality of truth could be represented as light. The development of the telescope and microscope also advanced the study of optics. This advancement was more the result of efforts by optical scientists to explain the behavior of the instruments than of efforts to advance the instruments by the direct application of optical theory.

Kepler's chief optical work, *Supplement to Witelo* (1604), took the form of a commentary on the medieval optical theorist Witelo (c. 1230–after 1275). Kepler came up with the theory of vision used today, which is based on the light-focusing role of the retina, and he also discovered the inverse-square law of the diffusion of light. A subsequent work, *Dioptrics* (1611), inspired by the telescope, concentrates on lenses and the theory of refraction. The law of refraction now known as Snel's law—that the sines of the angles of incidence and refraction always bear the same ratio in an interface between two media— was discovered independently at least three times. The first discovery was by Thomas Harriot, but like much of Harriot's work, it was unpublished and unpublicized. Willebrord Snel (1580–1626), a mathematics professor at the University of Leiden, discovered the law later but also left his work unpublished. Descartes later discovered it as well (he has been charged with plagiarism from Snel's manuscript, but there is no direct evidence of this), and he published his findings in

his *La Dioptrique* (1637). Descartes's influential theory of light held that light is the result of an instantaneously transmitted pressure on the eye emanating from the perceived object.

The Jesuit order, which liked to use the emblem of the light of the Sun—representing truth that dispels the darkness of error—was very active in optics and optical experimentation in the seventeenth century. The Jesuit professor of mathematics at the University of Bologna, Francesco Maria Grimaldi (1618– 1663), concentrated on light's diffraction and discovered that light does not travel in precisely straight lines. He was the first to elaborate a theory of light as a wave, viewing light as a wave in a fluid medium.

Both Descartes and Kepler (although not Grimaldi) had treated the propagation of light as instantaneous, with infinite speed. The Danish astronomer Ole Rømer (1644–1710) demonstrated in the 1670s the finite velocity of light through astronomical observation. Noting that the satellites of Jupiter seem to move more slowly when the Earth is moving away from them, he deduced that light moves at a large but finite speed. Christiaan Huygens, unlike those dogmatic Cartesians who continued to insist on light's infinite speed, incorporated Rømer's finding in a new mechanical wave theory of light expounded in his tour de force, *Treatise on Light,* published in 1690 but based on earlier work.

Huygens had to deal with a classic optical problem, the so-called double refraction of Iceland spar first described by the Danish scientist Erasmus Bartholin (1625–1698) in 1670. A ray of light refracted by Iceland spar generates two rays, one following Snel's law, the other not. Huygens solved some, but not all, of the problems with Iceland spar through a theory that describes light as a wave that is emitted from the particles of an object, setting off a series of additional waves. These waves are carried through an ether composed of particles (unlike Grimaldi's fluid).

Newton's contribution to optics resulted from the experiments with a prism he carried out in his "year of wonder" (1665–1666).

These experiments were devoted to the difficult problem of explaining color. Previous optical theorists such as Kepler or Newton's antagonist Robert Hooke had believed that color is caused by a mixture of light and darkness. Newton reversed their theory of colors by showing that white light is not the purest form of light that subsequently degenerates into colored light. Instead, white light is actually composed of a mixture of all forms of colored light. Newton established the traditional categorization of seven colors used in the English-speaking world—red, orange, yellow, green, blue, indigo, and violet. The results of his early studies were printed in an article in *Philosophical Transactions* in 1672, setting off a controversy with several leading scientists in England and Europe, including Hooke. Although at this time Newton was thinking of light as a wave, he never abandoned the idea of an ethereal medium, and in his later treatise *Opticks* (1704), he set forth a theory of light as composed of corpuscles. This uneasy combination of waves and particles would be the scientific revolution's legacy to the eighteenth century.

> *See also* Huygens, Christiaan; Kepler, Johannes; Newton, Isaac; Telescopes.
> **References**
> Westfall, Richard. *The Construction of Modern Science: Mechanisms and Mechanics.* New York: Wiley, 1971.
> Wolf, A., with the cooperation of F. Dannemann and A. Armitage. *A History of Science, Technology, and Philosophy in the Sixteenth and Seventeenth Centuries.* 2d ed., prepared by Douglas McKie. London: Allen and Unwin, 1950.

Oratorians
See Cartesianism; Education; Malebranche, Nicholas

Orta, García d' (c. 1500–c. 1568)
A physician, García d'Orta brought to European medical awareness the plants and diseases of India. Educated at the Universities of Alcalá and Salamanca in Spain, he was appointed a lecturer in natural philosophy at the University of Lisbon in 1530 and accompanied the Portuguese viceroy to the colony at Goa in India in 1534. D'Orta stayed on in Goa as a successful and wealthy physician and businessman for the rest of his life.

In 1563, d'Orta published at Goa the fruit of his study of Indian medicine, *Colloquies on the Herbs and Drugs of India,* the first scientific book printed in India. He described for a European audience Eastern plants such as aloes, camphor, ginger, sandalwood, and mangoes. D'Orta was familiar with the medical literature, both ancient and modern. Unlike many humanistic medical writers of his time, he had a high respect for nonclassical sources, particularly the Arab physicians, and he used his personal experiences with Indian remedies to correct what he viewed as the errors of the ancients. He claimed that were he still in Spain, he would not dare to say a word against Galen and the Greek physicians. He also drew medical and botanical knowledge from Indian physicians and other natives, as well as Persian and Chinese traders.

D'Orta's multicultural approach is not surprising, given that he was a *converso,* of Jewish descent. Externally conforming to Catholicism, D'Orta and his family members, many of whom followed him to India, secretly practiced Judaism. Goa provided an escape from the long arm of the Portuguese Inquisition, but an escape that proved temporary. After d'Orta's death, one of his sisters was burned by the Inquisition, and his own bones were dug up and burned.

> *See also* Apothecaries and Pharmacology; Indian Science; Jewish Culture.
> **Reference**
> Friedenwald, Harry. *The Jews and Medicine: Essays.* 2 vols. Baltimore: The Johns Hopkins University Press, 1944.

Oxford University
Oxford University was a leading center of British science in the seventeenth century. It had recovered from the damage caused by the

Christopher Wren designed this large sundial (c. 1640) at All Souls College, Oxford University. (Oxford University)

Protestant Reformation and a flirtation with Ramism in the sixteenth century. University education was relatively widespread in seventeenth-century England, and many young men of the aristocracy and gentry class received a grounding in natural philosophy as part of their undergraduate education. While Aristotle dominated the Oxford curriculum, as he did that of every European university, there was room for a variety of approaches and a tolerance for differences of opinion in natural philosophy. The Savilian professorships in astronomy and geometry, founded in 1619 by the mathematician Sir Henry Savile

(1549–1622), attracted a stream of distinguished incumbents. John Bainbridge (1582–1643), the first Savilian Professor of Astronomy from 1619 to 1643, set up a rooftop observatory and was one of the earliest in any university to teach Keplerian astronomy. Henry Briggs (1561–1630), the popularizer of logarithms, was the first Savilian Professor of Geometry, from 1619 to 1631. A flurry of benefactions accompanied and followed the Savilian professorships. Sir William Sedley, Savile's son-in-law, had founded the Sedleian Chair of Natural Philosophy in 1618, and Henry, Lord Danvers (1573–1644) established a botanical garden in 1621. Oxford also supported mathematics tutors for those undergraduates who wished to study mathematics beyond the basics.

The rise of the "new philosophy" at Oxford can be dated from the arrival of John Wilkins as Master of Wadham College in 1649. This followed the massive disruption of university life caused by the English Civil War, during which Oxford served as the Royalists's capital. Wilkins's formidable informal group of experimental philosophers included Robert Boyle, John Wallis (Savilian Professor of Geometry from 1649 to 1703), and Christopher Wren. Many left Oxford after the Restoration of Charles II in 1660 to form the nucleus of the Royal Society, but they established a tradition of experimental philosophy that would be carried on in the Oxford Philosophical Society, led by the natural historian Robert Plot (1640–1696) in the 1680s. Wilkins himself deflected Puritan attacks on the Oxford curriculum. There was also widespread interest in the mechanical philosophies of René Descartes and Pierre Gassendi. Despite Oxford's rudimentary medical faculty, it was also home to brilliant and innovative physicians such as Thomas Willis (1621–1675), Sedleian Professor of Natural Philosophy from 1660 to 1675, Richard Lower (1631–1691), and John Mayow (1641–1679), all of whom carried on active research in anatomy and physiology.

Although the center of English science shifted to London, Oxford continued to play an important role for a time. The Ashmolean Museum, including a chemical laboratory, was founded by a gift from the astrologer Elias Ashmole (1617–1692) in 1683, but it was plagued by lack of funding. The Savilian professorships were occupied by such luminaries as the Scottish Newtonian David Gregory (1659–1708), professor of astronomy from 1691 to 1708, and Edmond Halley, Wallis's successor as professor of geometry. Halley was often absent, though, and generally Oxford declined as a scientific center in the eighteenth century.

See also Physiology; Wilkins, John.
Reference
Tyacke, Nicholas, ed. *The History of the University of Oxford.* Vol. 4, *Seventeenth-Century Oxford.* Oxford: Oxford University Press, 1997.

P

Papacy

Until the middle of the seventeenth century, the papacy was the greatest institutional patron of science in Europe. However, the popes sometimes also tried to combat those scientific developments that they saw as dangerous to the church, the most notorious episode being the trial of Galileo. The role of the papacy in science emerged in several contexts. The papacy had a long concern with astronomy, as astronomical calculations were necessary to set the date of Easter. The most important example of papal patronage in this field is the calendrical reform sponsored by Pope Gregory XIII (pope, 1572–1585), which produced the calendar in use today. Gregory had a greater interest in science than was common among popes; he was also the patron of his distant relative Ulisse Aldrovandi, as well as Girolamo Cardano.

The popes influenced appointments in the universities in the papal territories, notably the Jesuit Collegio Romano and the University of Rome, known as "La Sapienza" or "Wisdom." Outside Rome, the popes had some influence on the University of Bologna, although the Bolognese civic authorities tried to keep control of the university for themselves. Medicine was of particular and personal interest to the popes, most of whom were old men when they attained the papacy.

The prestigious position of papal physician was often combined with a professorship in La Sapienza's medical school. Sapienza professors and papal physicians during the scientific revolution include such distinguished scientists as the ichthyologist Ippolito Salviani (1514–1572), the anatomist Andrea Cesalpino (1519–1603), and the pathologist and epidemiologist Giovanni Lancisi (1654–1720). Marcello Malpighi was also a papal physician, a position that did not prevent him from maintaining close connections with scientists in Protestant Europe. Scientists sometimes attained high position in the papal court; Pope Innocent XI (pope, 1676–1689) appointed the mathematician Michelangelo Ricci (1619–1682), a longtime papal official, as cardinal in 1682.

Men of education and often of wide intellectual interests, popes were not the foremost persecutors of scientific efforts in the Catholic Church. They sometimes intervened to protect scientists from the Inquisition, as did Gregory XIII in the case of the unorthodox Aristotelian natural philosopher Girolamo Borro (1512–1592), freeing him from the Inquisition in 1583. Likewise, Pope Clement VIII (pope, 1592–1605) protected Francesco Patrizi (1529–1597), a Platonic natural philosopher and Sapienza professor, from the Inquisition. There were limits to this clemency, of course; Pope Clement VIII did nothing for the heretic and magician Giordano Bruno, burned at Rome in 1600. But the most famous example of the interaction between the papacy and science in this period is that of the trial of Galileo in 1633, whereby Pope Urban VIII (pope, 1623–1644) successfully attacked Galileo for his Copernicanism, effectively silencing him and forcing him to publish outside Italy. However, this event was the culmination of a long struggle by Galileo to receive the patronage of the papal court, particularly that of Urban, who had been associated with Galileo even before he became pope. Urban VIII was more motivated by political and personal motives than by opposition to Galileo's science.

The trial of Galileo did not by any means end papal patronage of science. That same year, Urban's client the Jesuit Athanasius Kircher arrived in Rome as the new scientific star. But the papacy did play a less central role in fostering scientific innovation after 1633. In part this was simply because Italy was becoming a less important scientific center, eclipsed by Protestant Europe and France, where there was more money and intellectual freedom. The papacy was also a less central cultural institution generally, as Europe moved away from the age of the Wars of Religion after the Treaty of Westphalia in 1648. By the late seventeenth century, some of the lustre of papal patronage had been lost, as Louis XIV of France (r. 1643–1715) became Europe's premier patron, in the sciences as in many other fields.

> *See also* Courts; Gregorian Reform of the
>   Calendar; Malpighi, Marcello; Trial of Galileo.
> *Reference*
> Lindberg, David C., and Ronald L. Numbers, eds.
>   *God and Nature: Historical Essays on the Encounter*
>   *between Christianity and Science.* Berkeley and Los
>   Angeles: University of California Press, 1986.

Paracelsianism

Paracelsianism was the creation not of Paracelsus, but of the late-sixteenth-century writers who systematized his wildly unsystematic writings, along with pseudo-Paracelsian works widely thought to be his. Paracelsus had left most of his works in manuscript form during his life, but a number were published in both German and Latin translation after 1550, with a definitive ten-volume edition edited by the physician Johannes Huser (c. 1545–c. 1600) appearing from 1589 to 1591. The foremost synthesis of Paracelsianism in this early period was the innovative and influential work of the Dane Peter Severinus (1542–1602), *Idea Medicinae Philosophicae* (1571).

Physicians seeking an alternative to Galenism exalted Paracelsianism. Paracelsian physicians were found all over Europe north of the Mediterranean, from Poland to England, with Protestants having a particular affinity for the new doctrines. Paracelsians condemned ancient and Arabic medicine as anti-Christian. They envisioned the universe in terms of correspondences, the most important being the correspondence between the universe as macrocosm and humanity as microcosm. Paracelsians believed that diseases are cured by like rather than opposite medicines, as, for instance, when Oswald Croll claimed that a cancer could be cured by the application of a crab, shaped like a cancer. Paracelsian medical writers denied the Galenic theory of humors and proclaimed chemistry the foundation for all medical and natural knowledge, arguing that more effective medicines can be created by alchemical refining and distilling of the herbs used by conventional physicians. Divisions did exist within the Paracelsian movement, and many university-educated Paracelsians denounced unlearned Paracelsian practitioners as charlatans. There were also divisions between those who embraced the magical side of Paracelsus's legacy, including the authors of the Rosicrucian pamphlets, and those who spurned it.

Supporters of traditional Galenic medicine responded vigorously to Paracelsian attacks in a debate that was one of the most voluminous and bitter in sixteenth-century and early-seventeenth-century natural philosophy. Between 1572 and 1574, the Swiss Reformed theologian and physician Thomas Erastus (1524–1583) published a four-volume denunciation of Paracelsus as an ignorant, corrupt, and heretical magician, charging him with asserting that the world was created out of preexisting matter rather than from nothing. Nor was the debate limited to words. The Paris Medical Faculty, like the University of Paris always strongly anti-Paracelsian, succeeded in having a Paracelsian condemned by the Parlement of Paris, France's supreme law court, and expelled from the city in 1579. But not all traditional physicians simply opposed the new medical doctrines. Others, such as the influential Wittenberg medical professor Daniel Sennert (1572–1637), attempted to reconcile Paracelsian and Galenic medicine. Nor were all alchemists Paracelsians; Andreas Libavius denounced Paracelsus in *Neoparacelsica* (1594).

In the seventeenth century, the conservative and Catholic French medical establishment continued to exclude the mostly Protestant French Paracelsians, even though the formerly Protestant King Henry IV (r. 1589–1610) employed Protestant Paracelsians as his personal physicians. Elsewhere in central Europe and in England, however, Paracelsianism was largely accepted as useful for physicians. Seventeenth-century Paracelsianism lost much of its controversial nature as well as its identity as an independent movement, blending into new systems of chemical medicine and chemical philosophy, such as that of Johannes Baptista van Helmont.

See also Alchemy; Chemistry; Croll, Oswald; Helmont, Johannes Baptista van; Medicine; Paracelsus.

References

Debus, Allen G. *The Chemical Philosophy: Paracelsian Science and Medicine in the Sixteenth and Seventeenth Centuries.* 2 vols. New York: Science History Publications, 1977.

Debus, Allen G., and Michael T. Walton, eds. *Reading the Book of Nature: The Other Side of the Scientific Revolution.* Kirksville, MO: Sixteenth Century Journal Publishers, 1998.

Paracelsus (1493–1541)

As a young man, Philippus Theophrastus Bombast von Hohenheim, a German-Swiss physician and theologian, took the Latin name Paracelsus, meaning "greater than Celsus"; Aulus Cornelius Celsus was a first-century Roman physician whose works had recently been published. Paracelsus, the son of the municipal physician of the mining town of Villach, studied at the University of Ferrara and claimed to have a medical degree, although there is no record of this. He was briefly a professor of medicine at the University of Basel, receiving the post in 1527 and scandalizing the university by lecturing in German rather than Latin and publicly burning the *Canon* of the Arab physician Avicenna (980–1037), a basic medical text. In 1528, he was expelled from the university, and he spent the rest of his life wandering through Europe as far as Constantinople, dying in Salzburg.

Paracelsus was an extremely pugnacious personality who denounced professors and physicians as ignorant and greedy frauds and advocated learning from peasants and unlearned medical practitioners, including women. He broke with precedent by inviting unlearned people such as barber-surgeons as well as medical students to his lectures. Influenced by alchemy, Paracelsus attacked the dominant medical traditions handed down from the ancient and Arab physicians, particularly Galen, as inaccurate and useless, claiming that his shoe-buckles were more learned than Galen and Avicenna. Rather than the study of texts, Paracelsus advocated direct observation, and he urged physicians to study the knowledge of peasants and old women to learn the uses of local herbs.

Paracelsus was religiously unorthodox, remaining a Catholic while sharing many

This wood engraving of Paracelsus shows him with the Azoth, an alchemical universal medicine, and a book of the Kabbalah. (National Library of Medicine)

Protestant views. Some followers referred to him as the "Luther of Medicine," but he shunned the title. He saw medical reform as part of an overall project of reform based on religion: Physicians could complete Christ's mission of redemption by redeeming physical nature. Paracelsus argued that in addition to their ignorance of chemistry, the ancient Greeks should be rejected for their paganism, and the Arabs for their Islam, and that his own truly Christian medicine and natural philosophy should replace theirs.

Paracelsus's medical system was based on the relation between the human body as a microcosm and the universe as a macrocosm. In this view, the forces of the universe and the body interact, and thus it is vital for the physician to be an astrologer, to understand the cosmic forces that shape the individual body. Paracelsus's medicine relied heavily on the use of herbs, drugs, and metallic compounds, particularly taken internally. He had extensive experience with metals, thanks not only to his father's connections with mining but to his own experience; in 1537, he was employed at the Fugger family's mines in

Villach. Paracelsus wrote *On the Miner's Sickness and Other Miner's Diseases,* published in 1567, one of the first tracts on occupational diseases. He was also the first to recognize the difference between congenital and infectious syphilis. In alchemy, he discovered the method of concentrating alcohol by freezing it out of its solution, and he invented the "Paracelsian triad"—a doctrine that describes all matter as composed of salts, sulfurs, and mercuries—which would be very influential among later alchemists.

After his death, a Paracelsian movement of "chemical medicine" gained many followers as an alternative to Galenic medicine. Sometimes prevented from publishing by orthodox physicians, Paracelsus published little during his life. But his voluminous and sometimes wild writings were published posthumously, and his thought was systematized by his followers. He partially reoriented alchemy away from the quest for gold and toward medicine, and his work influenced scientists such as Johannes Baptista van Helmont.

See also Alchemy; Medicine; Paracelsianism.
References
Debus, Allen G. *The Chemical Philosophy: Paracelsian Science and Medicine in the Sixteenth and Seventeenth Centuries.* 2 vols. New York: Science History Publications, 1977.
Pagel, Walter. *Paracelsus: An Introduction to Philosophical Medicine in the Era of the Renaissance.* Basel, Switzerland, and New York: S. Karger, 1958.

Paris Observatory

The Paris Observatory, originally intended as a central building for the Royal Academy of Sciences, was founded in 1667 and completed in 1672. Situated in the southern outskirts of Paris, it was more than an astronomical observatory. It had a laboratory for physical experiments, as well as a staircase from which objects could be dropped to study the physics of falling bodies. It also held apartments, where some academicians lived, and it was a popular spot for visitors to Paris. Its leading personality was the Italian Gian Domenico Cassini, who arrived in 1669 and moved into the observatory in 1671. However, Cassini never held the title of director, and the observatory did not have the central organization of its English rival at Greenwich. Various astronomers carried out various projects with no central agenda.

There was some collaboration, however, notably between Cassini and others at the Paris Observatory and Jean Richer (1630–1696) at Cayenne to determine the parallax of Mars in 1672. Other astronomers who worked there included Christiaan Huygens, the Dane Ole Rømer (1644–1710), Adrian Auzout (1622–1671), and Jean Picard (1620–1682). The observatory was best known for its use of telescopes with micrometers for very precise measurements of celestial angles (Picard and Auzout were co-inventors of a successful micrometer based on moveable wires), pendulum clocks for increased accuracy in the measurement of time, and new telescopic sights for traditional astronomical instruments such as the quadrant.

See also Cassini, Gian Domenico; Royal Academy of Sciences.

Reference

Wolf, A., with the cooperation of F. Dannemann and A. Armitage. *A History of Science, Technology, and Philosophy in the Sixteenth and Seventeenth Centuries.* 2d ed., prepared by Douglas McKie. London: Allen and Unwin, 1950.

Pascal, Blaise (1623–1662)

Pascal was an eminent French mathematician, geometer, and physicist, as well as a great writer and religious thinker. The young Pascal, who did not attend a university, was introduced to Parisian salons and the scientific circle around Marin Mersenne by his father, the magistrate Etienne Pascal (1588–1651), himself a talented mathematician and musical theorist. Something of a child prodigy, Pascal wrote an impressive essay on conic sections at the age of 16, although it was not published until much later. This essay includes

"Pascal's theorem" on the properties of a hexagon inscribed in a conic section. Pascal engaged in barometric experiments, demonstrating the existence of a vacuum and the variability of air pressure in *New Experiments on the Void* (1647). (Descartes, who did not believe in the vacuum, described Pascal in a letter to a friend as having a vacuum in his head.) What took up a great deal of Pascal's time in the 1640s, though, was the invention and refinement of a calculating machine, originally intended to help his father collect taxes. (Pascal's labors on this early computer have inspired contemporary computer scientists to name a programming language for him.) Pascal eventually received a patent on the machine, but, unfortunately, few were sold and Pascal's patent never proved profitable. In 1653, he wrote *A Treatise on the Equilibrium of Liquids,* the first complete system of hydrostatics along with a discussion of the implications of his barometer experiments. Like much of Pascal's scientific and mathematical work, it was published after his death, in 1663.

Pascal moved in aristocratic French society. A conversation with a friend concerning the proper division of the stakes in a gambling game led Pascal to formulate the theory of probability in a correspondence with Pierre de Fermat in 1654. As a mathematician, Pascal also wrote another posthumously published treatise on the arrangement of integers now known as "Pascal's triangle," although it had already been known to the Arabs, Indians, and Chinese as well as some Europeans.

Pascal underwent two conversion experiences to Jansenism, a puritanical movement within the French Catholic Church emphasizing asceticism that was considered heretical by some Catholics. The first, in 1646, did not greatly affect his manner of life, but he largely abandoned mathematics and natural philosophy after his second and more profound conversion on the night of November 23, 1654. He became a theologian and a Jansenist polemicist, waging a particularly intense and

Only Blaise Pascal was able to attain greatness in both science and literature. He also denied natural theology, believing God's presence was shown only through miraculous violations of the laws of nature. (National Library of Medicine)

vicious struggle with the Jesuits, already an enemy because of Pascal's attacks on their Aristotelian natural philosophy. He now believed they were promoting moral laxness.

Unlike many of the pious natural philosophers of the scientific revolution such as Isaac Newton, Pascal did not view God as at all rationally comprehensible, instead finding the unknowability of God, the self, and the universe terrifying. Racked with disease and pain, Pascal rejected the Cartesian and Scholastic idea of attaining true knowledge of God through logic, proclaiming his loyalty to the "God of Abraham, Isaac, and Jacob," not the "God of the philosophers and scientists." For Pascal, God's presence is demonstrated not through the laws of nature, but through his occasional miraculous violations of those laws, such as the miraculous cure of Pascal's niece in 1656. Pascal's only involvement in science and mathematics in the last few years of his life was work on the properties of the

cycloid curve, originally undertaken as a distraction from the constant pain he endured. The mathematical challenges he issued attracted the attention of some of Europe's leading mathematicians.

> *See also* Cycloid; God; Mathematics; Probability.
> *Reference*
> Adamson, Donald. *Blaise Pascal: Mathematician, Physicist, and Thinker about God.* New York: St. Martin's, 1995.

Peiresc, Nicolas-Claude Fabri de (1580–1637)

Nicolas-Claude Fabri de Peiresc was one of the great patrons and organizers of early modern science as well as humanistic studies. From a wealthy family of French noble magistrates, he was educated at the Jesuit College at Tournon and the University of Montpellier. In 1610, he was one of the first Frenchmen to look at the sky through a telescope. He recorded the time of celestial phenomena such as the motions of the satellites of Jupiter, using the observations for geographic calculations. His extensive correspondence all over Europe and the Mediterranean enabled him to coordinate observations over a wide area. Peiresc coordinated a widely dispersed series of observations of the solar eclipse of 1635 through French embassies, using the results to correct the exaggerated length estimate of the Mediterranean. He traveled to Switzerland, Italy, and England and spent a few years in Paris, but most of his career was spent at his home in Provence, where he sponsored human and animal dissections and had the third largest botanical garden in France.

Peiresc's intellectual activities and his extensive international patronage raised his social standing. His most illustrious clients and allies included Tommaso Campanella, who stayed at Peiresc's house temporarily after his release from jail, Marin Mersenne, whose propensity to quarrel with Peiresc's other clients led to much tension in their relationship, and Galileo Galilei, on whose behalf Peiresc unsuccessfully intervened at

Rome in 1634 and the publication of whose works in France he promoted. A particularly close ally was Pierre Gassendi, who lived with Peiresc from 1634 to 1637. Gassendi took up astronomy at Peiresc's behest and wrote his biography.

See also Correspondence; Gassendi, Pierre.
Reference
Sarasohn, Lisa. "Nicolas-Claude Fabri de Peiresc and the Patronage of the New Science in the Seventeenth Century." *Isis* 84 (1993): 70–90.

Periodicals

The scientific periodical was a relatively late bloomer in the scientific revolution, emerging over 200 years after the printing press. Its precursors include the annual almanacs that had appeared in Europe for centuries and sometimes dealt with scientific subjects, and the published weekly records of Théophraste Renaudot's conferences at the Office of Address in the early seventeenth century. But periodicals systematically reporting on natural philosophy had to wait until a sufficiently large readership had formed. The first major periodicals to appear were the French *Journal des scavans,* founded in 1665, and the English *Philosophical Transactions,* which followed a few months later. Both were private ventures, although *Philosophical Transactions* benefited from the association of its publisher, Henry Oldenburg, with the Royal Society, and the *Journal des scavans* reported on the activities of the Royal Academy of Sciences. *Philosophical Transactions* was primarily devoted to natural philosophy and mainly published original work. In contrast, the *Journal des scavans* and the German but Latin-language *Acta Eruditorum,* founded in 1683, were general reviews of learning whose primary mission was to review new books. They published reviews of new natural-philosophical books, along with books of theology or humanistic subjects like history and philology.

With the exception of *Philosophical Transactions,* most early journals that published original work concentrated on medicine.

Miscellanea Curiosa, published by the College of the Curiosities of Nature, a group of German physicians and natural philosophers, first appeared in 1670. Like the *Acta Eruditorum,* it was published in Latin, enabling physicians from all over Europe, few of whom understood German, to read and contribute to the journal. A similar Latin medical journal, *Acta Medica et Philosophica Hafniensa,* was published at Copenhagen from 1673 to 1680. Dominated by the Danish physician and anatomist Thomas Bartholin (1616–1680), it published works on medicine, zoology, and botany from Danish and foreign physicians. The medical journals usually published reports of interesting cases rather than general works of medical science.

A number of French-language journals were published outside France in the late seventeenth century, beginning with Pierre Bayle's (1647–1706) Amsterdam-based *News from the Republic of Letters* in 1684. These journals, mostly published by French Protestants like Bayle who had been forced from France by increasing persecution, were designed to provide summaries, reviews, and extracts of books from different countries, presenting this material in French, the language of polite European society. This broad dissemination was possible partly because the journals were printed in the Dutch Republic, a place relatively free from censorship. Science was only one of many subjects these journals covered, but given the general ignorance of the English language in Europe at the time, the periodical press played an important role in introducing English science to a Continental audience. There were also a number of short-lived English journals that tried to reciprocate by reporting in English on Continental work. The Dutch and Italians also published such journals, but unlike the French journals, their circulation was limited to their own linguistic area. (Antoni van Leeuwenhoek published his microscopic results in one such journal, *De Boeksall van Europe,* for his Dutch countrypeople, but he relied for international transmission on *Philosophical Transactions.*) In a time when the production of scholarly and scientific literature in Europe was expanding rapidly, these publications made it possible for both the working natural philosopher and the amateur of science to keep up with what was going on.

Scientific societies in the seventeenth century usually supported the publication of books rather than periodicals. Although *Philosophical Transactions* sometimes included papers presented at the Royal Society, the first major scientific society to regularly publish its own proceedings in a periodical form was the French Royal Academy of Sciences. After several abortive efforts, the academy succeeded in publishing annual reports of its proceedings beginning with the 1702 publication of the proceedings for 1699, the year of the academy's reorganization. This annual publication, *History and Memoirs,* combined a narrative of the academy's work for the year with a selection of the outstanding papers read at academy meetings. The academy had also informally taken over the *Journal des scavans* by this time.

By the end of the seventeenth century, periodicals were presenting science in a popularized form for the benefit of a public interested in keeping up with the new science without being natural philosophers themselves. One of the earliest examples is John Dunton's (1659–1733) *Athenian Mercury,* a London publication devoted to answering questions sent in by readers that ran, with some gaps, from 1691 to 1697. Scientific questions included such subjects as the tides, the existence of a vacuum, and how the floating positions assumed by drowned men differ from those of drowned women. Dunton employed the mathematics teacher Richard Sault (d. 1702), a contributor to *Philosophical Transactions,* to answer the mathematical and scientific questions. Following in Dunton's wake were a number of other short-lived publications that focused on the entertainment value of science. One successful annual publication presenting mathematical and scientific problems was *The Ladies' Diary, or the* *Woman's Almanack,* which ran from 1704 to 1840. Founded by a mathematical schoolmaster, John Tipper (d. 1713), its purpose was to introduce educated women to mathematics and science.

See also Journal des scavans; *Philosophical Transactions;* Printing.

References

Goldgar, Anne. *Impolite Learning: Conduct and Community in the Republic of Letters, 1680–1750.* New Haven, CT: Yale University Press, 1995.

Kronick, David A. *A History of Scientific and Technical Periodicals: The Origins and Development of the Scientific and Technical Press, 1665–1790.* 2d ed. Metuchen, NJ: Scarecrow Press, 1976.

McEwen, Gilbert D. *The Oracle of the Coffee House: John Dunton's Athenian Mercury.* San Marino, CA: Huntington Library, 1972.

Personifications and Images of Nature

Early modern natural philosophers used a variety of metaphors to understand the natural world, its role in a supernatural universe, and their own relation to it. Some had classical and Christian roots, while others, such as the metaphor of nature as a vast machine or clock, emerged during the scientific revolution itself.

Nature was most often personified as a goddess or god, having either a cooperative or conflictual relationship to science. In metaphors for both types of relationship, nature could be gendered as either male or female. One of the most basic metaphors, rooted in the classical tradition, was that of nature as the goddess Natura. This metaphor suggested fecundity and multiplicity, and it was allied to a metaphor of the Earth as a fertile womb from which living things emerge. Visually, nature was often represented as a nude woman bearing emblems of fertility such as ears of wheat. The fertility of nature could also be represented by a woman with supernumerary breasts. Despite Natura's bounty, however, she could also be conceived as decaying, growing old or senescent as the end of nature and the world approached, signaling the Apocalypse. This idea, which implied a steady process of deterioration in history, was opposed by those who emphasized progress and the advancement of knowledge, such as Francis Bacon.

Nature was also pictured as an adversary of the natural investigator, someone whose secrets needed to be exposed, an image expressed most memorably, although not most typically, in Francis Bacon's call for the "vexing" or tormenting of nature by art to force her to reveal her secrets. Although this metaphor of conflict was often gendered as a conflict between a goddess and male investigators, it was not dependent on this gendering. Nature could also be portrayed as a male god, Pan, from the Greek word meaning "all," as in Bacon's *Of the Wisdom of the Ancients* (1609). Visually, Pan was presented holding a pipe with seven tubes, representing the seven planets. The conflict between the investigator and nature was also expressed by the myth of the binding of the shape-changing god Proteus in ancient epics such as the *Odyssey*, whose heroes forced Proteus to reveal secrets. Pan and Proteus were linked as embodiments of nature's unity and multiplicity respectively. The metaphor of the binding of Proteus was also used by Bacon, as by other writers of the scientific revolution.

The ultimate extension of the personification of nature was to treat the personification not as a metaphor, but as actually existing. This could be traced back to the medieval distinction between *natura naturans,* nature acting, and *natura naturata,* nature being acted upon. *Natura naturata* is the nature that people see and perceive, *natura naturans* is that which acts upon nature to produce change. In the early modern period, those who believed in a self-acting natural universe, often influenced by alchemical or magical thinking, sometimes spoke of a "plastic spirit" that shaped the material universe. This view, which did not imply that the spirit had intelligence or volition, was common among the late-seventeenth-century Cambridge Platonists.

Despite the Platonists' piety, all views of

The frontispiece to an English translation of Giambattista della Porta's Natural Magick *(1658) shows nature in the form of a woman with multiple pairs of breasts.*

nature as a deity or spirit, but particularly the image of Natura, trailed a slightly suspect air of paganism. Many claimed that *natura naturans* is God's subordinate or handmaid, but such a being seemed to others incompatible with Christianity. Mechanical philosophers such as Robert Boyle, author of *A Free Enquiry into the Vulgarly Received Doctrine of Nature* (1686), were troubled by the tendency to make nature an independent being, which seemed to derogate from the power of God. The idea of a self-acting universe seemed inconsistent with the mechanical philosophy as it developed, and the mechanical philosophy's superior piety could be asserted by claiming that it took all agency from the material universe, vesting it solely in God. The overall tendency in the scientific revolution was away from the metaphors that personify nature to those that make it more mechanical, such as the image of a vast machine or a clock made by God. Although the "clockwork universe" image is often associated with Newtonian physics, Newton himself never employed it. The idea of the universe simply running without the need for the involvement of divine providence was abhorrent to him. However, Newton's great rival, Gottfried Wilhelm Leibniz, had himself worked on creating a mechanical calculating engine and often used clock metaphors for nature.

Nature could also be treated as analogous to a human body, and conversely, the body as analogous to nature, in the so-called microcosm-macrocosm analogy. This idea, which also had ancient roots, held that humanity and nature are linked as a network of correspondences, so that the head or the heart in the body corresponds to the Sun in the heavens. Naturally, this idea was easily adaptable to astrology, where each planet in the heavens could be seen as corresponding to a part of the body or organ that it governed. Society was also brought into the system of correspondences, a system surviving in our term "body politic." The fullest development of the microcosm-macrocosm analogy in the scientific revolution was that of Leibniz. In Leibniz's physical system, each monad, the spiritual atom of which the universe is composed, reflects all the other monads, being simultaneously macrocosm and microcosm.

> **See also** Baconianism; Book of Nature; Great Chain of Being.
> **References**
> Merchant, Carolyn. *The Death of Nature: Women, Ecology, and the Scientific Revolution.* San Francisco: Harper & Row, 1980.
> Tillyard, E. M. W. *The Elizabethan World Picture.* New York: Macmillan, 1944.

Perspective
See Art.

Pharmacology
See Apothecaries and Pharmacology.

Philosophical Transactions
Founded in 1665 by Henry Oldenburg, corresponding secretary to the Royal Society, *Philosophical Transactions* was the first periodical to be devoted exclusively to natural philosophy, as distinguished from general journals of the world of learning, such as the *Journal des scavans,* which was founded slightly earlier. The first issue of the *Transactions,* 16 pages long, appeared on March 6, 1665. For the period of Oldenburg's editorship, ending with his death in 1677, the monthly issues were mostly filled with extracts from Oldenburg's foreign and English correspondence describing experiments, natural wonders, observations, instruments, and mathematical innovations. In the early issues, Oldenburg translated foreign contributions into English (excepting those in Latin), and he wrote most of the book reviews, which began in the second number. Although everyone knew of the connection between Oldenburg and the Royal Society, *Philosophical Transactions* was not originally an official journal of the society but Oldenburg's private venture, and its first incarnation ended with his death. Under Oldenburg, *Philosophical Transactions* played an important role in disseminating the society's work to the world, and the early reputation of the Royal Society on the Continent was largely based on this publication.

Widely circulated during Oldenburg's time, *Philosophical Transactions* languished after 1677, with only a few issues appearing under the editorship of Nehemiah Grew. Robert Hooke, now secretary to the society, attempted to replace it with a journal called *Philosophical Collections,* which ran from 1679 to 1683, when the society's new secretary, Dr. Robert Plot (1640–1696), revived the old title. *Philosophical Transactions* again suspended publication between 1687 and 1691, and it was not restored to full health until the energetic and internationally minded Hans Sloane (1660–1753) took over as secretary to the society in 1695. Sloane expanded the journal and revived its foreign correspondence, which had severely declined since Oldenburg's time.

Philosophical Transactions tended to reflect the interests of its editors. Under Sloane, a physician and natural historian, it printed many accounts of oddities and curiosities, which led to friction between Sloane and the society's Newtonian leadership. The Newtonian Edmond Halley, who edited the journal as clerk to the Royal Society in 1686 and took it over from Sloane in 1714, printed much astronomy and physics and little natural history. In 1752, *Philosophical Transactions* became an official journal of the Royal Society.

See also Grew, Nehemiah; Halley, Edmond; Oldenburg, Henry; Periodicals; Royal Society.

References

Atkinson, Dwight. *Scientific Discourse in Historical Context: The Philosophical Transactions of the Royal Society of London, 1675–1975*. Mahwah, NJ: L. Erlbaum Associates, 1999.

Hall, Marie B. *Promoting Experimental Learning: Experiment and the Royal Society, 1660–1727*. Cambridge: Cambridge University Press, 1991.

Physicians

In early modern Europe, physicians were medical practitioners with university training. They were associated with the other learned professions that required training by graduate-level university faculties—law and the ministry. Physicians, all male, were economically and socially of higher status than other medical professionals—first below physicians were the apothecaries and surgeons whose practices physicians often tried to control, and below them, midwives and the freelance medical entrepreneurs with secret medical formulas the Italians called *ciarlatani,* from which the English word "charlatan" is derived.

Learned medicine as taught at medieval and Renaissance universities was a complex tradition, developed over the centuries and dominated by commentaries on the ancient texts of the Greek physicians Hippocrates (c. 460–377 B.C.) and Galen (A.D. 129–c. 199) and their Arab commentators. Learned Galenic physicians took what we now call a holistic approach, emphasizing diet and proper care for the body rather than specific cures for specific diseases, an approach associated with the medical heresy of Paracelsianism. However, Paracelsus himself and many Paracelsians were also physicians, and physicians were famously disputatious. A wide variety of medical and scientific thinking could be found in the community of European physicians.

Physician training was a growth industry in the early modern period, particularly north of the Alps, where Dutch universities, with Leiden in the lead, had passed the University of Padua and the notoriously conservative University of Paris as Europe's best centers for medical study by the late seventeenth century. University medicine remained text based, although dissections were practiced in some university medical schools, particularly in Italy, and the practice spread widely in the seventeenth century. University medical study began to incorporate studies of human health that had been excluded from medieval curricula, such as surgery and pharmacology, and medicine, as opposed to law or theology, was the obvious graduate course of study for men interested in the physical universe.

Even before beginning specialized medical training, aspiring physicians had taken an undergraduate degree, usually with some study of Aristotelian natural philosophy. Many medical educators believed that a strong grounding in natural philosophy was necessary for medicine and that physicians were thereby qualified to comment on natural philosophical questions. (The converse was also true; nonphysician natural philosophers such as René Descartes proclaimed the relevance of their work for medicine.) Physician training was also transformed by humanism, which emphasized Greek over

Physicians were often viewed as egotistical. This early-eighteenth-century illustration represents a physician as God, surrounded by books, instruments, and scenes of medical procedures. (National Library of Medicine)

Latin or Arabic medical writing. Many physicians were themselves humanists. By the eighteenth century, however, medicine had become the least text and lecture based of the major academic disciplines, with lectures at many universities devoted to specific medical topics rather than classic texts.

Many advances and innovations in medicine were associated with university-based physicians, particularly Andreas Vesalius and the anatomists who followed him at Padua, or the mechanical physician Herman Boerhaave (1668–1738) at Leiden. Outside the universities, practicing physicians were organized into colleges, which were dedicated to the protection of physicians' interests from rival medical professionals, patients, church, and state. In some of these colleges, for example the London College of Physicians, medical research was carried out and medical knowledge was advanced through lectures and pub-

lic dissections. Physicians working independently or at courts also made significant medical innovations.

The scientific interests of many physicians were not restricted to medicine. Medical school professors and graduates researched in many scientific fields, such as botany, chemistry, and mathematics, all of which were part of the curriculum in early modern medical schools. Botanical gardens were founded at universities as part of medical schools, and physicians such as García d'Orta who sought remedies outside Europe were in the vanguard of scientific investigation in European colonies and outposts the world over. Many people known for scientific achievements outside medicine originally received some medical education and practiced as physicians, most notably Nicolaus Copernicus, who briefly studied medicine at Padua, the greatest nursery of physicians and scientists in the sixteenth century, and William Gilbert, the experimenter with magnets, who was an eminent London physician.

Physicians constituted an international and interconfessional community, as medicine was the only learned profession in which Christian Europe allowed Jewish participation. This led to some suspicion on the part of the Catholic Inquisition, both of *converso* physicians—Jews or descendants of Jews who had accepted Christianity—and of Catholic physicians who owned medical books by Protestants. Paracelsianism especially aroused inquisitorial wrath. But the widespread communication between physicians throughout Europe, mostly in Latin, also meant that medicine and natural philosophy as practiced by physicians had a European rather than a narrowly national quality.

See also Cardano, Girolamo; Fludd, Robert; Gilbert, William; Grew, Nehemiah; Harvey, William; Hernández, Francisco; Malpighi, Marcello; Medicine; Orta, García d'; Paracelsus; Steno, Nicolaus; Sydenham, Thomas; University of Leiden; University of Padua; Vesalius, Andreas.

References

Conrad, Lawrence I., Michael Neve, Vivian Nutton, Roy Porter, and Andrew Wear. *The Western Medical Tradition: 800 B.C. to A.D. 1800.* Cambridge: Cambridge University Press, 1995.

Porter, Roy. *The Greatest Benefit to Mankind: A Medical History of Humanity.* New York: W. W. Norton, 1997.

Physics

Physics underwent some of the most dramatic changes of any science during the scientific revolution. The main movement was from a qualitative Aristotelian science to a quantitative and mechanical one, culminating in the Newtonian system.

The fundamental intellectual challenge presented to Aristotelian physics was that of Copernican, Sun-centered astronomy. Aristotelianism explained terrestrial motion as a natural motion to or away from the center of the Earth, the center of the universe. Thus

heavy objects fall, and light substances such as smoke rise. This explanation could not be adapted to a Sun-centered system. Neither did the Aristotelian motion theory fit with a moving and rotating Earth, as it could not explain why a dropped object falls in a straight line rather than landing at a point to the west of where it was dropped, as would seem necessary if the Earth were moving. Thus, the acceptance of Copernicanism required a new physical system that did not depend on Earth's centrality and immobility.

The most important Copernican to attack Aristotelian physics was Galileo Galilei. Galileo initially built on a late medieval avant-garde Aristotelian concept of "impetus," the product of a long struggle among Aristotelians with the inadequacies of Aristotle's theory of "violent" motion, motion that is not natural motion to or away from the center of the Earth. This concept described impetus as a force inherent in a moving body that keeps it moving, then gradually leaks out as the object stops moving. This was a departure from Aristotle, who asserted that a body moves from a cause outside it.

The problem of violent motion had been much on people's minds since the recent introduction of cannon in battle, and figuring out the course of a cannonball was a question of great practical importance. The projectile's motion was traditionally a combination of the violent motion of the explosion and the natural motion that eventually brought it to the ground. But Aristotle gave no clear guidance on how the two motions combined. The Italian mathematician Niccolò Tartaglia (c. 1499–1557), among the first to apply sophisticated mathematics to a physical problem, had succeeded in calculating the trajectory of a cannonball using the impetus theory. He demonstrated that a cannon at a 45-degree angle to the Earth would send its missile farthest.

Galileo later abandoned the impetus theory, however, in favor of something resembling the modern concept of inertia, although he did not use the term. He surmised that because a dropped body already moves along with the rotating Earth, it does not move relative to the Earth's surface while falling. But Galileo's greatest influence on subsequent physics was his insistence on mathematizing physical principles. Aristotelians had usually viewed natural philosophy and mathematics as separate disciplines, with mathematics the inferior of the two. "Mixed sciences"—aspects of physics that were expressed in mathematical terms, like optics, were of relatively low status. Galileo, influenced by Platonism, broke with this tradition by claiming that the book of nature was written in the language of mathematics. Galileo's assertions about the nature of motion differed from Aristotle's in that they assumed a mathematically perfect universe, rather than the one we actually live in. Galileo's famous law of falling bodies, which states that the distance covered by a falling body varies with the square of the time of the fall, ignores the factors of friction, wind, and air resistance. And despite its mathematical power, Galileo's physics lacked what Aristotelianism, and at this time Aristotelianism alone, had—a system of explanation that put all physics, and indeed all science, in a single intellectual framework.

One possibility for an alternative physical system was the so-called magnetic philosophy, which built on the work of William Gilbert. This subsumed a number of forces, including gravity, as varieties of magnetism. This idea attracted the interest of Galileo and more seriously of Johannes Kepler, who theorized that the movements of the planets could be explained by magnetic influences. If the Earth is a magnet, as Gilbert had demonstrated, couldn't all planets be magnets? But Kepler's physical approach, which rested on magical and Platonic ideas of celestial harmony, was not the direction physics was going to take.

The most successful physical alternative to Aristotelianism was not the magnetic but the mechanical philosophy. The most important theoretical physicists of the first half of the seventeenth century were two Frenchmen, René Descartes and Pierre Gassendi, champions of the two main positions of mechanical philosophy. Gassendi built on one aspect of the revival of ancient philosophy by the humanists, the new availability of alternative, non-Aristotelian ancient physical systems. He rejected Aristotelianism in favor of a revived and Christianized Epicurean atomism, reducing the universe to the movement of atoms—small, indivisible particles—in the void.

Descartes was more radical in that he openly proclaimed a new philosophy rather than reviving an old one. Descartes's solution also relied on the mechanical principles of matter and motion, but unlike the Epicureans and Gassendi he followed Aristotle in denying the existence of a vacuum. Cartesianism was the first rival to Aristotelianism to be a complete system of nature; Descartes claimed that he could explain not just the motion of bodies but optical, magnetic, and other physical phenomena as well. Descartes, like Gassendi, clearly formulated a principle of inertia. This meant that what had to be explained was not motion itself, as Aristotelians and impetus theorists had been assuming, but rather the changes in a body's state of motion or rest. Since the Cartesian universe, like the Aristotelian, is a "plenum," perfectly full of matter with no empty space, motion occurs only by the displacement of matter by other matter. Since the empty space left by the moving object has to be instantly filled, Cartesian physics is dominated by "vortices," circular whirlpools of matter generated by motion.

The Aristotelian idea of heavenly motion as circular died extraordinarily hard. Despite his awareness of Kepler's work on elliptical orbits, Galileo had held to circular motion, and although Descartes claimed that all natural motion occurs in a straight line and that the heavens are governed by the same principles as the Earth, his vortices still carried the planets around the Sun in circles. For Descartes, gravity is the pushing of things to the center of the vortex, just as solid objects are forced to the center of a whirlpool.

Likewise, light is a pressure emanating from an object and transmitted along a line of particles, and magnetism is caused by corkscrew-shaped particles, generated by the turning of the vortex, that fit into pores in iron. Magnetic polarity, then, is a matter of left-handed versus right-handed screws. Although Descartes was a great mathematician, he did not fully mathematize his physical system, which was basically one of qualitative description.

The Cartesian system was very successful, particularly in France and Holland, but problems did emerge. The existence of vacuums was demonstrated by the barometer and air-pump experiments of Galileo's disciple Evangelista Torricelli (1608–1647), the German Otto von Guericke (1602–1686), Blaise Pascal, and Robert Boyle. But Cartesian physics remained intellectually fruitful for decades, as can be seen in the work of the Dutch physicist Christiaan Huygens, a somewhat unorthodox Cartesian who worked out the theory of the pendulum and devised a wave theory of light.

The most important physical developments of the late seventeenth century occurred in England, where Cartesianism never became an orthodoxy. This was partly simply because Cartesianism was French, partly because it seemed suspiciously materialistic, and partly because the Baconians of the Royal Society distrusted grand theoretical schemes. However, English physicists felt free to draw from Cartesianism, as well as from Gassendi's revived Epicureanism, the physics of the ancient Stoics, the magnetical philosophy, and even Aristotelianism.

The culmination of the physics of the scientific revolution came with Isaac Newton, whose *Mathematical Principles of Natural Philosophy* (1687) set forth an elegant system of mathematical laws built on the work of Galileo and Kepler. The three laws of motion include the first law, a clear statement of the principle of inertia, that a moving body tends to remain in motion and a body at rest tends to remain at rest, the second law, that changes of motion are proportional to the force im-

pressed, and the famous third law, that for every action there is an equal and opposite reaction. The work also set forth Newton's theory of universal gravity, which holds that between any two bodies in the universe there exists a force directly proportional to the product of the masses of the two bodies and inversely proportional to the square of their distance. Although Newton's was an immensely powerful system, it marked a retreat from the project of the mechanical philosophy in describing all physical action in terms of matter and motion, as it did not explain universal gravity in mechanical terms— indeed, it did not explain universal gravity at all. Both Newton's great rival, Gottfried Wilhelm Leibniz, and European Cartesians like Huygens picked up on this deficiency as an objection. But Newtonianism would not be denied, and in the eighteenth century, Newtonian physics was accepted not only as a true picture of the workings of the universe but as the model for all other sciences.

See also Aristotelianism; Atomism; Cartesianism; Descartes, René; Force; Galilei, Galileo; Gravity; Mechanical Philosophy; Mechanics; Newton, Isaac; Optics; Stoicism; Vacuum.

References

Cohen, I. Bernard. *The Birth of a New Physics.* Revised and expanded. New York: W. W. Norton, 1985.

Hall, A. Rupert. *The Scientific Revolution, 1500–1800: The Formation of the Modern Scientific Attitude.* London: Longmans, 1954.

Westfall, Richard. *The Construction of Modern Science: Mechanisms and Mechanics.* New York: Wiley, 1977.

Physiology

The question of the living body's functioning was fiercely disputed during the scientific revolution by Galenists, Paracelsians, and mechanical philosophers. Their explanations were not purely material, as all living bodies were considered to possess "souls" or spirits enabling them to function. Galenists viewed the body as balanced between the four humors of blood, phlegm, choler, and black bile. From the humors the "spirit" was thought to emerge. Galenist physiology divides the body into three systems centered on the three major organs—liver, brain, and heart. Blood, thought to be derived from food digested in the stomach by heat, originates in the liver, as does the system of veins. Blood is carried to the right ventricle of the heart, where it mixes with the spirit taken in by the lungs. The arteries carry enriched blood in a system originating in the heart. The third system, centered on the brain, distributes vital spirits through the nerves. The key problem for Galenic physiology was how blood gets from the right to the left ventricle of the heart. Galen had claimed that the blood seeps through tiny pores in the "septum," or wall, between the ventricles, but most anatomists could find no physical evidence of this.

The materialism of Galenic physiology was challenged in the sixteenth century. Jean-Francois Fernel (1497–1558), one of the first to use the word "physiology," modified Galenism by identifying the nutritive, vital, and animal spirits as emerging not out of the humors, but from the celestial regions. Fernel, a medical professor at the notoriously conservative University of Paris, continued to consider himself a Galenist. A more radical challenge was mounted by Paracelsus, who rejected Galenism totally and put forth a system based on medical alchemy. Paracelsus was not a very consistent or systematic thinker, and his physiology seems to have derived from his use of the microcosm-macrocosm analogy. Under this analogy, the processes of the human body are governed by immaterial spiritual forces called *archei*. Paracelsus denied the humors, but they crept back into his theories in different guises, such as the four "tastes"—bitter, salt, acid, and sweet—which correlate with different temperaments in much the same way as humors.

The most important developments in physiology in the early seventeenth century were William Harvey's discovery of the circulatory system and Johannes Baptista van

Helmont's innovations within the Paracelsian tradition, which dominated later "iatrochemical" (medical-chemical) thought. Van Helmont rejected the macrocosm-microcosm analogy and identified living processes as chemical. For example, he described the digestive processes of the stomach as based on chemical interactions like that of fermenting wine, as opposed to the Galenic theory of digestion by heat. But van Helmont's chemistry did not make him a materialist. Like Paracelsus, he asserted that physiological processes are carried on by *archei*, spiritual entities that govern the body.

The theoretical implications of the circulation of the blood were unclear. Harvey himself believed that blood carries with it a spirit endowing the body with life. The circulation would be interpreted in an Aristotelian way by Harvey, in a mechanistic way by René Descartes, and in a spiritual way by Robert Fludd, but it was definitely a crippling blow to Galenic physiology, which quickly began to recede from consideration, although it continued to be taught in some medical schools.

One influential school of thought, the "iatromechanists" (medical mechanists), extended the Cartesian mechanical interpretation of the circulation to other physiological processes. Iatromechanists included the scientist Giovanni Alfonso Borelli (1608–1679), who studied the workings of the muscles, and the physician Giorgio Baglivi (1668–1707), who minimized the importance of fluids and compared the body to a windup mechanical doll. Other late-seventeenth-century scientists investigated particular living systems with a less theoretical and more empirical agenda. In England, Richard Lower (1631–1691), John Mayow (1641–1679), and other "Oxford physiologists" carried on Harvey's work, investigating blood circulation and respiration and identifying the redness of blood with its exposure in the lungs to "nitro-aerial" particles in the air. Influenced by both van Helmont and Harvey, Franciscus Sylvius (1614–1672), a medical professor at the University of Leiden, continued Van Helmont's study of the stomach, and several investigators studied the glands, a focus of great interest in the late seventeenth century. The English physician Thomas Wharton's (1614–1673) *Adenographia* (1656) was the first complete treatment of the human glandular system.

Although their theory was influential, the iatromechanists did not have everything their own way. The German physician and chemist Georg Stahl (1660–1734) put forth an influential theory of "vitalism," the belief in an intangible and nonmaterial life-substance that animates the bodies of living things. The rivalry of mechanists and vitalists would continue throughout the eighteenth century.

See also Circulation of the Blood; Galenism; Harvey, William; Helmont, Johannes Baptista van; Paracelsianism; Paracelsus.

References

Foster, Sir Michael. *Lectures on the History of Physiology during the Sixteenth, Seventeenth, and Eighteenth Centuries.* Cambridge: Cambridge University Press, 1901.

Hall, Thomas S. *Ideas of Life and Matter: Studies in the History of General Physiology, 600 B.C.–1900 A.D.* Vol. 1, *From Pre-Socratic Times to the Enlightenment.* Chicago: University of Chicago Press, 1969.

Planetary Spheres and Orbits

During the scientific revolution, the dominant idea of a planet's path in space changed from a perfect circle around the Earth, in which the planet was embedded in an enormous sphere or orb, to a Sun-centered ellipse derived from the gravitational relationship between the planet and the Sun. This complex change involved nearly every prominent astronomer and physicist of the period. The system of spheres was ultimately derived from Aristotle, who had conceived of the planets as carried in spheres centering on the Earth and exhibiting the most perfect motion, which was circular. It was substantially modified in ancient times by Claudius Ptolemy (90–168 A.D.), who unlike Aristotle was a technical astronomer. Realizing that the planets do not move as if they are carried at a

uniform rate on Earth-centered spheres, Ptolemy introduced the idea that the spheres that carried the planets are centered on a point near but not identical with the Earth—that they are "eccentric" to the Earth. The picture became even more complex with the addition of epicycles—whereby the planet combines the larger orbit with a smaller one around a point on the larger eccentric orbit—and equants—points other than the center of the sphere relative to which the sphere's motion is uniform. The introduction of Greek and Islamic astronomy and cosmogony to Latin Europe in the Middle Ages ensured that this picture of spheres, although not the only ancient system, was intellectually dominant.

Copernicus adjusted this received picture by centering it on the Sun rather than the Earth and getting rid of the equants. However, the Copernican system otherwise closely resembles the traditional one, down to the spheres carrying the planets, the epicycles, and the circular orbits; indeed, Copernicus viewed the greater circularity of his system as an advantage over Ptolemy's. The gravedigger of the celestial spheres was not Copernicus but Tycho Brahe. Tycho argued that comets, specifically the comet of 1577, are heavenly bodies rather than disturbances in the atmosphere, as traditional Aristotelians would have it. Heavenly comets had to cross the celestial spheres in which the planets were supposedly embedded to reach the Sun, and thus were clearly incompatible with the traditional picture. Tycho's fully developed view of the planetary system, the "Tychonic system," put the Earth at the center, with the Sun and Moon orbiting it and the other planets orbiting the Sun. This was mathematically equivalent to the Copernican system, but Tycho abolished the celestial spheres. There was no single center of the system for them to revolve around—the spheres of Mars, which in Tycho's system would have been centered on the Sun, would have had to cross the sphere of the Sun, centered on the Earth.

The Tychonic system and variations such as

Nova Mundani Systematis Hypotyposis ab Authore nuper adinuenta, qua tum vetus illa Ptolemaica redundantia & inconcinnitas, tum etiam recens Coperniana in motu Terra Physica abfurditas, excluduntur, omniaq, Apparentiis Cœleſtibus apt:ßime correfponent.

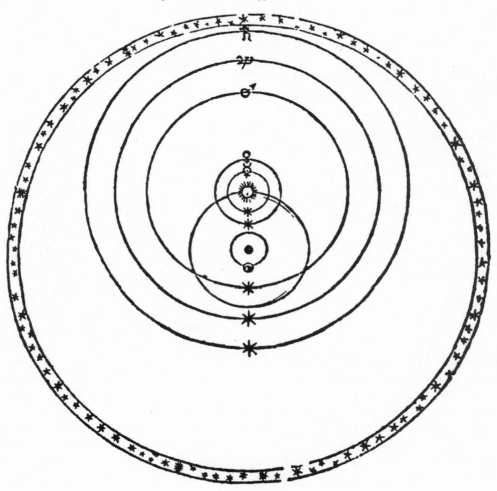

A diagram of the Tychonic system of the cosmos from Tycho Brahe's Progymnasmata *(1610).*

the "semi-Tychonic" system, which combined a Tychonic planetary system with a rotating Earth, became very popular, supplanting the Ptolemaic system as the chief rival to Copernicanism. (Ironically, following the church's condemnation of Copernicanism, Catholic astronomers tended to follow the Protestant Tycho, while Protestants followed the Catholic Copernicus. Some Catholic astronomers continued to defend the Tychonic system into the eighteenth century.) The late sixteenth century also saw a revival of interest, even among Ptolemaic astronomers, in ancient Stoic and Epicurean theories that held that the planets are not embedded in spheres but travel like fish in the sea or birds in the air. Conservative Scholastic natural philosophers responded to these challenges at first by putting greater emphasis on the hardness and rigidity of the spheres, although after around 1630 they too largely abandoned hard spheres.

The next great blow to the traditional picture was that dealt by Johannes Kepler. Using Tycho's excellent data and his own outstanding mathematical skills, Kepler produced a modified Copernican system, abandoning the celestial spheres and circular orbits and replacing them with ellipses, with the Sun at one focus. This involved the abandonment of epicycles and the expression of planetary motion as a smooth and continuous curve. However, the prejudice in favor of circularity was ingrained; even Kepler thought the universe as a whole was spherical. For this reason, neither of the other leading Copernicans of the first half of the seventeenth century, Galileo Galilei and René Descartes, accepted Kepler's innovation. Neither of them was a technical astronomer, and so the mathematical power of Kepler's theories was not their central consideration. Conversely, technical astronomers, who viewed Kepler's astronomical tables, the *Rudolphine Tables* (1627), as the best available, were not necessarily interested in the theory behind them.

Kepler had viewed the planets as alternately pushed and pulled in their orbits by a magnetic force emanating from the Sun. This explanation proved unconvincing, particularly as experiments with actual magnets showed that magnets cannot make other magnets rotate around them. Descartes provided a physical explanation for nonspherical planetary orbits by assimilating the planets to his idea of the universe as full of "vortices," or whirlpools of matter. He postulated that these circular whirlpools carry the planets along in their journeys round the Sun.

This Cartesian model was never completely dominant in England, where Keplerian astronomy was widely accepted and where the influence of the "magnetic philosophy" of William Gilbert worked against pure mechanism. Magnetic philosophy, as mediated by Robert Hooke, influenced Isaac Newton. It was Hooke's question about planetary orbits that inspired Newton to formulate the theory of universal gravitation, under which the planets' motions are determined by the gravity of the Sun. Newton's *Mathematical Principles of Natural Philosophy* (1687) ended the period of transition by deducing Kepler's laws from the law of universal gravitation.

See also Astronomy; Brahe, Tycho; Copernicanism; Gravity; Kepler, Johannes.

References

Grant, Edward. *Planets, Stars, and Orbs: The Medieval Cosmos, 1200–1687*. Cambridge: Cambridge University Press, 1994.

Taton, Rene, and Curtis Wilson, eds. *Planetary Astronomy from the Renaissance to the Rise of Astrophysics. Part A: Tycho Brahe to Newton. The General History of Astronomy*, ed. M. Hoskin, vol. 2. Cambridge: Cambridge University Press, 1984.

Platonism

One characteristic of early modern thought in many areas was a revival of interest in the ancient Greek philosopher Plato, the vast majority of whose writings were unavailable to Westerners in the Middle Ages. The fifteenth century, with increased knowledge of Greek, had seen a massive effort to publish Plato, both in Greek editions and in Latin translations (and eventually in vernacular translations as well). The first complete translation of Plato's works into Latin, by the Florentine philosopher Marsilio Ficino (1433–1499), was published in 1484. Ficino, leader of Florence's Platonic Academy, interpreted Plato in the Neoplatonic tradition, stemming from the late Roman Empire. He also translated and published the work of the foremost Neoplatonic Greek philosopher, Plotinus (c. 205–270).

Neoplatonists presented Plato as a philosopher of the divine and its emanations in the universe. The Neoplatonic Plato was a divinely inspired teacher of dogma rather than a philosopher in the modern sense. Neoplatonism had influenced Christian theology in the first Christian centuries and was easy to present as compatible with Christianity, as Ficino did in his *Platonic Theology* (1482). Neoplatonism was also easy to combine with magical traditions such as Hermeticism and the Kabbalah, both of which it had originally inspired, and Plato was sometimes included in the chain of ancient sages who were thought to have passed down the divine holy wisdom derived from God.

Plato never seriously rivaled Aristotle in the university curriculum, as his works, all in dialogue form, did not lend themselves to be arranged in a course of study the way Aristotle's did. Nor did they cover Aristotle's enormous range of topics. However, there were professors of Plato in Italian universities by the 1570s, the foremost being Francesco Patrizi (1529–1597), whose views of space and mathematics anticipate those of Isaac Newton. Although many tried to combine Plato and Aristotle (and to combine both with Christianity), the banner of Plato was frequently waved by opponents of Aristotelianism such as Petrus Ramus (1515–1572). For their part, anti-Platonists were suspicious of Plato's paganism and his endorsement of homosexual love. Catholic authorities often looked on Platonism with suspicion as related to magic and heresy, and Patrizi's works were condemned after his death. However, many Platonists were loyal Catholics; Patrizi, for example, taught at La Sapienza, the University of Papal Rome, and Pope Clement VIII (pope, 1592–1605) protected him from the Inquisition.

The Platonic view of the actual world as an imperfect embodiment of the "real" world of archetypes, or Forms, has affinities with the scientific view of the physical world as imperfectly exemplifying "real" mathematical laws. Galileo, for example, sometimes proclaimed himself a Platonist for this reason. According to legend, inscribed above the gate of the original Platonic Academy at Athens were the words "Let none enter who are ignorant of geometry," and Plato's description of God as a geometer was frequently referred to during the scientific revolution. And Platonism did award mathematics a higher intellectual status than did Aristotelianism. Galileo's contemporary Johannes Kepler was also influenced by Neoplatonic ideas of the divine geometrical harmony of the universe. The five regular solids with equal faces and angles that Kepler believed organized the universe—the tetrahedron, the cube, the octahedron, the dodecahedron, and the icosahedron—had been discussed by Plato and were known as the "Platonic" solids. Gottfried Wilhelm Leibniz was also interested in Plato, who contributed substantially to his metaphysical system. Leibniz was among the earliest to attempt to disentangle the original Plato from the Neoplatonists.

The most significant Platonic philosophers for the sciences in the seventeenth century were the so-called Cambridge Platonists, whose intellectual leaders were Henry More (1614–1687) and Ralph Cudworth (1617–1688). They essentially viewed Platonic philosophy as a way to spread a nondogmatic version of Christianity that would not lead to conflicts such as the English Civil War. Although neither participated in scientific inquiry, both had an interest in it and were fellows of the Royal Society. The most important aspect of Cambridge Platonism for late-seventeenth-century English science was its opposition to Cartesianism. After an initial attraction, the Cambridge Platonists found Cartesianism to be materialistic and conducive to atheism and Hobbesianism. The philosophy of the Cambridge Platonists influenced that of Isaac Newton, whose conception of absolute space and time has Platonic affinities.

See also Aristotelianism; Humanism.
Reference
Burtt, E. A. *The Metaphysical Foundations of Modern Physical Science.* 2d ed., rev. Garden City, NY: Doubleday, 1954.

Political Arithmetic

The systematic application of mathematical and quantitative techniques to social issues flourished during the seventeenth century, principally in England. Its earliest practitioners included the close friends John Graunt (1620–1674) and Sir William Petty (1623–1687). Graunt's *Natural and Political Observations Mentioned in a Following Index, and Made upon the Bills of Mortality* (1662) was a sensation, attracting the notice of the king and allowing Graunt to become one of the founders of the Royal Society, unusual for a tradesman.

The London Bills of Mortality, dating from the late sixteenth century, were weekly tabulations of christenings and deaths, with some attempt made to identify causes of death, particularly plague. In *Natural and Political Observations,* Graunt used this information and other sources to estimate the total population of London as around 380,000, debunking contemporary estimates that ran into the millions. He also attempted to aggregate the statistics and compensate for their unreliable aspects to draw conclusions about the medical condition and average lifespan of the London populace, making him one of the first to show an awareness of what is now called public health.

Petty, a physician and an extraordinarily creative mind, got his start in applying mathematics to social questions as physician-general to the English Army occupying Ireland. In 1654, he proposed to take over the surveying of land forfeited by Irish rebels against English authority. Paid in land for his services, he became a rich Irish landowner and survived the end of the English Commonwealth and the Restoration of King Charles II (r. 1660–1685) in 1660 to become a founding member of the Royal Society.

Petty's first publication in political arithmetic, *A Treatise of Taxes and Contributions* (1662), was followed by a series of works attempting to estimate national wealth, including the posthumously published *Polit-*

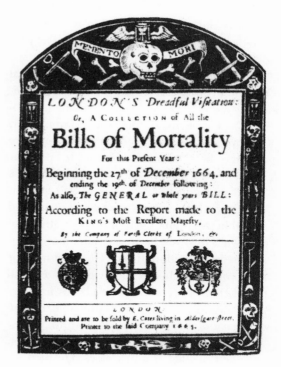

The Bills of Mortality, which collected death totals for London, were an important primary source for political arithmetic. (London's Dreadful Visitations, *1665*)

ical Arithmetick (1690), which gave its name to the field. (Much of Petty's writing circulated in manuscript and was published posthumously if at all.) Petty's works were the first detailed quantitative efforts to estimate national capital, income, and expenditure. He always saw his work in the context of policy recommendations, a particular concern in Ireland, where the ravages of war and English oppression had produced a poverty-stricken population. In both England and Ireland, Petty suggested public works schemes as the best approach to the problems of poverty. His suggestions were often more feasible in theory than in practice. For example, in his last years he put forth an ambitious plan, far beyond the reach of any seventeenth-century government, to depopulate Ireland, make it a huge cattle farm with a population of a few hundred thousand, and settle the excess population in England. Petty also attempted to estimate the world's population based on the assumed doubling rate of the eight passengers on Noah's Ark.

In the next generation, English political arithmetic was carried on by Charles Davenant (1656–1714) and Gregory King (1648–1712). Davenant, a political pamphleteer, attempted to use political arithmetic to figure out the best taxation scheme to pay for England's wars with France. King, Davenant's friend and collaborator, was a civil servant who served on several commissions related to public accounts but published nothing of his own. However, he was a much more original political arithmetician. He is known for an ambitious scheme to divide the English population into a number of classes for purposes of analysis. First, he estimated the size of the national population, the resulting number being close to that of modern scholars, and then divided it into classes. He then set forth in tabular form the income and expenditure of each class. King also systematically compared the English economy with that of its principal rivals, France and the Dutch Republic, and he estimated global population based on the distribution of people over land.

Another impressive work in political arithmetic was Edmond Halley's study of the demography of the German city of Breslau, a useful object for analysis due to its isolation. The information had been compiled by a clergyman, Caspar Neumann, who sent it on to Halley. In a paper presented to the Royal Society in 1693, Halley analyzed the numbers to produce the first modern table of life expectancy at different ages, meant to be of use to the infant life insurance industry and the sellers of annuities.

The English dominated political arithmetic but did not monopolize it; the eminent French military engineer Sébastien Vauban (1633–1707) published several works in the field in the late seventeenth and early eighteenth centuries. Although the study of political arithmetic declined in the eighteenth century, it remained to influence the development of economics.

See also Mathematics; Politics and Science.
References
Grattan-Guiness, Ivor. *The Norton History of the Mathematical Sciences: The Rainbow of Mathematics.* New York: W. W. Norton, 1998.
Stone, Richard. *Some British Empiricists in the Social Sciences, 1650–1900.* Cambridge: Cambridge University Press, 1997.

Politics and Science

Early modern science had always to adapt itself to the European political context in which it functioned, and by the seventeenth century the methods of science were themselves influencing European political theory and practice as well. Scientists offered governments specific solutions to different problems, ranging from navigation to gunnery. They also offered rulers the more intangible but nonetheless real asset of prestige, the product of an association with scientific advance. The classic case here occurred when Galileo Galilei named the newly discovered satellites of Jupiter the "Medicean stars," for his patron in the Medici family, the duke of Tuscany. Scientific associations also served both tangible and intangible political functions. The Royal Academy of Sciences expressed the glory of the Sun King, Louis XIV (r. 1643–1715) and fueled French pretensions to European intellectual hegemony, as well as serving the more mundane political function of examining devices that the French king was asked to patent. In Britain, the Royal Society offered a politically stabilizing form of natural knowledge to a country recovering from civil war and regicide.

The hierarchical world picture of Aristotelian cosmology did play an important ideological role in justifying the political hierarchy. In the sixteenth century, the state was often viewed as a microcosm, with the same order and subordination as the macrocosm, or universe. God, with his hierarchy of angels, was seen as working through the celestial spheres, each sphere imparting its motion to the lower. Similarly, the king was depicted as the prime mover of his kingdom, working through the social hierarchy. Copernicanism forced some rearrangement of this neat picture, although it did not entirely overthrow it. The mid-seventeenth-century English "Leveller," or radical egalitarian, Richard Overton (fl. 1642–1663) argued against the traditional notion that God lives in the empyrean heaven outside the universe, from where he was alleged to govern the cosmos through the celestial hierarchy. Instead, Overton proposed that God must live in the Sun, at the center of the universe. Overton's antihierarchical and materialist interpretation of Copernicanism even went so far as to deny the hierarchy existing between man and beast, both equally material. But solar metaphors could also be adapted for conservative purposes, and Copernicanism contributed to an increased tendency to compare the monarch ruling his kingdom to the Sun ruling the planets.

Despite the prominence of political arguments based on natural-philosophical theories of the macrocosm, the dominant ways of understanding political power in the sixteenth century remained theological, legal, and humanistic rather than scientific. For instance, although Niccolò Machiavelli's (1469–1527) value-neutral attitude toward politics has led some to hail him as a predecessor of modern political science, he remained a traditional humanist in his sources and method. The support of science played only a minor role in governance, and scientists were dependent on the whim of individual patrons more than on the continuity of state policies. Tycho Brahe's observatory, the greatest example of political patronage of science in the sixteenth century, had to be abandoned when a new Danish king decided it was no longer worth maintaining.

Then, in the first half of the seventeenth century, a succession of writers, the utopians Tommaso Campanella, Johann Valentin Andreae (1586–1654), Francis Bacon, and the activists of the Hartlib circle, set forth programs for the closer integration of innovative science and political authority. Bacon,

the most successful politician among scientists and the greatest scientist among politicians in the early modern period, set forth a program both to buttress the state with the support of science and technology and to foster and discipline science with the support of the state. Parts of this program were actually put into place during the time of Puritan rule in Britain in the 1650s, and it was eventually embodied, however inadequately, in the Royal Society and the Royal Academy of Sciences.

The mid-seventeenth century also saw the methods and practices of science directly applied to political questions and problems. This movement took many, sometimes conflicting, forms. Thomas Hobbes took an abstract approach, deducing politics from first principles on the model of geometry, for him the paradigm of all true knowledge. Humanistic thought, he asserted, had demonstrated its utter uselessness in constructing a true political philosophy, and only the method of geometry offered hope. Hobbes's picture of a society composed of individuals all alike in their desire for self-preservation was a political equivalent of the mechanical philosophy. The organic metaphor for society, which subordinated the individual to the group and justified the traditional hierarchies of Europe, faced a powerful rival in the mechanical view of society as a great clock or engine, in which authority was described as a social convention. Paradoxically, Hobbes's philosophy, intended to support authority, became identified with political radicalism due to its materialism and anticlericalism.

Hobbes's disciple and rival James Harrington (1611–1677), although much more humanistic and less enamored of mathematics than Hobbes, also employed mechanical metaphors. He described the ideal society in terms of a vast machine he claimed to have seen in Rome, in which the struggles of imprisoned cats served to turn the spit and baste the meat in a mock-kitchen. Harrington's favorite science was not geometry but medicine, and he argued that a true politics could only be arrived at through the study of particular political bodies rather than the abstract philosophical principles of Hobbes.

Hobbes, as a mechanical philosopher, was an associate and to some degree a follower of René Descartes. Although Descartes himself and the early Cartesians appear to have had little interest in politics, as a rival to the traditional hierarchical worldview, Cartesianism might be seen to have politically unsettling implications. The Cartesian (or for that matter the atomist) universe, with its myriad of tiny bits of matter bumping against each other like Hobbesian individuals, clearly lacked a cosmic hierarchy. However, French Cartesians like Bernard Le Bouvier de Fontenelle gave Cartesianism a conservative spin, comparing the Cartesian solar system, where natural laws and forces keep each planet in its destined space, with the world, where the laws of society keep individuals in their proper "sphere." This kind of conservative interpretation, emphasizing the laws that govern the universe and the providential care of God, was also applied to Newtonianism.

Sir William Petty (1623–1687), founder of "political arithmetic," took a methodological approach to political questions that is directly opposed to that of Hobbes. Petty relied on empirical quantitative data and estimates rather than on deductions from first principles, and he eschewed discussion of the questions of legitimacy and right that were characteristic of both traditional and Hobbesian political philosophy. This kind of political empiricism grew increasingly popular in the later seventeenth century, whether it was the work of individuals, as was usually the case in England, or of a powerful state, as in the gathering of statistics in France.

See also Courts; Hobbes, Thomas; Political Arithmetic.

References
Olson, Richard S. Science Deified and Science Defied: The Historical Significance of Science in Western Culture. Vol. 1, From the Bronze Age to the Beginnings of the Modern Era, ca. 3500 B.C. to ca. A.D. 1640. Berkeley, Los Angeles, and London: University of California Press, 1982.

———. Science Deified and Science Defied: The Historical Significance of Science in Western Culture. Vol. 2, From the Early Modern Age through the Early Romantic Era, ca. 1640 to ca. 1820. Berkeley, Los Angeles, and Oxford: University of California Press, 1990.

Popularization of Science

An important aspect of the creation of science as a cultural phenomenon during the scientific revolution was a vastly increased effort to communicate science to nonscientists. Popularization was aimed at a number of different audiences, but it excluded the poorest elements of European society, peasants and illiterate laborers. Instead scientific popularization was directed at men and to a lesser degree women educated at least to the level of literacy and having some disposable income. Popularization operated for different motives and through a number of vehicles, increasing in scope during the course of the scientific revolution.

One of the earliest forms of scientific popularization to emerge was intended to teach people specific skills for use in their work. Vernacular treatises on practical mathematics for merchants and businessmen went back to fifteenth-century Italy. In the sixteenth century, accessible works of science and mathematics were aimed at others who had use for these skills in their professions, particularly navigators and agriculturalists. This tradition was especially strong in the Dutch Republic and England, where it involved such leading natural philosophers as John Dee. As an institutional program, this effort was embodied, with limited success, in Gresham College.

The literate lower and middle classes were not blank slates for natural knowledge; they had long found sources of scientific and magical information in such popular reading matter as almanacs, cheap medical manuals, and books of secrets. Printing vastly increased the availability of such material. By the late sixteenth century, much of this popular knowledge had been characterized in learned culture as "vulgar errors," but these demotic forms were possible vehicles for the popularization of elite science. As early as the beginning of the seventeenth century, some almanacs endorsed the Copernican system, and later some communicated the new discoveries and, in a simple form, the theoretical innovations of Galileo Galilei, Johannes Kepler, and Isaac Newton. There was also a growing tendency to use natural-philosophical language and explanations in such popular entertainments as monster shows and exhibits of curiosities.

Despite Francis Bacon's own elitist concern with the possible dangers of extending natural knowledge beyond a state-aligned group of wise men, some of the earliest attempts to put Baconianism to work involved the spreading of scientific knowledge to larger groups of people. These included Théophraste Renaudot's Office of Address, active in Paris during the 1630s, and the Baconian projects of the Puritans during the English Revolution. Renaudot brought together noblemen and middle-class Parisians to discuss a number of subjects, including natural philosophy, and then published the proceedings. The Puritans, motivated by a millenarian belief in the increase of knowledge before the Last Days, attempted to spread natural knowledge through educational reform and cooperative projects such as the founding of an office to circulate useful information, although few of these efforts came to fruition.

Scientific popularization increased during the second half of the seventeenth century, as science assumed a higher cultural profile with the founding of the Royal Society and the Royal Academy of Sciences. It became fashionable among the upper classes to know something of the current developments in science. The idea of science as a civilized diversion was particularly strong in the salon culture of France and the Italian courts, where the ideal of the civilized person began to include the ability to discourse knowledgeably, or at least with an appearance of knowledge, on the science of the day.

Few of the leading natural philosophers themselves wrote for a popular audience, although there were exceptions, such as Robert Boyle's book of simple medical recipes, *Medicinal Experiments* (1692). Instead, many effective popularizers had a foot in the camps of both natural philosophy and literature or journalism, often presenting their works in dialogue form. The Frenchman Bernard Le Bouvier de Fontenelle's *Conversations on the Plurality of Worlds* (1686) was the most popular and successful work of popularization of the time, going through five editions in the four years after its publication and translated into several languages. Like a number of later popularizations, such as Francesco Algarotti's (1712–1764) *Newtonianism for Ladies* (1737), Fontenelle's work was aimed at upper-class women, with the interlocutors being a learned male Cartesian natural philosopher and a noblewoman. A competing work of Cartesian popularization was Christiaan Huygens's *The Celestial Worlds Discovered* (1698). Written in Latin, it originally addressed a much narrower and more male audience, but it was soon translated into English and French.

By the late seventeenth century, English periodicals such as the London-based *Athenian Mercury* (1691–1697) had become vehicles of scientific popularization. On an international level, the French-language periodicals published in the Netherlands by Huguenot exiles also circulated scientific information to nonscientists, playing a particularly important role in disseminating Newtonianism to the Continent. Even more original natural-philosophical periodicals, such as *Philosophical Transactions* and the various publications of the French Royal Academy of Sciences, addressed a broad literate public as well as natural philosophers.

In addition to printed texts, science was also exhibited and demonstrated in a variety of public and semipublic venues such as coffeehouses and, by the early eighteenth century, Masonic lodges. Some eminent English scientists supported themselves in part by charging admission to coffeehouse lectures, one notable example being the physicist William Whiston (1667–1752), who was barred from the established institutions of English science due to his brazenly heretical religious views. Other lecturers, such as the Huguenot-Anglican clergyman and Freemason John Desaguliers (1683–1744), active in London, the English provinces, and the Continent, were essentially professional popularizers. The popularization of science and mathematics in England led to an increased tendency to think in terms of Newtonian mechanics among people of all classes above the poorest, a development that contributed to the Industrial Revolution.

Science was often popularized for theological reasons, particularly during the British vogue for natural theology in the late seventeenth and early eighteenth centuries. It was believed that scientific popularization would be an effective technique for combating atheism because it could demonstrate a design in nature and thereby establish the existence of a Designer. The printed Boyle Lectures of the Reverend William Derham (1657–1735), *Physico-Theology* (1713) and its sequel *Astro-Theology* (1715), were bestsellers of the early eighteenth century, not just in Britain but, through translated versions, on the European Continent as well. Motivations for popularizing natural-philosophical explanations also included the desire to combat potentially destabilizing "enthusiastic" or superstitious understandings of natural phenomena with politically harmless scientific ones. Thus Edmond Halley's broadside illustrating the cause of a great eclipse in 1715 as the Moon's blocking the light of the Sun had the stated purpose of refuting interpretations of the eclipse as a divine condemnation of the legitimacy of the recently crowned British King George I (r. 1714–1727).

See also Books of Secrets; Demonstrations and Public Lectures; Fontenelle, Bernard Le Bouvier de; Monsters; Museums and Collections; Periodicals; Salons; Wilkins, John.

References
Jacob, Margaret. *Scientific Culture and the Making of the Industrial West.* New York: Oxford University Press, 1997.
Sutton, Geoffrey V. *Science for a Polite Society: Gender, Culture, and the Demonstration of Enlightenment.* Boulder, CO: Westview Press, 1995.

Porta, Giambattista della (1535–1615)

Giambattista della Porta was the leading natural magician in the late sixteenth century. Della Porta was the son of a wealthy Neapolitan official and, as was the custom of his class, he did not attend a university. His voluminous *Natural Magic* (1558, expanded 1589) was the most influential work in the field. Della Porta wrote in Latin for a learned audience, but his works were translated into the major European languages.

Della Porta argued that superstitious practices, such as touching the flesh with a white magnet to produce love, could be explained in natural, nondemonic terms drawing on properties of objects and the universe's structure of attractions, repulsions, and correspondences. He believed that natural secrets could be actively sought through the reading of texts, the observation of artisans, and the performance of experiments, and that art could duplicate the marvels of nature or even create new ones. For instance, he gave recipes and procedures for creating wonderful and monstrous things, like cucumbers in the shapes of dragons.

Like many natural magicians, della Porta was interested in technology, particularly spectacular technology, and he perfected the camera obscura. He also published works on hydraulics and military engineering. In the 1550s, della Porta established an Academy of Secrets at his home in Naples to perform experiments and study natural secrets, but it was shut down by the Inquisition, concerned about both popular and learned magic. Despite his loyal Catholicism (he became a lay brother of the Jesuits by 1585) the Inquisition also banned his philosophical works in 1592, lifting the ban in 1598.

Della Porta's science fit easily with the interest in wonders and curiosities characteristic of Renaissance courts, and many princes and prelates of the church visited him, supported him financially, or invited him to their courts, among them the Emperor Rudolf II. Della Porta himself formed a museum of natural history, and he was a member of the Accademia dei Lincei from 1610 until his death. He also wrote on cryptography and the meanings of the structures of the human face.

See also Accademia dei Lincei; Natural Magic.
Reference
Eamon, William. *Science and the Secrets of Nature: Books of Secrets in Medieval and Early Modern Culture.* Princeton: Princeton University Press, 1994.

Preformationism
See Embryology.

Printing
The introduction of printing with movable type to Europe by Johannes Gutenburg (c. 1390–1468) in the 1450s had a profound impact on European science. Many scientists, such as Paracelsus, spoke with scorn of book learning as compared to the direct study of nature, but books, and particularly printed books, in practice remained indispensable for the circulation of knowledge. Few of the major works of the scientific revolution were bestsellers, but the ability of print to circulate and stabilize knowledge enabled science to become fully cumulative for the first time. When combined with the humanist effort to recover ancient texts, print allowed an unprecedented circulation of ancient Greek and Latin texts, both original and in translation, on natural and mathematical subjects.

Print also allowed for the circulation of images as well as words. Before print, images and maps had been even more vulnerable than words to corruption at the hands of generations of copyists. Printing had a particularly great impact in the field of cartography, which had advanced little since the ancients, and on anatomy, where the detailed illustrations of Andreas Vesalius's *Of the Fabric of the Human Body* (1543) had an even greater effect than did the text. Printing made possible the distribution of much more precise and standardized illustrations of plants, animals, rocks, and machines. In astronomy, the printing press made it relatively easy for the aspiring astronomer to acquire the body of available ancient, Arabic, and medieval stellar and planetary observations and theory. It also facilitated the dissemination of new calculations based on new astronomical systems, such as those of Nicolaus Copernicus or Tycho Brahe. Erasmus Reinhold's (1511–1553) Copernican *Prutenic Tables* (1551), Johannes Kepler's *Rudolphine Tables* (1627), and the stellar catalogues of Johannes Hevelius and John Flamsteed could all be distributed throughout Europe and to European outposts abroad. Brahe included a printing press in his observatory complex at Uraniborg and owned a papermill to keep it supplied.

Printing played a role in the rise of Protestant Europe and France as scientific powers in the seventeenth century. Although no European country had what would now be considered freedom of the press, the most thorough and efficient censorships in Europe were Catholic. Censorship in Spain virtually removed that country from the mainstream of European intellectual development. The famous Index of Forbidden Books, set up in Rome in the mid-sixteenth century as part of the Counter-Reformation, was originally directed at Protestant and other heretical works. However, it quickly laid claim over all areas of printing, including science, banning the publication of Copernicus's *On the*

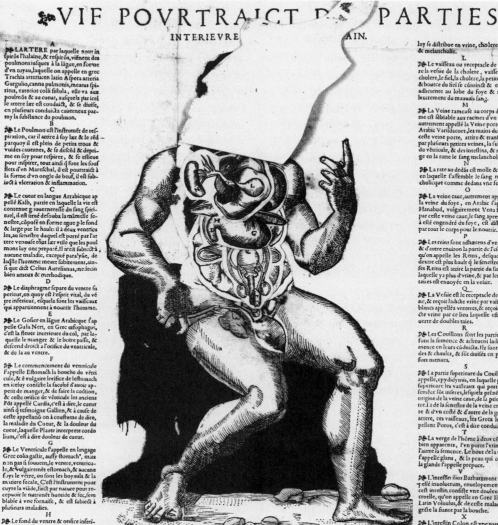

This anatomical broadsheet had a flap that could be lifted to reveal the internal organs. (Wellcome Library)

Revolutions of the Celestial Spheres (1543) "until corrected" in 1616. The work of Copernicus, a Catholic, was banned for its content, but in theory all Protestant works on any subject, including the scientific, were banned.

The establishment of the Index accompanied a movement away from the major centers of printing in Italy, particularly Venice, to those outside the control of the church, either in Protestant countries or in France, whose Catholic Church preserved a great deal of independence from Rome. Major Copernican works, whether by Catholics or Protestants, issued from Protestant presses. Galileo's later works issued not from Italian presses, but from those of the Dutch city of Amsterdam, whose dynamic capitalist economy, skilled printers, and relatively weak Protestant state church made it Europe's greatest seventeenth-century printing center.

In England, the Royal Society possessed the right to license books for printing and sponsored scientific publishing, including works of Catholic scientists such as Marcello Malpighi. The official printer to the Royal Society, John Martyn (d. 1679), specialized in scientific publishing. Although the Royal Society was involved in printing Isaac Newton's *Mathematical Principles of Natural Philosophy* (1687), it was published not by Martyn but by Joseph Streater, best known as England's first large-scale pornographer. (Other works Streater published included *The School of Venus* and *Sodom, or the Quintessence of Debauchery.)*

The growth of science affected printing, both in the increased volume of scientific works published and in the creation of a new genre of publishing, the printed scientific periodical, of which the most influential was the Royal Society's *Philosophical Transactions.* In the course of the scientific revolution, printing became more important to a scientist's career. Although some leading scientists shunned print, those who did often had less impact on the scientific community than their talents warranted. The French mathematician Pierre de Fermat and the English mathematician and astronomer Thomas Harriot are examples. Harriot began to study the heavens with a telescope at the same time as Galileo, but the Florentine's willingness to print his observations associated his name, not Harriot's, with telescopic astronomy forever. By the end of the seventeenth century, it was possible for a scientist to derive a significant portion of his income from printed works. Since publication on topics not devotional or domestic was often considered disgraceful for women, printing also supported the male domination of science.

Science carried on below the level of the major thinkers, observers, and experimenters of the scientific revolution also benefited from print. Among the most popular and profitable printed genres of early modern Europe was the almanac, a table of celestial events for the coming year that frequently included astrological prognostications. Books of secrets were also widely distributed, and medical charlatans took to print to advertise their wares.

See also Illustration; Libraries; Periodicals.

References

Eisenstein, Elizabeth. *The Printing Press as an Agent of Change: Communications and Cultural Transformations in Early Modern Europe.* 2 vols. Cambridge: Cambridge University Press, 1979.

Johns, Adrian. *The Nature of the Book: Print and Knowledge in the Making.* Chicago: University of Chicago Press, 1998.

Priority

The scientific revolution was marked both by "priority" disputes—fierce disputes over who was the first to make an original scientific discovery, theory, or observation—and by changes in the concept of priority itself. One of the earliest disputes over an intellectual claim was that between the Italian mathematicians Niccolò Tartaglia (c. 1499–1557) and Girolamo Cardano. Tartaglia accused Cardano of stealing his technique for solving equations of the form $x^3 + ax = b$, which he had told Cardano in confidence. This was not exactly a priority dispute, as Cardano did not claim to have originated the formula himself but claimed to have heard of it from a source independent of Tartaglia and thus to be justified in treating it as common knowledge. For a professional mathematician like Tartaglia, this was the equivalent of theft, in that the formulas that he and only he knew were the source of his income. Natural philosophers shared in the Renaissance tendency to treat fame and glory as legitimate motivations for action, and subsequent disputes were over the fame and reputation that come with making new discoveries as well as assertions of property in the discoveries themselves.

The new philosophy was intertwined with questions of priority from the beginning. The very model of knowledge as progressive, rather than as the steady recovery and adaptation of ancient knowledge, meant that the glory of new discovery was always potentially in dispute. The printing press, which made it possible to get ideas in circulation much more quickly, also made priority more of an issue. Priority disputes were usually bitter and personal; few were merely about the question of which independent investigator had first happened on a discovery. They usually revolved around accusations of plagiarism and intellectual theft.

Tycho Brahe had a bitter dispute with Nicholas Reimers (1551–1600), known as Ursus, over the credit for inventing the Tychonic system. Reimers, a brilliant but unstable and unscrupulous man of peasant origin, copied some of Tycho's writings surreptitiously when visiting his observatory and then tried to pass the system off as his own. Tycho, with the rage of a great aristocrat insulted by a peasant (he referred to Reimers as a "shrewmouse" and hoped that he would be hanged), succeeded in removing Reimers from the position as mathematician to the Holy Roman Emperor. He himself took the position and procured the destruction of Reimers's books. Johannes Kepler, originally an admirer or at least a flatterer of Reimers, supported Tycho. His *A Defence of Tycho against Ursus,* left incomplete and not published until the nineteenth century, was one of the first works of the history of science. Kepler's contemporary Galileo Galilei, a shrewd self-publicizer passionately concerned with his own reputation, was also passionate about asserting his own priority in discoveries. He announced his initial discovery of the phases of Venus in a Latin anagram until he could confirm his observations and publish the full story.

One way of attacking a discoverer's reputation was to assert that not he but another had made the discovery and therefore had priority. William Harvey was frequently asserted to have stolen the credit for originating the idea of the circulation of the blood, and the Englishman John Wallis, partly inspired by national pride, accused René Descartes of having stolen the idea for analytic geometry from another Englishman, Thomas Harriot. Ironically, Wallis himself had a bad reputation as a plagiarist.

Some of the nastiest priority disputes of the scientific revolution involved Robert Hooke and Isaac Newton. Hooke was particularly prone to priority disputes due to the fertility of his ideas, which led him to put them forth and never really follow up on them. He would then claim priority when the ideas were developed by others. This poisoned his relations with Henry Oldenburg, whose interest as an "intelligencer" was to communicate new scientific ideas, not keep them the preserve of their originator. Hooke's quarrel with Christiaan Huygens in 1675 over the credit for the invention of the spring balance for clocks and watches involved not only honor but potential earnings, as a patent was at stake. Hooke also accused Nicolaus Steno of plagiarism, and Steno may indeed have been influenced by Hooke when he put forth his theory of geographical strata. But Hooke's great quarrel, which shaped English science for years, was that with Isaac Newton over gravity. Both Hooke and Newton were men from obscure backgrounds whose social ascent in large part depended on their reputations for "ingenuity." (Things were easier for a great aristocrat like their contemporary Robert Boyle, who received credit for Boyle's law without originating it. But even Boyle often rushed his work into print to forestall plagiarists.) Hooke claimed to have originated the idea of gravitational attraction as the inverse square of the distance between two objects, and he disdained Newton's (true) claim to have come up with the idea independently. Newton viewed this as Hooke's attempt to seize credit for the idea without doing the hard mathematical work. As a result of this feud, Newton partially withdrew from the Royal Society, only reentering on Hooke's death in 1703.

The greatest priority dispute of the entire scientific revolution was that between Newton and Gottfried Wilhelm Leibniz over the discovery of the calculus. The dispute shows Newton at his worst; he pursued Leibniz, who had discovered the calculus independently of Newton's earlier work, with unrelenting vindictiveness. Newton did not display himself openly. He worked through front men like the mathematician John Keill (1671–1721), who accused Leibniz of plagiarism, and the clergyman Samuel Clarke (1675–1729), who defended Newtonianism in the Leibniz-Clarke correspondence. Newton used his institutional position as president of the Royal Society against Leibniz, stacking the society committee that investigated the matter. Their report was published in 1712, and in 1715 the society's journal, *Philosophical Transactions,* devoted the January-February issue to another attack on Leibniz. Newton's use of the society contrasts with the efforts of Hooke, who had never been able to get the society to support his claims fully. (By the end of the scientific revolution, institutional bodies generally had a larger role in certifying priority and originality.) The feud over the calculus had a disastrous impact on British mathematics, making Newton's version of the calculus a matter of national pride and delaying for a century British acceptance of the superior Leibnizian calculus, already used on the Continent.

Struggles to be first with a dramatic discovery, priority disputes, and accusations of plagiarism continue to be integral parts of the culture of Western science, as can be seen in such modern cases as the discovery of DNA or of the human immunodeficiency virus that causes AIDS.

See also Brahe, Tycho; Cardano, Girolamo; Clitoris; Gravity; Hooke, Robert; Leibniz, Gottfried Wilhelm; Newton, Isaac.
References
Boorstin, Daniel. *The Discoverers: A History of Man's Search to Know His World and Himself.* New York: Random House, 1983.
Rosen, Edward. *Three Imperial Mathematicians: Kepler Trapped between Tycho Brahe and Ursus.* New York: Abaris Books, 1986.

Probability

The emergence of probability theory in the scientific revolution was a consequence of bigger changes, notably the abandonment of the Aristotelian idea that the goal of inquiry is certainty. Specifically, mathematical probability emerged from the culture of upper-class gambling, extremely popular during the period. There was an early false start by the inveterate gambler Girolamo Cardano and some work on dice by Galileo Galilei, but the first extended considerations of mathematical probability occurred in a letter of Blaise Pascal to Pierre de Fermat regarding a problem concerning the frequency with which double sixes would emerge from the throw of two dice in a given number of trials. Another gambling problem, discussed by Pascal and Fermat in 1654, was the "problem of points," concerning the division of the stakes in a dice game that had ended suddenly and unexpectedly. Fermat dealt with the problem by aggregating the possible outcomes; Pascal, by multiplying the probabilities of each outcome by the associated winnings.

Pascal's early work on probability was expanded by Christiaan Huygens. Huygens corresponded with Fermat, who posed probability problems for him. Huygens's *Treatise on Reasoning in Games of Chance* (1657) includes his solutions to Fermat's problems. Huygens expanded the study of games to situations involving more than one player, and he discussed the problem of the "gambler's ruin," the possibility of a gambler winning the entire stake in a given multiplayer game. In 1662, Pascal's fellow Jansenists, Antoine

Arnauld (1612–1694) and Pierre Nicole (1625–1695), published a logic textbook titled *Logic, or the Art of Thought*. Later well-known and influential, the textbook was widely translated as the *Port Royal Logic,* after the community of Port Royal, where the authors lived. Their program included an attempt to use mathematical notions of probability to construct a system of logical inference—part of a movement to broaden probability theory beyond gambling.

Legal authorities became interested in the possibilities of probability theory for the settlement of complex cases. The application of probability to legal questions was pioneered by Gottfried Wilhelm Leibniz. Probability was also applied to actuarial statistics by political arithmeticians and by the burgeoning late-seventeenth-century annuity industry. Probability was a tool not merely for understanding random events like the results of a throw of the dice, but for estimating the degree of belief that should be given to assertions the veracity of which is unknown. This offered a new model of knowledge, one more immune from skeptical attack than the Aristotelian (or Cartesian) quest for certainty had been. This form of probabilism was especially popular in England, where it became virtually the official epistemology of the Royal Society.

The next major contribution after Huygens to the mathematics of probability was made by Jakob Bernoulli (1655–1705). Bernoulli corresponded with Huygens on probability theory and solved the problems Huygens had posed to stimulate further work in *Treatise on Reasoning in Games of Chance*. Bernoulli's *Art of Conjecture,* published eight years after his death and intended as a sequel to *Logic, or the Art of Thought,* extended probability from the consideration of single trials to problems involving the distribution of outcomes over a large number of trials. Bernoulli's "law of large numbers," which he referred to as the "golden theorem," showed that the distribution of the results of a large number of trials would continually approach the ratio of probabilities. Thus, given a sufficiently large number of flips of an honest coin, the ratio of heads to tails will approach 1:1.

Bernoulli was most responsible for lifting probability theory from an avocation of mathematicians and scientists whose principal interests lay elsewhere, such as Fermat and Huygens, to an issue of central mathematical concern. Building on Bernoulli's work, mathematicians such as Abraham de Moivre (1667–1754) would make probability theory one of the most active branches of mathematics in the eighteenth century.

See also Bernoulli Family; Fermat, Pierre de; Huygens, Christiaan; Pascal, Blaise.

Reference
Grattan-Guiness, Ivor. *The Norton History of the Mathematical Sciences: The Rainbow of Mathematics.* New York: W. W. Norton, 1998.
Hacking, Ian. *The Emergence of Probability: A Philosophical Study of Early Ideas about Probability, Induction, and Statistical Inference.* Cambridge: Cambridge University Press, 1975.

Prodigies

Strange and aberrant events outside the usual order of nature, such as comets, monstrous births, apparitions of armies in the sky, or rains of blood, have often been considered to bear meaning for humans. During the scientific revolution, prodigies in learned culture moved from the religious realm to the realm of natural philosophy. The idea that marvels and wonders can illuminate the workings of nature was an old one. Francis Bacon called for the compilation of accounts of prodigies to illuminate the regular order of nature. Prodigies were seen as evidence of the infinite fecundity and variety of nature, or alternatively, of nature's exhaustion and decay.

But the dominant way of looking at prodigies in the sixteenth century was as signs of God's will. This idea had both Christian and classical roots. Classical historians spoke of prodigies foretelling great events, such as the assassination of Julius Caesar. Christian tradition treated prodigies as divine signs from God, the most famous example being the star of Bethlehem. Many believed that an increase in prodigies will herald the Second Coming of Christ. Prodigy stories were used by all sides in the political and religious controversies of early modern Europe, such as the Protestant Reformation and the English Civil War. For example, during the English Civil War, the Royalist supporters of King Charles I (r. 1625–1649) told prodigy stories about the beheading of the king in 1649. They associated his death with monstrous births and unusual tides, and claimed that the king had bled twice as much as an ordinary man. Parliamentarians pointed to prodigies to argue that God was punishing his enemies; for example, mothers who wished that they might have children with no heads rather than Roundheads, the insulting nickname for Parliamentarians, would be cursed with headless babies.

Several natural philosophers, including Pierre Gassendi, wrote against the idea that prodigies are divine signs. The most thorough attack on providential interpretation came from John Spencer (1630–1693), a Cambridge professor and later friend of Isaac Newton. Spencer's *Discourse Concerning Prodigies* (1663) was inspired by widely circulated recent prodigy stories, such as that of an invasion of mice, that were presented as signs of God's displeasure with the English government. Spencer's *Discourse* worked on three levels of argument. Politically, he argued that belief in providential prodigies fosters political unrest. Religiously, Spencer argued that most prodigies are trivial and unworthy of the divine majesty, mocking as ludicrous the idea of looking for the "Jewel" of divine wisdom on the "dunghill of obscene and monstrous births, apparitions of lying spirits, strange voices in the air, mighty winds, alterations in the face of heaven, &c" (Spencer 1663, 10). The magnificent regularities of nature are better evidence of God's glory than are prodigies.

Scientifically, Spencer argued that the proper people to evaluate prodigies are natural philosophers. But these natural philosophers must eschew the magical worldview that sees the world in terms of similitudes, and they must be centrally organized, like the recently founded Royal Society.

It is to be wisht that there was a kind of Philosophy office; wherein all such unusual occurences were registered; not in such fabulous and antick circumstances wherein they stand recorded in the writers of Natural Magick (designing nothing but wonder in their Readers) nor with a superstitious observation of any such dreadfull events which such relations are usually stain'd, in the writers which intend a service to religion in them: But in such faithfull noticies of their severall circumstances, as might assist the understanding to make a true judgement of their Natures and Occasions. (Spencer 1663, 104)

Spencer's *Discourse* became a standard work on the subject for natural philosophers in the English-speaking world, although some found Spencer's willingness to deny providence any role too radical. In French-speaking culture, Pierre Bayle's (1647–1706) *Thoughts on the Comet* (1682) made similar arguments.

Natural philosophers, growing more suspicious of the wonder that prodigies often provoked, continued to present themselves as the people best qualified to interpret them. Edmond Halley, discoverer of the periodicity of Halley's comet, published a broadside before a total eclipse of the Sun in 1715. The broadside included a diagram illustrating how the eclipse was produced: the Sun's light was being blocked by the Moon. The caption said that the picture and explanation would prevent people from looking on the eclipse as a bad omen for King George I (r. 1714–1727), whose Hanoverian dynasty had just succeeded to the British throne.

See also Comets; Monsters.

References

Burns, William E. "'Our Lot Is Fallen into an Age of Wonders': John Spencer and the Controversy over Prodigies in the Early Restoration." *Albion* (1995): 237–252.

Daston, Lorraine, and Katharine Park. *Wonders and the Order of Nature, 1150–1750.* New York: Zone Books, 1998.

Spencer, John. *Discourse Concerning Prodigies.* Cambridge: Cambridge, 1663.

Ptolemaicism

See Astronomy; Planetary Spheres and Orbits.

Puritanism and Science

One of the most controversial issues in the historiography of the scientific revolution is the relationship between English Puritanism and science. One widely held hypothesis was originated by the American sociologist Robert Merton (b. 1910), who analyzed the religious positions of prominent seventeenth-century English scientists. The "Merton thesis" claims that Puritanism, an extreme form of Protestantism, was characterized by a reforming drive, engagement with the world, and emphasis on education, all of which played a central role in the flowering of English science in the seventeenth century. There are problems with this argument, particularly in that it is often impossible to identify the precise religious allegiances of particular figures, as nearly all English Protestants before the Civil War were regarded as members of the Established Church of England, and not everyone's individual religious position is well documented. It is also difficult to rigorously define Puritanism, particularly as Puritans themselves did not use the term, which was invented by their enemies. The most significant figures in English science in the first half of the seventeenth century, Francis Bacon, William Gilbert, and William Harvey, were definitely not Puritans, and many Puritan leaders either took no interest in natural philosophy or regarded it with suspicion as a distraction from religious issues.

Nor were identifiable Puritans disproportionately active in the scientific movement in late sixteenth and early seventeenth century England.

The key moment for the connection between Puritanism and science is the mid-seventeenth century. The Civil War in the 1640s pitted a largely Puritan Parliamentary party against a religiously conservative monarchist party, the latter of which supported the established order of the Church of England. The victorious Puritans ruled England in the 1650s, the time of the domination of Oliver Cromwell (1599–1658). This was a time of great innovation and many projects, as the destruction of the monarchy, the House of Lords, and the Established Church gave people in England and elsewhere a great sense of possibility as well as encouraging millenarian hopes, voiced by the Hartlib circle, among others.

The ruling Puritans opposed the mostly Aristotelian and religiously conservative English universities that had supported the Royalists during the Civil War. They promoted projects for educational reform and broadened the educational base by teaching natural philosophy, medicine, astrology, and magic at the universities on a non-Aristotelian basis. This period also saw the spread of Baconian ideology in England, closely identified with Puritan movements for educational reform and an emphasis on technological advance that, it was hoped, would summon the millennial restoration of the control of nature exercised by Adam.

The science promoted by the Puritan regime emphasized pragmatic and technological application over abstract theorizing. However, Puritanism did not possess a unified scientific ideology, whether Baconian or millenarian, and different tendencies within the Puritan movement took different scientific positions. Some radicals, such as the communist "Digger" Gerard Winstanley (c. 1609–c. 1660), who was deeply influenced by alchemy, took a pantheistic-magical approach, identifying God with nature.

Others, usually more conservative, opposed magical reform plans, such as the plan to add courses in astrology to the university curriculum, suggesting instead plans for institutionalizing the mechanical philosophy. Some of the most vigorous promoters of scientific activity in the period, such as John Wilkins, were involved in the defense of the universities from Puritan radicals, and in fact, Puritan reforms had little impact on the university curriculum.

After the fall of the Puritan regime and the Restoration of Charles II (r. 1660–1685) in 1660 with widespread popular support, institutionalized science became identified with the critique of post-Puritan religious extremism, also known as "enthusiasm." Active natural philosophers of the previous decade, such as Wilkins and Robert Boyle, conformed to the restored Church of England, indicating their shallow commitment to Puritanism. (Wilkins even became a bishop.) They promoted the belief that the study of natural philosophy contributes to social stability by diverting people's interest from religious controversy.

See also Boyle, Robert; Hartlib Circle; Millenarianism; Religion and Science; Wilkins, John.

References

Cohen, I. Bernard, ed. *Puritanism and the Rise of Modern Science: The Merton Thesis.* New Brunswick, NJ: Rutgers University Press, 1990.

Merton, Robert K. *Science, Technology, and Society in Seventeenth-Century England.* New York: Howard Fertig, 1970.

Webster, Charles. *The Great Instauration: Science, Medicine, and Reform, 1626–1670.* New York: Holmes and Meier, 1976.

R

Race

The great age of European expansion brought with it an influx of new information about the peoples of the world, and this new knowledge inspired particularly urgent questions about the differences among various peoples and cultures. The accuracy of travelers' accounts varied widely, but these accounts, sometimes compiled into massive ethnographies, were the most important sources of information on other societies available. All this new information had to be fitted into a body of knowledge derived from the Bible and the Greek and Roman classics. Writers traced the descent of foreign peoples from Noah, or argued that the natives of America were descendants of the ten lost tribes of Israel. Negative information as well as positive affected people's thought. As more of the world became known, it became clear that the explorers had not encountered the traditional monstrous races thought to inhabit areas remote from Europe—the one-legged and headless races, among others. Thus, such phenomena lost credibility, although some continued to believe in their existence throughout the period.

The category of "race" did not have the centrality for European thought about human divisions that it attained later. The term was not in common use, and when used to refer to a people, it could just as easily refer to the social class of the European elite as to a race in the modern sense. Religion, not race, remained the most important way of categorizing the world's peoples for most early moderns. A roughly fourfold division of the world into Christians, Jews, Muslims, and "idolaters" was common.

A simpler division was twofold, between the "civil" European elite and everyone else. This division was unstable, as some distinguished non-European and non-Christian peoples dwelling in urbanized societies such as the Chinese, Japanese, and Muslims were seen as civil. The uncivil could include European peasants as well as non-Europeans, and comparisons between the natives of the Americas and the European lower classes were quite common.

Another common way of categorizing differences was the climatic theory, which held that the character of different societies is determined by their natural environments—a theory with classical roots. Thus, the darker pigments of Africans and Americans was thought to be due to their exposure to the heat of the Sun, although some objected to this theory by pointing out that children retained their parents' coloration no matter where they were born. The dominant planet and astrological sign of a given area was also believed to determine or influence the nature of its people.

Most Europeans believed in the superiority of their cultures over those they encountered elsewhere in the world, although a few, such as the French essayist Michel de Montaigne (1533–1592), used the experience of other cultures to argue against assumptions of European or Christian superiority. As Europeans established a more dominant position in the world, beliefs about the relative status of different groups reflected European practices. For instance, African slavery was often explained and legitimated by referring to Africans' non-Christian beliefs or the Biblical curse on the descendants of Ham. Partly because many Africans had already been enslaved when they fell into European-Christian hands, the question of African slavery was not intellectually as important in European culture as that of the status of the peoples of the New World.

The most important early modern intellectual debate on the subject of race was provoked by the Spanish conquests in America. The question of whether Native Americans had souls, and were therefore both human and potentially Christian, a position supported by the missionary orders of the Catholic Church, was settled in the affirmative by a papal bull, *Sublimi Deus,* in 1537. Vigorous debate on whether Indians were barbarians in the Aristotelian sense, and therefore were naturally slaves, was carried on in Scholastic terms in the Spanish universities during the sixteenth century, where the justification of the Spanish conquests remained highly controversial. Both sides produced voluminous treatises, one side stigmatizing the differences between Native American and European society, the other arguing that since the Aztecs and Incas had lived in urbanized and political societies, they were not barbarous, merely idolaters in need of the Christian revelation. Champions of the "non-barbarous" interpretation, who located the Spanish right to rule in the need to bring Christianity to pagans, included the Dominican Bartolomé de Las Casas (1474–1556) and the Jesuit José de Acosta, both of whom had direct experience of the New World. The greatest champion of the "natural slave" interpretation was the humanist Aristotelian Juan Ginés de Sepúlveda (1490–c. 1573), who had not been to the Americas. The great debate between Las Casas and Sepúlveda staged at Valladolid in 1550 was inconclusive, but the opponents of natural slavery won the debate in the Spanish intellectual world as a whole. However, their influence on actual Spanish practice in the New World, which remained extremely harsh, was slight.

Schemes for classification of human races, along the lines of classifications of plants and animals, emerged relatively late in the scientific revolution. One of the first was by the French traveler and Gassendist Francois Bernier (1620–1688), whose *New Division of the World among the Different Species or Races of Men That Inhabit It* (1684) was one of the earliest works to use "race" in the modern sense. Bernier divided humanity into a small number of groups based on skin color, physiognomy, and areas of habitation. "Whites" included Native Americans, Middle Easterners, and Indians as well as Europeans, as the darkness of these peoples was thought to be caused by exposure to the Sun. Africans were "Blacks," as their darkness was innate. Chinese and Japanese were another race, as were the Laplanders. The full development of biological theories of the innate and immutable characteristics of different human groups, called "scientific racism," would occur in the eighteenth and nineteenth centuries.

See also Acosta, José de; Exploration, Discovery, and Colonization.

References

Hannaford, Ivan. *Race: The History of an Idea in the West.* Washington, DC: Woodrow Wilson Center Press, 1996.

Pagden, Anthony. *The Fall of Natural Man: The American Indian and the Origins of Comparative Ethnology.* Cambridge: Cambridge University Press, 1982.

Ramism

Pierre de La Ramée (1515–1572), better known by the Latin form of his name, Petrus Ramus, was a French Platonist and mathematician who put forth a logical method that he hoped would replace Aristotle's as the basis for pedagogy at both the school and university level. His early attacks on Aristotle aroused great hostility among university professors, and he was ineffectively forbidden to teach philosophy by royal decree in 1544. Despite the professors' hatred, Ramus was a popular teacher whose Aristotle-bashing went over well with undergraduates. Ramus became a Calvinist Protestant, despite his friendliness with Catholic cardinals who had protected him from the wrath of the Aristotelian university professors. He was killed in the great massacre of Protestants in Paris in August 1572, the Saint Bartholomew's Day massacre. His death helped make his system popular among Protestants, many of whom were suspicious of Aristotle and the traditional university curriculum as "pagan" and Catholic. In the closing decades of the sixteenth century, universities throughout Protestant Europe were invaded by dogmatic Ramists, certain that the Ramist "method" offered a surefire way of attaining truth in all fields. The obnoxiousness of the Ramists was compounded by the fact that Ramist logic emphasized argument, naturally appealing to argumentative people.

Ramus claimed that logic should follow the geometric method of Euclid rather than the syllogistic method of Aristotle. This method applied a simple set of logical relations (usually unacknowledged borrowings from Aristotle) to all knowledge, and indeed all reality. These relations were often represented in the form of a diagram, a device Ramus did not originate but did much to popularize. The Ramist method aroused opposition not only from university Aristotelians, but from humanists who found it simplistic and intellectually arrogant. But it was useful in arranging textbooks, and non-Ramist textbooks in many fields, including natural philosophy, adopted Ramist principles of organization by charts and diagrams.

Ramus did not replace Aristotle as the supreme logician, and Ramism as a doctrine as opposed to a method had been beaten back in the universities by the beginning of the seventeenth century. It did retain some popularity in the intellectual backwater of Puritan New England. Although the new anti-Aristotelian doctrines of the seventeenth century, Baconianism and Cartesianism, were not directly based on Ramism, Ramus did arouse a greater interest in logical method.

See also Aristotelianism; Education; Universities.
Reference
Ong, Walter J. *Ramus: Method and the Decay of Dialogue; from the Art of Discourse to the Art of Reason.* Cambridge: Harvard University Press, 1958.

Ray, John (1627–1705)

The most influential natural historian of seventeenth-century Britain, John Ray was also a creator of the English tradition of natural theology. The son of an Essex blacksmith, Ray entered Trinity College, Cambridge, in 1646, receiving a B.A. in 1648 and an M.A. in 1651. He held various posts at Trinity until 1662, when the government of Charles II (r. 1660–1685) required an oath from college officials that Ray, who held Puritan opinions, was unwilling to swear. At Cambridge, Ray acquired his interest in natural history and formed an alliance with the gentleman and amateur of natural history Francis Willughby (1635–1672), who became his collaborator and patron. Ray's first published work was a catalog of the plants of Cambridge and the surrounding country in 1660, and botany absorbed the bulk of his labors. He and Willughby planned an exhaustive and systematic natural history of all existing living things, Ray to handle the plants and Willughby the animals.

After his resignation from Trinity, Ray spent three years on the European Continent, studying anatomy, botany, and zoology, after which he returned to England. He was

Ray's Botanical Method *(1682) was one of the most systematic treatments of botanical classification during the scientific revolution. (National Library of Medicine)*

admitted as a fellow of the Royal Society in 1667 and published an exhaustive catalog of English plants in 1670, along with a catalog of English proverbs. Willughby's sudden death ended their partnership, but Willughby left Ray an annuity in his will, and Willughby's mother, Cassandra, continued to be Ray's patron until her death in 1675.

Ray published catalogs of birds and fishes under Willughby's name, although much of the work that went into them was Ray's. Ray's *Botanical Method* (Methodus Plantarum) (1682) set forth a scheme for arranging knowledge of plants by species, linked by various natural characteristics. *History of Plants* (Historia Plantarum), published in three volumes from 1686 to 1704, attempted to describe all plants known to European science at the time. Ray also published books on "serpents and quadrupeds" and on insects, which like his botanical works were packed with startling observations and precise descriptions. In addition to providing the most important work on the concept of species during the scientific revolution, Ray

was the most influential classifier of species before the work of Carl Linnaeus (1707–1778) established the modern system of biological classification in the eighteenth century. Although he believed in the fixity of species, Ray concluded that some had become extinct, with some lost species preserved as fossils.

Ray was a devout Christian, who despite his Puritanical background remained a member of the Church of England. He was offered positions in the church several times, turning them down, as he did an offer to succeed Henry Oldenburg as secretary to the Royal Society on Oldenburg's death in 1677. Ray integrated modern geology with the biblical narrative of the creation, flooding, and destruction of the world in *Miscellaneous Discourses Concerning the Dissolution and Changes of the World* (1692), reprinted and expanded the next year as *Three Physico-Theological Discourses*. His other work on science and religion was the extremely popular and frequently reprinted *The Wisdom of God Manifested in the Works of Creation* (1691). This work inaugurated a British tradition of biologically based natural theology that examined the adaptation of living species to their environment. This tradition culminated in Darwinism, although Ray saw God rather than natural selection as the ultimate designer.

> *See also* Botany; Natural History; Natural Theology; Royal Society.
> *Reference*
> Raven, Charles. *John Ray, Naturalist: His Life and Works.* 2d ed., with an introduction by S. M. Walters. Cambridge and New York: Cambridge University Press, 1986.

Religion and Science

During the scientific revolution, different forms of religion and science interacted in complex ways that cannot be simply reduced to conflict or cooperation. At a time when religion was central to cultural and intellectual life, many important natural philosophers with strong religious commitments, such as Isaac Newton, saw their work as primarily religious, giving glory to God through exploring his creation. Religion inspired scientists to challenge traditional wisdom, as Paracelsus challenged medical tradition in the name of Christianity. In other cases, developments in science, notably the mechanical philosophy, could be exploited as religiously beneficial or combated as religiously dangerous.

The scientific revolution emerged in a Europe divided by the conflicts of the Protestant and Catholic Reformations. Reasserting itself after the Council of Trent in the mid-sixteenth century, Catholicism became much less tolerant of diversity of opinion in many areas, not just science, than it had been before the Reformation. The Catholic Counter-Reformation set up a number of controls over intellectual life, notably an elaborate system applying censorship and granting permission to publish. This system culminated in the Index of Forbidden Books, established in 1557—a list of books that Roman Catholics were forbidden to own or read. Although it was not consistently enforced, the Index was eventually expanded to include all works by Protestants, whatever their subject, including books of natural philosophy. However, religious works were the primary target of the Index.

Before its controversy with Galileo Galilei, the Catholic Church tended to focus its attacks on magicians such as Giordano Bruno, burned for heresy in 1600, or Tommaso Campanella, imprisoned and tortured, rather than natural philosophers. The church was not opposed to natural science; indeed, the papacy promoted science through the University of Rome and the patronage of the papal court. The popes were particularly involved in astronomy because the calendar and timekeeping in general were important to religious issues, such as calculating the date of Easter. The Gregorian Calendar in use today was instituted by Pope Gregory XIII (pope, 1572–1585) (and many Protestant countries delayed its adoption as an expression of their anti-Catholicism, retaining instead the less accurate Julian Calendar).

The idea that the Catholic Church was opposed to science really began with the trial of Galileo in 1633. The trial became the supreme symbol of conflict between religion and science, and it was used by Protestants and anticlericals to argue that Catholicism was opposed to science specifically and intellectual development in general. This criticism was ironic since the biblical literalism eventually used to condemn Galileo was more characteristic of Protestantism than Catholicism. Although the validity of the church's judgment condemning heliocentrism outside Italy was questionable, Galileo's trial and his condemnation to house arrest for life did cast a chill over scientific activity in the Catholic world, and by the late seventeenth century, Galileo was looked on as a martyr to freedom of thought against papal despotism in Protestant countries.

René Descartes, a Catholic who spent much of his career in the Protestant Dutch Republic, despite extreme cultural and religious alienation, was quite worried about the fate of Galileo. He wrote to a friend that the judgment had made him consider burning his papers. Descartes was safe from the church in the Netherlands, and he probably would have been safe had he gone back to France. However, despite his cordial relations with many Protestants, he was a loyal Catholic who did not want to publish anything against the church or anything that could be perceived as giving aid and comfort to its enemies. He suppressed a book he was on the verge of publishing because it was Copernican, and he largely abandoned natural philosophy in favor of metaphysics. The work of the priest Pierre Gassendi, another important French natural philosopher, was also religiously problematic as materialist and atomist. Gassendi lessened his risk but also his readership by publishing in Latin rather than French.

The leading Catholic country in science in the late seventeenth and eighteenth centuries was France, which did not have an Inquisition and whose Catholic Church retained great independence from the papacy. Copernicanism was accepted quite early in France and was supported by many churchmen who wanted to preserve some independence from Rome. But even outside France, the post-Galileo Catholic world was far from scientifically sterile, and no effort was ever made to suppress science per se. Church authorities were always much more concerned with religious heresy than with scientific error. Much Catholic science was carried on by the Society of Jesus, a new religious order founded in 1540 by Ignatius of Loyola (1495–1556) and designed as an elite group whose mission was to combat sin and heresy. Although Aristotelian, the Jesuits were less bound to intellectual tradition than some of the older orders.

The mechanical philosophy was developed in a religious direction by Marin Mersenne, a natural philosopher and correspondent of Descartes and Gassendi. Mersenne, a priest of the Franciscan Minims, a particularly ascetic order, feared magic and popular superstition but not natural philosophy. He saw magic, and particularly the belief that nature is alive, as the root of all heresy, and he was particularly horrified by the heresies of Bruno. Mersenne seized upon the idea that nature is not alive, but mechanical—that it acts in a mechanical way at all times, except when God acts directly on it. In this view, God is not the soul of the world, but its master or governor. (The Protestant Robert Boyle made a similar argument for similar reasons in his 1686 *A Free Enquiry into the Vulgarly Receiv'd Doctrine of Nature.*)

Mersenne was not opposed to Galileo or Copernicanism, and he played an important role in circulating Galileo's works in France. The mechanical philosophy enabled a clear distinction between nature in its ordinary course and miraculous actions. Catholics were particularly eager to allow for the occurrence of miracles because the dominant school of thought in the Protestant world held that miracles had ceased to occur in the time since the Apostles. Catholics believed that miracles were continuing, and that the persistence of miracles in the Catholic Church was strong evidence that Catholicism was true. This made it necessary to have a rigorous definition of a miracle, and much of the church bureaucracy for evaluating miracles was put into place at this time.

However, there was a terrible trap in the mechanical philosophy from the Catholic point of view. It closed the doors on magicians, but it opened the floodgates to Protestants. The problem was transubstantiation. Catholic doctrine held that the bread and wine of the Mass are transmuted into the body and blood of Jesus Christ, which Protestants denied. Transubstantiation had been defined in the Middle Ages through the Aristotelian distinction between substance and accidents, the substance of the bread and wine being changed to flesh and blood while the accidents remain the same. Mechanical physics, not acknowledging this distinction, was incompatible with transubstantiation. This issue was far more important in the Catholic world than heliocentrism, and Protestant polemicists did use natural-philosophical arguments to attack transubstantiation. Those in the Catholic world who supported Epicurean or Gassendist atomism were always running a particularly great risk.

Protestant propagandists, then as now, linked the Protestant challenges to Catholic religious authority with challenges to natural philosophical authority, such as Aristotelianism. However, in practice things were more complicated. Some Protestant areas, such as Lutheran Germany or Scotland in the early seventeenth century, were intellectually conservative and formed an alliance with Scholastic Aristotelianism that historians call Protestant Scholasticism. Other areas, such as England and Holland, were more progressive. By the late seventeenth century, there was also some participation in science from the frontiers of the Protestant world in the

English and Dutch colonies of North America and the Caribbean.

There were few Protestant countries in which the established church was as powerful as the Catholic Church was in most Catholic countries. Censorship, for example, existed on a much more ad hoc basis in Protestant countries. They had no equivalent of the Index, and although governments exerted some controls, there was relative freedom from censorship. This was particularly true in the Dutch Republic, where the combination of a weak state church, a commercial orientation, and a thriving printing industry made Amsterdam the major center for publishing things that couldn't be published elsewhere, including many works of natural philosophy. Some of Galileo's works were published for the first time in Amsterdam. As a general rule, it was much easier to obtain Catholic works, including works of natural philosophy, in Protestant environments than vice-versa.

Protestants also had different problems with the mechanical philosophy than did Catholics. Many Protestant intellectuals viewed the mechanical philosophy as useful in combating superstition but thought that such consistent materialism could be seen as a disputation of divine action in the world. This, they feared, might lead to "atheism," which in the seventeenth century was defined less as a lack of belief in God than as a lack of belief in God's actions in the universe. The belief that God is indifferent to human beings and never bothers to interfere in the workings of the cosmos, which were mechanical, was called atheist.

Descartes himself got into trouble in the Netherlands with conservative Calvinist-Aristotelian authorities because of the supposed atheism of his mechanical philosophy. Baruch Spinoza, a pantheist and determinist of Jewish heritage, was also called an atheist; he had his troubles with Jewish authorities as well as Protestant and was expelled from the Jewish community. Spinoza and Spinozism, identified with freedom of thought and even outright opposition to Christianity, became great bugbears among Protestants. It was difficult, however, to find anyone who identified him or herself as Spinozist, partly because Spinozists could lose their jobs and be thrown in jail. Some Spinozist writings were too religiously dangerous even to print in the Netherlands, and they survived only in manuscript form. For example, the anonymous work called *Book of the Three Impostors,* which argued that Moses, Jesus, and Mohammed were all frauds, was not printed.

The Cambridge Platonists and others in Restoration England, fearing materialism and atheism, moved away from the pure mechanism identified with Descartes and Thomas Hobbes. Some asserted that there are spiritual forces at work in the universe as well as matter in motion, and others invoked a "plastic spirit" intermediary between God and the universe, which shapes things. This helped set the stage for Newton's theory of gravity.

See also Bible; Boyle Lectures; God; Jesuits; Jewish Culture; Natural Theology; Newtonianism; Papacy; Pascal, Blaise; Spinoza, Baruch; Trial of Galileo.

References
Brooke, John Hedley. *Science and Religion: Some Historical Perspectives.* Cambridge: Cambridge University Press, 1991.
Hooykaas, Reijer. *Religion and the Rise of Modern Science.* Edinburgh: Scottish Academic, 1972.
Westfall, Richard S. *Science and Religion in Seventeenth-Century England.* New Haven, CT: Yale University Press, 1958.

Renaudot, Theóphraste (c. 1584–1653)

Theóphraste Renaudot had a finger in every pie in the Paris of Cardinal Richelieu (1585–1642). A physician trained at the University of Montpellier, Renaudot was appointed Intendant of the Poor in 1618. He ran pawnshops, a free medical clinic, and an Office of Address—an agency of employment, housing, and information. He also operated the first French newspaper, the *Gazette,* a mouthpiece for his patron, Richelieu.

Renaudot held conferences at the warehouse of the Office of Address every Monday afternoon from August 22, 1633, to September 1, 1642. These conferences were open and averaged about 100 attendees, including both nobles and bourgeoisie, and many physicians who worked at Renaudot's clinic participated. (There is no evidence that women attended, although they were not formally barred.) Each conference was devoted to a particular topic. Religion and politics were excluded, and many conferences were devoted to medical or natural-philosophical topics, ranging from the existence of unicorns to the thought of Paracelsus. A few conferences were devoted to such innovations as the work of Nicolaus Copernicus on the motion of the Earth or that of Galileo Galilei on sunspots. (Unfortunately for Renaudot, the Copernican conference was published shortly before the condemnation of Galileo, and Renaudot had to quickly publish an abjuration and promise the conferences would not discuss the question again.) Conference speakers argued from a wide variety of intellectual positions, from Aristotelian to Paracelsian, although many were not exclusively committed to a particular school and were willing to use a variety of concepts. Renaudot published both weekly proceedings and massive collections of 100 conferences each. These went through a number of editions and repackagings, including a two-volume English translation published in 1664 and 1665.

Renaudot promoted a scheme for diagnosing and treating patients by mail. He also proposed the establishment of state-supported medical laboratories, arousing the ire of the physicians of the University of Paris. His plans, partially inspired by Bacon, had to be abandoned because of the civil discord that wracked France in the late 1640s.

See also Baconianism.

Reference

Solomon, Howard M. *Public Welfare, Science, and Propaganda in Seventeenth-Century France: The Innovation of Théophraste Renaudot.* Princeton: Princeton University Press, 1972.

Rhetoric

Rhetoric, the art and science of persuasive speech, was a vastly more important and active area of knowledge in the early modern period than it is now. The relation of rhetoric to science was ambivalent. Rhetoric was central to humanist education, and most early modern scientists would have been exposed to it. Some, like Robert Boyle, acquired substantial technical knowledge of rhetorical terms and categories or skillfully employed rhetorical devices in their scientific writings. The use of metaphor and analogy in scientific argument had more roots in rhetoric than in the rival tradition of logic. The move from the Aristotelian and Cartesian quest for absolute certainty to the claim that scientific assertions are valid when they carry a very high degree of probability also brought science closer to the rhetorical than the logical tradition. But early modern scientists also proclaimed their rejection of rhetoric as "mere words."

Two of the most gifted rhetoricians of the scientific revolution lived at the same time in the early seventeenth century—Francis Bacon and Galileo Galilei. Galileo's *Dialogue on the Two Chief Systems of the World* (1632) employed the form of the dialogue, often associated with rhetorical display. This work also used many devices of epideictic rhetoric, the rhetoric of praise and blame, to praise the Copernican and blame the Ptolemaic system. Bacon, whose work was rich in rhetorical devices of all sorts, was also suspicious of the power of words to persuade, or even worse, their power to perpetuate unending contention. Calling for the close study of things rather than of words, he set forth a program that was at least implicitly antirhetorical.

The scientific academies of the late seventeenth century were aware of the importance of rhetoric and literary style. In France, the matter of science and the matter of language were the portfolios of separate academies, the Royal Academy of Science and the French Academy. But Britain had no official body dedicated to the oversight of language, and the Royal Society filled this position by default, setting up a committee for language reform in 1664. When leading members of the Royal Society talked about language and language reform, they often adopted an antirhetorical stance—contrasting "Rhetorical Flourishes" with sober scientific discourse restricted to the recounting of facts. But the committee produced few real results, and in practice, members did not avoid the use of rhetoric. The Royal Society employed a skilled writer and rhetorician with little interest in science, Thomas Sprat (1635–1713), rather than one of its own natural philosophers to write the most prominent book representing it. Sprat's denunciation of rhetoric in favor of a plain and unadorned style in his *History of the Royal Society* (1667) was itself a rhetorical device, designed to encourage the reader's confidence in the writer's veracity.

See also Literature.

References
Slawinski, Maurice. "Rhetoric and
    Science/Rhetoric of Science/Rhetoric as
    Science." In Stephen Pumphrey, Paolo L. Rossi,
    and Maurice Slawinski, eds., *Science, Culture,
    and Popular Belief in Renaissance Europe.*
    Manchester, England: Manchester University
    Press, 1991: 71–99.
Vickers, Brian, ed. *English Science, Bacon to
    Newton.* New York: Cambridge University
    Press, 1987.

Rosicrucianism

The Rosicrucian movement, aimed at uniting spiritual alchemy and religious and social reform, emerged in early-seventeenth-century Germany and was associated with three manifestos. The first, written anonymously in German, was *The Discovery of the Fraternity of the Most Noble Order of the Rosy Cross* (Fama Fraternitatis), which was first printed along with related writings in 1614 but circulated in manuscript earlier. The second, also anonymous, was *The Confession of the Laudable Fraternity of the Most Honorable Order of the Rosy Cross, Written to All the Learned of Europe* (Confessio Fraternitatis), published in 1615. These manifestos told the story of a mysterious brotherhood of healers and spiritual adepts founded by the legendary fifteenth-century German knight Christian Rosenkreutz (or "Rosycross"). The third manifesto, *The Chemical Wedding of Christian Rosenkreutz* (1616), was an alchemical romance. It also was published anonymously but is known to be the work of the Lutheran theologian Johann Valentin Andreae (1586–1654), who was probably also the author of *The Discovery.* Andreae emphasized those elements of Rosicrucianism that were based on religious renewal and later turned against magical Rosicrucianism. Andreae's subsequent *Christianopolis* (1619) described an ideal community devoted to the worship of God and the acquisition of natural knowledge and was an influence on Francis Bacon's *New Atlantis* (1627). This early Rosicrucianism was associated with Protestant politics in the early stages of the Thirty Years War (1618–1648).

Rosicrucian discourse was incredibly eclectic, drawing from alchemy, astrology, Paracelsianism, Hermeticism, Christian mysticism, millenarianism, the lore of chivalric orders, and the Kabbalah. The Rosicrucian order was itself a myth, although many readers of the Rosicrucian tracts believed that such an order existed; the English magician Robert Fludd, who used Rosicrucian language in his writings, was even irked that the order had not contacted him. The imaginary order attracted the enmity of the arch antimagician Marin Mersenne, and, briefly, the interest of René Descartes, who on his return to Paris from Germany in 1623 was accused of being a Rosicrucian. Rosicrucian language and

This seventeenth-century engraving of the mythical Temple of the Rosy Cross teems with Rosicrucian imagery, emphasizing the importance of the spiritual to the Rosicrucian worldview. (North Wind Picture Archive)

imagery was adopted by many individuals and groups in the seventeenth century and strongly influenced Freemasonry.

See also Alchemy; Hermeticism; Kabbalah; Magic; Millenarianism; Paracelsianism.

References

Akerman, Susanna. *Rose Cross over the Baltic: The Spread of Rosicrucianism in Northern Europe.* Leiden, the Netherlands: Brill, 1998.

Yates, Frances. *The Rosicrucian Enlightenment.* London: Routledge and Kegan Paul, 1972.

Royal Academy of Sciences

The Academie Royale des Sciences, or Royal Academy of Sciences, was founded in 1666 by King Louis XIV of France (r. 1643–1715) with the enthusiastic support of his principal minister, Jean-Baptiste Colbert (1619–1683). The opening meeting took place in the King's Library on December 22, 1666, and was commemorated by a medal with an image of the king's head on one side and the goddess Minerva, patroness of wisdom, on the other. The academy's subsequent career was to be balanced between the services of these two masters.

The Royal Academy of Sciences was part of a project to center the intellectual life of France on a series of Parisian institutions closely identified with the king and serving to spread his glory. As an extension of the power of a king who valued harmony, the academy should not be a place of conflict, and dogmatic adherents of one or another school, whether Cartesian or Aristotelian, were excluded. Foreign scholars and even Protestants were not excluded, however, and the star of the academy in its early days was the Dutch Protestant Christiaan Huygens. Another foreigner, the Italian Gian Domenico Cassini, was the head of the Paris Observatory.

Physically, the academy was built around the King's Library, the Royal Garden, and the Paris Observatory, which was built to serve the academy's needs. The library included a dissecting room and some apartments, where Huygens and some other academicians lived. Members met on Wednesdays and Saturdays at the King's Library, with the Saturday meetings for natural philosophy and the Wednesday meetings for mathematics, although the same people attended both meetings. The academy was a small group in the seventeenth century, with a total membership of 34 at the most. Students of academicians were admitted to these meetings, but they were closed to the public. Since not all members resided in Paris, meetings were quite small, ranging up to two dozen. Minutes of the meetings remained the academy's property and were not published, although reports of the academy's work were published in the *Journal des scavans* and elsewhere. In its early days, the academy's publications were identified with the academy itself rather than with individual scientists. If an academician published something under his own name not previously approved by the academy, he was expected not to refer to himself as an academician in the work. However, as the academy developed, it gave more recognition to individual contributions and emphasized corporate identity less.

Although the academy was in part inspired by the desire to emulate and outdo the Royal Society across the English Channel, it was a fundamentally different sort of institution. For the most part, it was composed of working scientists rather than a mixed group that included dilettantes, and it was funded by the king, rather than by the dues of its members. In fact, members of the academy were paid salaries by the king, although these salaries were not enough to live on and usually supplemented members' income from other sources. The salaried academicians provided a model for the professional scientist that countered that of the independent virtuoso.

With access to the Royal Treasury, controlled by Colbert, the academy launched an ambitious and expensive program of expanding natural knowledge, financing expeditions to the Canary Islands and also to the Island of Hven, where members pinpointed the exact location of Tycho Brahe's Uraniborg, which was needed for astronomical purposes. Although the academy included no Jesuits, it made an agreement with Jesuit missionaries in the Far East to send back scientific information, some of which was published by the academy. The academy had an ambitious publishing program, producing elaborate folio volumes such as the impressive *Memoirs of the Natural History of Animals* (1671) and *Memoirs of the Natural History of Plants* (1676). The academy also established a position of international leadership in the dissection and

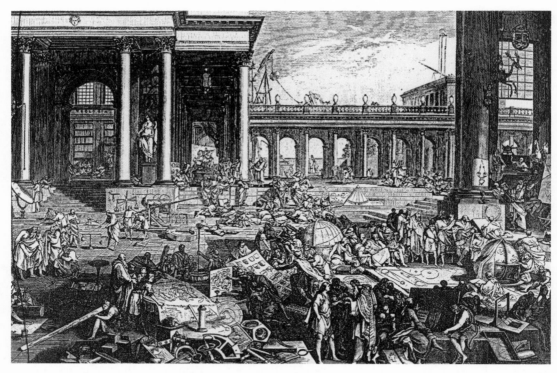

This engraving of the Royal Academy of Sciences shows a busy workplace, with construction, mapmaking, astronomy, and many other activities. (Ronan Picture Library and Royal Astronomical Society)

analysis of animals and plants, and it was expected to serve the utilitarian needs of the French state, working on navigation problems, the creation of an accurate map of France, and plans for the water supply of the king's new palace at Versailles.

The academy had to cut back its operations in the 1670s due to the fiscal demands of the wars. It also suffered after Colbert's death in 1683, when he was succeeded as ministerial sponsor of the academy by the Marquis de Louvois, Francois-Michel Le Tellier (1641–1691). Although Louvois was a powerful figure, he did not share Colbert's interest in science, and the academy suffered from cuts in funding, the abandonment of expeditions and publishing projects, and a shrinking membership. The academy also lost Protestant talent due to Louis XIV's increasingly repressive religious policies, which caused Huygens to return to the Netherlands in 1681, a serious blow. Louvois's death transferred the supervision of the academy to Louis Phelypeaux de Ponchartrain (1643–

1727), under whose sponsorship the academy underwent a modest revival.

In 1699, the academy was reorganized under the supervision of Ponchartrain's nephew the Abbe Bignon (1662–1743), who in 1691 became the first person formally recognized as president of the academy. Its internal organization, until then somewhat loose, was now laid out in great detail. This institutionalization had the effect of giving the academy an existence somewhat more independent of the crown. The revitalized academy would go on to become Europe's leading scientific institution in the eighteenth century.

See also Academies and Scientific Societies; Cassini, Gian Domenico; Fontenelle, Bernard Le Bouvier de; Huygens, Christiaan; National Differences in Science; Paris Observatory; Royal Society.

References

Hahn, Roger. *The Anatomy of a Scientific Institution: The Paris Academy of Sciences, 1666–1803.* Berkeley, Los Angeles, and London: University of California Press, 1971.

Stroup, Alice. *A Company of Scientists: Botany, Patronage, and Community at the Seventeenth-Century Parisian Royal Academy of Sciences.* Berkeley and Los Angeles: University of California Press, 1990.

Royal Society

The Royal Society of London for Improving Natural Knowledge (the Royal Society) was the first enduring public organization devoted to scientific research, internationally recognized as a leading scientific body in the late seventeenth and eighteenth centuries. In 1660, the year of the restoration of the monarchy in England, a group of natural philosophers met to establish a society for advancing natural knowledge. King Charles II (r. 1660–1685) chartered their group as the Royal Society on July 15, 1662. The society, which met at Gresham College, followed along the same lines and involved many of the same people as had the informal groups of men interested in natural philosophy that met in London and Oxford during the English Interregnum. But the society was the first attempt to institutionalize nonsectarian science, in that membership was theoretically open to all adult men, regardless of religious affiliation and even including Catholics, such as Sir Kenelm Digby. Many people who had been supporters of the Puritan regime during the 1650s, such as John Wilkins, were also included.

One reason for establishing the Royal Society was the belief that diverting people's attention from religious and political strife to scientific disputes, seen as politically harmless, would lessen the possibility of another civil war. In practice, the society was dominated religiously by members of the Church of England, and socially by men of gentlemanly status. Initially the members, or fellows, of the Royal Society included magicians such as Elias Ashmole (1617–1692) and Digby and natural philosophers representing a variety of viewpoints and disciplines, most notably Robert Boyle. It also included a number of gentlemen interested in natural philosophy

but not practicing it themselves, a group headed by Charles II and his brother James Stuart, the Duke of York and later James II (1633–1701). Balancing the needs of natural philosophers with those of amateurs was a problem for the society for many decades.

Its charter gave the Royal Society certain rights under English law, including the right to license books for publication, to correspond with foreigners, and to hold property. The society exercised this last right to acquire a museum, which became a leading tourist attraction in London, as well as a collection of instruments, an elaborate archive, and a library. The society was governed by an annually elected president and a council of 21 fellows. New members were elected by a vote of the fellows. Society meetings were initially held on Wednesday afternoons, for the convenience of members who had business affairs to attend to earlier in the day, and sociability was an important feature of the society's allure. However, not all fellows lived in London or even in England—there were a number of foreign fellows, including luminaries such as Christiaan Huygens and Antoni van Leeuwenhoek. Nor did all members who were London residents attend meetings—many became fellows simply for the social prestige.

There were few precedents for this form of scientific organization, so the society was largely in uncharted organizational territory. Unlike its French rival founded in 1666, the Royal Academy of Sciences, it was not funded by its monarch, and its strategy of admitting great aristocrats as fellows in hopes of gaining their patronage was usually unsuccessful. Meetings often involved complaints about money and efforts to get fellows to pay their subscriptions. The first decades of its history were full of plans for reform and reorganization, most of which were ineffectual. For example, in 1664, the society established standing specialist committees for different branches of natural knowledge, but these met for only a year before collapsing because of poor attendance.

Ideologically, the society was united behind the banner of Baconian empiricism, which was particularly useful in establishing the group's English character. Baconianism, with its emphasis on collaborative fact-gathering, was also less intellectually divisive than more sharply defined philosophical positions such as Cartesianism or Aristotelianism. The society extended beyond mere Baconian fact-gathering, however, proclaiming itself devoted to experimental knowledge. Meetings generally involved experiments, demonstrations, and dissections followed by discussion, as well as the reading of papers describing natural phenomena and experiments held elsewhere. Experiments performed at society meetings both demonstrated natural-philosophical principles and provided entertainment. From 1662 to 1677, Robert Hooke served as curator, a paid position with the duty of presenting experiments to the society meetings. He continued to present experiments until his death in 1703, although he was less active at the end of this period.

Another manifestation of the society's Baconianism was its emphasis on the economic usefulness of its activities. There were projects for the gathering and publication of information about various crafts and professions, the so-called histories of trades, and there were occasional drives to recruit merchants and tradesmen as fellows. The early society was sometimes involved in development projects, such as that of a Somerset gentleman who wished to abolish famine in England by spreading potato cultivation, but in the end these contributed little to English economic development.

The society was a frequent target for criticism and satire from its early days, and it was defended vigorously, with vast and somewhat inflated claims of its contribution to natural knowledge put forth in the works of fellows, such as Thomas Sprat's (1635–1713) *History of the Royal Society* (1667) and Joseph Glanvill's (1636–1680) *Plus Ultra* (1668). The society's publishing endeavors also included sponsoring the publication of several works of natural philosophy, such as Hooke's *Micrographia* (1665), and the licensing for publication of Isaac Newton's *Mathematical Principles of Natural Philosophy* in 1687 (the society was financially unable to publish the latter itself). The first successful scientific periodical, Henry Oldenburg's *Philosophical Transactions,* was independent of the society but often associated with it. Oldenburg was the society's secretary, and his elaborate network of Europe-wide correspondence was important for establishing the society's profile. His death in 1677 was followed by the temporary abandonment of *Philosophical Transactions* and a shrinking of the society's European network, and the society was in the doldrums by the 1680s. Another zealous and internationally minded intelligencer, Hans Sloane (1660–1753), became secretary in 1693, a position he was to hold for 20 years. Sloane restored much of the society's international correspondence.

The society's most dominant personality of the early eighteenth century was Isaac Newton, who was elected president in 1703 and every year thereafter until his death in 1727. Unlike many society presidents, Newton was a constant attendant at society meetings, and he restored their emphasis on experiment, which had been marginalized in favor of the reading of papers.

See also Academies and Scientific Societies; Boyle, Robert; Digby, Sir Kenelm; Gresham College; Grew, Nehemiah; Hooke, Robert; Malpighi, Marcello; Newton, Isaac; Oldenburg, Henry; *Philosophical Transactions;* Ray, John; Royal Academy of Sciences; Wilkins, John; Wren, Christopher.

References

Hall, Marie B. *Promoting Experimental Learning: Experiment and the Royal Society, 1660–1727.* Cambridge: Cambridge University Press, 1991.

Hunter, Michael. *The Royal Society and Its Fellows: The Morphology of an Early Scientific Institution.* Chalfont St. Giles, England: British Society for the History of Science, 1982.

———. *Establishing the New Science: The Experience of the Early Royal Society.* Woodbridge, England: Boydell, 1989.

Royal Touch

The belief in the power of certain individuals, notably the kings of France and England, to heal by their touch was seldom directly challenged during the scientific revolution. The belief that the French and English kings could heal scrofula, a disease caused by poor nutrition and characterized by unsightly growths, dated back to the Middle Ages, and by the seventeenth century this belief was an important part of monarchist ideology in both countries. Charles II of England (r. 1660–1685), the same monarch who first chartered the Royal Society, was a particularly avid toucher, using the touch to help reassert the divine status of the monarch after the English Revolution.

Whatever skepticism was voiced about this miracle was not that of natural philosophers, but of those Protestants reluctant to ascribe divine powers to a human being (or for that matter to the coins given out by the king to the sufferers at the touching ceremony, often kept as talismans). Some natural philosophers accepted the phenomenon as real but attempted to explain it by nonsupernatural means as a piece of natural magic—a power belonging to the family of the monarch but no more divine than the ability to heal possessed by certain jewels and herbs. Girolamo Cardano believed that the king of France actually carried out his cures by means of herbs concealed on his person.

The situation in Restoration England was complicated by the extraordinary popularity of an Irish gentleman, Valentine Greatrakes (1629–1683), who also claimed the ability to heal by his touch, attracting the support of Robert Boyle (who also believed in the royal touch) and Anne Conway, among others. The overthrow of the Stuart dynasty in the revolution of 1688 brought in King William III (r. 1689–1702), who did not practice the rite, and the subsequent accession of the German Hanoverian dynasty in 1714 marked the end of the king's touch for scrofula in England. The practice persisted in France until the French Revolution of 1789. Despite the ambivalent attitude of many early modern scientists and physicians, disbelief in the royal touch and other forms of touching for healing became widespread among Europe's educated elite in the eighteenth century.

See also Natural Magic; Politics and Science.
Reference
Bloch, Marc. *The Royal Touch: Sacred Monarchy and Scrofula in England and France.* Translated by J. E. Anderson. London: Routledge and Kegan Paul, 1989.

Rudbeck, Olof (1630–1702)

Sweden's leading contributor to the scientific revolution, Olof Rudbeck was an original scientist, an engineer, and an administrator. Rudbeck was a student at the University of Uppsala, Sweden's leading university, where his father was a professor of mathematics and theology. While at Uppsala, Rudbeck made his scientific reputation with the discovery of the lymphatic system and the description of the function of the lymph glands in 1650, publishing his discovery as *New Anatomical Study* in 1653. (He later disputed the priority of this discovery with the Danish anatomist Thomas Bartholin [1616–1680].)

On the strength of his discovery, Rudbeck received a grant from Queen Christina (r. 1632–1654) to study at the University of Leiden, where he received an M.D. He returned to Uppsala as a professor of medicine in 1655. Rather than pursue anatomy, Rudbeck devoted himself to botany, establishing Uppsala's botanical garden in 1657 and publishing a catalog with descriptions of over 1,000 plants in 1658. An influential university politician, Rudbeck was the head of the university in the 1660s, building an elaborate anatomy theater and chemical laboratory and promoting scientific and technological studies. Rudbeck was involved in bitter struggles over the teaching of Cartesian and Copernican science, which was opposed both by conservative Swedish Lutheran ministers and Aristotelian professors. Although not a Cartesian himself, Rudbeck took the side of the Cartesians, who won by royal decree in 1689 the right to teach as they saw fit as long as they did not contradict the Bible. In 1665, Rudbeck was also appointed by the government as "commissioner of the country's culture," with the job of promoting the development of Sweden's natural resources.

Rudbeck resigned his chair at Uppsala in 1691. In addition to engineering, two massive projects filled the last two decades of his life. The four-volume *Atlantica* (1679–1702) is a history of Sweden in which Rudbeck employed a great deal of ingenuity and archeological knowledge to argue that Sweden was the main source of classical culture and in fact was the Atlantis described by Plato. He also planned an illustrated book containing descriptions of all known plants. Sadly, much of his work on this project was destroyed in the great fire at Uppsala in 1702, and the two-volume work resulting from the surviving research contained only 1,811 plants. Rudbeck's son, also named Olof Rudbeck (1660–1740), followed in his footsteps at Uppsala as a botanist and zoologist.

See also Anatomy; Botany.
Reference
Eriksson, Gunnar. *The Atlantic Vision: Olaus Rudbeck and Baroque Science.* Canton, MA: Science History Publications, 1994.

Rudolf II (1552–1612)

Rudolf von Hapsburg, Holy Roman Emperor from 1576 to 1611, was the greatest royal patron of science and magic in the late sixteenth century. The eccentric emperor's capital at Prague attracted leading alchemists and magicians from all over Europe. Oswald Croll, Giordano Bruno, the German Paracelsian court physician Michael Maier (1568–1622), the English alchemists John Dee and Edward Kelly, the Pole Michael Sendivogius (1566–1636), and hundreds of other foreigners visited or were associated with the imperial court and, along with natives such as the imperial physician and astronomer Tadeas Hajek (1525–1600), made Prague a Euro-

pean center of magical and natural philosophical activities. Rudolf was a skilled magical practitioner himself, believed to perform alchemical experiments in a secret laboratory and sometimes charged with devil worship. Although Catholic, he was personally tolerant and hoped for religious reconciliation on the basis of the universal harmony to be found in nature, and his court was religiously mixed. He employed the two leading astronomers of his time, the Lutherans Tycho Brahe and Johannes Kepler, in succession as imperial mathematicians, although much of the service he valued them for was astrological. Kepler spent the most productive period of his life at Rudolf's court, from 1600 to 1612, although he had great difficulty collecting his salary. He published his treatise *Astronomia Nova* (1609) at Prague with a dedication to the emperor. He also named his stellar tables the *Rudolphine Tables* (1627), after Rudolf.

See also Brahe, Tycho; Courts; Dee, John; Kepler, Johannes.
Reference
Evans, R. J. W. *Rudolf II and His World: A Study in Intellectual History, 1576–1612.* Oxford: Clarendon Press, 1973.

Rumph, Georg Eberhard (1627–1702)

Georg Eberhard Rumph (or Rumphius), a German in the employ of the Dutch East India Company, wrote exhaustive works on the plants, animals, shellfish, and minerals of Indonesia. Amazingly, he accomplished much of this work while blind. The son of a German military engineer, Rumph left his home town of Hanau at the age of 18. He spent time as a mercenary soldier in Portugal and as an architect in Germany. In 1652, he departed Europe, never to return. He spent most of his colonial career on the island of Ambon in the Banda Sea, a position ideally suited to his linguistic skills.

As a natural historian, Rumph was inspired not by Baconianism but by Aristotle and the encyclopedic works of the ancient Roman

In his The Ambonese Curiosity Cabinet *(1741), Georg Eberhard Rumph described the* Pagurus reidjugan *as the most common edible crab of Ambon.*

Pliny the Elder (A.D. 23–79). He studied the flora, fauna, and minerals of Ambon, both observing them himself and consulting with the local inhabitants, for whose knowledge he had great respect. He lost his sight in 1670, and he and his manuscripts endured a number of disasters, including fire and earthquake, in addition to the problems posed by the difficulty of communications with Europe. He published little during his lifetime, mainly letters addressed to the German scientific society, the College of the Curiosities of Nature. The college had admitted him in 1681 under the name "Plinius," which pleased Rumphius, a great admirer of the ancient natural historian.

Rumph's *Ambonese Herbal* was suppressed for many years by the paranoia of the Dutch East India Company, which wanted to control access to information on Indonesia, and it was not published until 1741. The *Ambonese Curiosity Cabinet,* a work on shellfish and minerals, was published shortly after his death in 1704. It contained thousands of items, many described for the first time to a European audience. Its extensively illustrated chapter on shells is presented in the form of a curiosity cabinet, reflecting Rumph's experience as a member of a network of collectors stretching back to Europe. His works also contain a great deal of information on Indonesian customs and history. His other manuscripts, on animals, have been lost.

See also Exploration, Discovery, and
    Colonization; Merian, Anna Maria Sibylla;
    Natural History.
Reference
Beekman, E. M. Introduction to *The Ambonese*
    Curiosity Cabinet, by Georgius Eberhardus
    Rumphius, E. M. Beekman, trans. New
    Haven, CT: Yale University Press, 1999.

S

Salons

The salon became an important area for the performance and dissemination of science in the late seventeenth century. Salons originated in France in the early seventeenth century as weekly gatherings hosted by aristocratic women in special rooms dedicated to the purpose. Their original mission was the refinement of manners, speech, and literature, and science did not originally play a large role in salon culture. But by midcentury, many salon hostesses became interested in natural philosophy, particularly Cartesianism.

Cartesianism dominated the salons for several reasons. Descartes's works were written in elegant French, accessible to salon hostesses who could not read Latin. (Descartes was aware of this and chose French partly because it could be read by women. His rival Gassendi, who wrote a difficult Latin, had little impact on salon culture.) Cartesianism provided a platform from which to oppose the Latin Aristotelianism of the exclusively male French universities, which salon culture stigmatized as "pedantic." Occultism was also considered intellectually suspect and somewhat degrading in comparison with a mechanist Cartesianism. The Cartesian idea of the separation of the mind from the body appealed to women, as it meant their minds were not affected by their "inferior" female bodies.

Salon hostesses and members valued interesting and entertaining natural phenomena, such as the chameleons one salon hostess kept in a heated cage. They also valued ingenious explanations for puzzling phenomena, although it was considered rude to insist too strongly that one possessed the only correct explanation. To be welcomed in leading salons was an important qualification for aspiring Parisian natural philosophers such as Christiaan Huygens or Bernard Le Bouvier de Fontenelle, both of whom used salon contacts to build their careers. Salons also served to introduce science into the polite culture of upper-class Parisian men and women.

See also Cartesianism; Popularization of Science; Women.

Reference
Sutton, Geoffrey V. *Science for a Polite Society: Gender, Culture, and the Demonstration of Enlightenment.* Boulder, CO: Westview Press, 1995.

Scholasticism

See Aristotelianism; Medieval Science.

Sexual Difference

The scientific revolution inherited age-old Western traditions of thought about the relationship between male and female. Ancient and medieval medicine and natural philosophy, emerging in a male-dominated society, had elaborated theories of sexual difference and female subordination that continued to be accepted into the seventeenth century. The new medicine and natural philosophy of the scientific revolution dismantled some of these theories, although they made little practical difference in the gendered society of early modern Europe.

All agreed that the physical differences between males and females and their respective roles in reproduction were capable of scientific and medical explanation. Early modern theories of sexual difference covered a wide intellectual ground. The basic difference was between those models that emphasized the similarities of the sexes, and those that emphasized their differences. The similarity school, drawing on Aristotelian and Galenic thought, believed women to be incomplete men, whose organs had not been fully pushed out due to insufficient heat. This "one-sex" school emphasized the degree to which men's and women's organs were analogous—the male nipples to the female breasts, for instance. Thus the vagina was thought to be an inside-out penis, and the ovaries were often referred to as female testes. (The increased attention given the clitoris in the late sixteenth century complicated this picture, as it competed with the vagina for the title of "female penis.") Biological processes as well as organs were seen as common to male and female, with menstruation being the equivalent of hemorrhoidal bleeding. The similarity school believed in the possibility of spontaneous sex-changes caused by a sudden access of heat. Sex changes were always from female to male, as nature always aims at perfection. Some alleged cases of sexual transformation, such as the late-sixteenth-century French girl-turned-boy Marie Germain, received extensive publicity in the early modern period. Another consequence of one-sex thinking was the emphasis placed on the female orgasm in conception. Since a man must climax to beget a child, it was thought that a woman must also climax to conceive one.

The difference school, which had less ancient textual authority, conceived of men and women as radically different and complementary, at least in their sexual and reproductive roles. Difference thinkers minimized the importance of structural similarities and denied the possibility of spontaneous sex-changes. Those who underwent sex changes were thought to be merely hermaphrodites whose male organs had been concealed, or possibly women with enlarged clitorises, or simply frauds. The "two-sex" school also made a religious argument, claiming that since God created woman, and God created all things perfect, women are equally as perfect as men. Although some two-sex thinkers also argued that women should receive more equal treatment in society, this was not a necessary corollary, any more than the perfection of animals required that they should be treated as equal to humans. Beginning in the late sixteenth century, the two-sex thinkers were an increasingly insistent presence in European anatomy.

The similarity approach was linked to an Aristotelian hierarchical cosmology. Women, exhibiting the qualities of coldness and moistness as opposed to the heat and dryness of men, were hierarchically subordinate. (Many difference thinkers also retained the categories of cold and heat to explain sexual difference, even though they were severed from the Aristotelian and Galenic context.) The legacy of Aristotelian ideas of female subordination was a powerful one; the non-Aristotelian Margaret Cavendish, while strongly asserting her right as an individual woman to philosophize about nature, accepted in many of her works the intellectual inferiority of women as a class—perhaps due to the softness of their brains.

Seventeenth-century natural philosophers displayed a wide range of attitudes on gender questions. Francis Bacon and many Baconians, including the founders of the Royal Society, argued that true science is a masculine endeavor. However, this claim of difference did not always correlate with the physical differences between the sexes; for Baconians, the science of the ancients, created by males like Aristotle, was quintessentially "feminine," being passive and weak in its relation to nature. A more masculine science takes a more active and dominant role in regard to nature.

The seventeenth-century natural philosophy most congenial to challenges to male domination was Cartesianism. This was not because Descartes himself was a feminist or even particularly interested in gender issues; he did have women intellectual friends and correspondents, but he accepted a similarity model of the sexes. The reason Cartesianism was available for feminist use was that it assumed a disconnection between the mind and the body. Descartes's dualism, coupled with the assumption that "souls have no sex," common in Christian theology, meant that advocates of women's equality could now assert that any physical differences existing between men and women are irrelevant to their spiritual and intellectual capacities.

The denigration of Aristotle also offered champions of gender equality the chance to attack the Aristotelian theories of subordination. The Cartesian Francois Poullain de La Barre (1647–1725) wrote an influential feminist tract, translated into English as *The Woman as Good as the Man* (1677). John Locke's natural philosophy also placed little emphasis on differences of gender, and he advocated the same education for upper-class girls as for upper-class boys. An anonymous Englishwoman published *An Essay in Defence of the Female Sex* (1696), which combined Cartesian and Lockean ideas to assert the intellectual equality of men and women. Of course, on the level of practice, male natural philosophers, whatever their philosophy, excluded women from scientific institutions.

The principal intellectual justification for female subordination in the early modern period remained religious. The creation of an intellectually powerful ideology of female subordination based on post-Aristotelian scientific assertions about biological differences would be the work of the eighteenth and nineteenth centuries.

See also Clitoris; Embryology; Women.
References
Laqueur, Thomas. *Making Sex: Body and Gender from the Greeks to Freud*. Cambridge: Harvard University Press, 1990.
Schiebinger, Londa. *The Mind Has No Sex? Women in the Origins of Modern Science*. Cambridge: Harvard University Press, 1989.
Schleiner, Winfried. "Early Modern Controversies about the One-Sex Model." *Renaissance Quarterly* 53 (spring, 2000): 180–191.

Skepticism

Skepticism in early modern Europe meant philosophical belief in the impossibility of certain knowledge. A thoroughgoing skepticism required that all knowledge, including scientific knowledge, be open to question. (People who doubted the truth of religion—skeptics in the modern sense—would be identified with Epicureanism rather than skepticism.) Early modern thought faced a skeptical crisis similar to the crisis of postmodernism in contemporary Western culture. Skepticism had played little role in European thought during the Middle Ages, but the rediscovery and reinterpretation of ancient skeptical texts, most importantly the writings of the third-century Greek philosopher Sextus Empiricus, led to a rebirth of skepticism in the sixteenth and seventeenth centuries. Greek manuscripts of Sextus's work began to circulate in the late fifteenth century, and in 1562 these were published in Latin translation in France. This edition was published mainly for humanist reasons, as a source of ancient thought and as a way of combating "dogmatists," the skeptical term for those who believed in certain knowledge, particularly Aristotelians.

Skepticism spread through its use in religious conflicts, as Protestants and Catholics used skeptical techniques to attack the foundations of the other side's beliefs. Catholics attacked Protestant biblicism—How could

Protestants be sure they had interpreted the Bible correctly?—and Protestants attacked Catholic reliance on papal authority—Even if the pope could be sure of the right answer to a theological question, how could any believer be sure that the person claiming to be the pope was indeed the pope? On the other hand, some argued that skepticism destroys all possibility of attaining truth through reason and that therefore the only way to know truth is through religious faith.

Skeptical arguments were also used in natural philosophy to attack the claims of certainty made by Aristotelian natural philosophers, alchemists, and astrologers, as well as the science of Copernicus and Galileo. The diversity of opinion within science made it particularly vulnerable to skeptical attack. By the early seventeenth century, a number of philosophers and scientists, particularly in France and England, were attempting to found scientific knowledge on a basis immune from skeptical attack. Francis Bacon believed that his empirical approach to natural knowledge would overcome skeptical objections, and René Descartes in his *Meditations on First Philosophy* (1641) put forth a logical structure based on the famous formula "I think, therefore I am," which he hoped would be unassailable by skeptical objections. Cartesians continued to view certainty as the goal of natural-philosophical inquiry, as did Thomas Hobbes in his universal application of the methods of Euclidean geometry.

The ultimate solution to the problem posed by the skeptical assault on the possibility of certainty was mitigated skepticism, which defined the scientific project not as a search for certainty, but as an attempt to make statements that have a very high degree of probability. In the English context, this was called "moral certainty"—a sufficiently high degree of probability that an assertion is true to justify acting as if the assertion were indeed certain. This was often allied with a view of the universe as a manifestation of God's power rather than his intellect, a view characteristic

of the tradition of Descartes's opponent Pierre Gassendi rather than Descartes. An emphasis on God's power makes the universe more arbitrary in its characteristics, with a logical structure less open to our observation than one that emphasizes God's intellect. Marin Mersenne, a contemporary of Descartes and Gassendi, suggested that rather than seeking the unknowable true nature of things, we seek the truth of appearances.

This modified skepticism, often identified with a proclaimed intellectual modesty that refused to go beyond the observed facts, was most fully developed in late-seventeenth-century England, where it was the characteristic approach of the natural philosophers of the Royal Society. English modified skepticism developed in opposition to both the dogmatic Aristotelianism characteristic of the English (and European) university curriculum and the "enthusiastic" claim to certain knowledge by direct revelation from God, which many blamed for the English Civil War of the midcentury. This modified view is put forth in writings identified with the Royal Society and its program, such as Joseph Glanvill's (1636–1680) *The Vanity of Dogmatizing* (1661). (Glanvill became a fellow of the Royal Society in 1664 on the presentation to the society of a substantially altered edition of *The Vanity of Dogmatizing* published in 1664 as *Scepsis Scientifica* with a dedication to the society.) Modified skepticism was also useful in distinguishing English natural philosophy from Cartesianism.

The skeptical crisis dissolved the quest for absolute certainty that had been characteristic of medieval Scholastic natural philosophy. Modern science, whose methodology denies the possibility of absolute certainty, emerged in part out of skepticism.

See also Descartes, René; Hobbes, Thomas; Humanism; Mersenne, Marin; Probability; Religion and Science; Royal Society.
References
Popkin, Richard H. *The History of Scepticism from Erasmus to Spinoza.* Rev. ed. Berkeley and Los Angeles: University of California Press, 1979.

Shapiro, Barbara J. *Probability and Certainty in Seventeenth-Century England: A Study of the Relationships between Natural Science, Religion, History, Law, and Literature.* Princeton: Princeton University Press, 1983.

Slide Rules

The invention of calculating instruments based on converting the multiplication of numbers into the addition of their logarithms followed soon after the invention of logarithms themselves. It was the English who first developed these instruments. Edmund Gunter (1581–1626), a Gresham College professor of astronomy and instrument designer, created the calculating device based on logarithms, combining a line of numbers with a compass to measure distances. The true slide rule with moving parts, however, was invented in both its rectangular and circular forms by the mathematician and clergyman William Oughtred (1575–1660). Oughtred's pupil, William Forster, described these instruments in *The Circles of Proportion and the Horizontal Instrument* (1632). This work was followed by a number of improvements, and the modern rectangular slide rule with a moving part between two fixed parts was described by surveyor Seth Partridge (1603–1686) in *The Description and Use of an Instrument Called the Double Scale of Proportion* (1672), although the instrument had been devised about two decades earlier. The slide rule did not come into common use in the seventeenth century.

See also Logarithms.
Reference
Wolf, A., with the cooperation of F. Dannemann and A. Armitage. *A History of Science, Technology, and Philosophy in the Sixteenth and Seventeenth Centuries.* 2d ed., prepared by Douglas McKie. London: Allen and Unwin, 1950.

Spinoza, Baruch (1632–1677)

Although not a significantly original natural philosopher, Baruch Spinoza helped create the climate in which many late-seventeenth-century scientists worked, and his own phi-

By considering human beings as part of nature, rather than concluding that humans are separate from nature because they possess of immortal souls, Baruch Spinoza has been considered a founder of scientific psychology but also shunned as an atheist. (National Library of Medicine)

losophy was deeply marked by contemporary scientific developments. Born into the prosperous Portuguese Jewish community of Amsterdam, he was educated in Amsterdam's excellent Jewish school. He followed his father into commerce, but in 1656 he was expelled from the Jewish community for his heretical opinions. This was the culmination of Spinoza's growing dissatisfaction with traditional Jewish learning and his interest in the new sciences of Christian Europe, particularly Cartesianism. He learned Latin and took up the profession of lens grinding, which led to a study of optics.

Spinoza engaged in science in the 1660s, performing experiments and commenting on Robert Boyle's discoveries in a correspondence with Henry Oldenburg. He was also acquainted with Christiaan Huygens, Nicolaus Steno (who tried to convert him to Catholicism), and Gottfried Wilhelm Leibniz. Spinoza's *Descartes's Principles of Philosophy* (1663) presented Descartes's thought with some Spinozist modifications in the form of a strictly logical and deductive system modeled on Euclidean geometry. He remained basically Cartesian as a natural philosopher. Although at times he endorsed the Baconian ideal of beginning with fact gathering in investigation, he preferred logical and systematic organization when it came to drawing conclusions and presenting ideas.

Spinoza saw all knowledge as interrelated, and thus, his theological, philosophical, and political work as part of a project shared with the natural and mathematical scientists of his time. He was a pantheist who saw God as immanent in nature. Since in this view God is nature, nothing is supernatural, including people. For his consideration of human beings as part of nature, rather than as separated from nature by virtue of their immortal souls, Spinoza has been called a founder of scientific psychology. He viewed body and soul as different aspects of a single reality, avoiding Cartesian dualism, and he was a rigid determinist who denied that human beings are possessed of free will.

Spinoza's ideas, which advocated religious liberty and attacked the supernatural origin of the Bible, horrified both the Christian and Jewish intellectual leaders of Europe. Of particular concern was his anonymously published work *Tractatus Theologico-Politicus* (1670). Spinoza was fortunate to live in the Dutch Republic, the most tolerant European state, but even so he published nothing else in his life. His bad reputation as an atheist clung to him for over a century after his death.

See also Cartesianism; God; Jewish Culture; Religion and Science.

Reference
Garrett, Don, ed. *The Cambridge Companion to Spinoza*. Cambridge: Cambridge University Press, 1996.

Spontaneous Generation

The belief that small creatures or "lower animals," such as insects or mice, can emerge spontaneously from the earth or from decaying organic matter was an old one by the time of the scientific revolution. The emergence of maggots from rotting meat was one frequently invoked example. Belief in this phenomenon dates back to the ancient world and was expressed by classical natural historians such as Aristotle and Pliny the Elder (A.D. 23–79). The phenomenon was explained in many ways. Occultists who believed in the hidden power of matter to generate life believed in spontaneous generation, but so did Aristotelians such as William Harvey and mechanical philosophers including René Descartes. Mechanical philosophers believed that living beings are only matter arranged in a different way than nonliving beings, and so the fact that living beings might spontaneously arise from nonliving matter was no more difficult to explain in principle than the sudden transformation of living into nonliving material through death.

What threatened belief in spontaneous generation in the late seventeenth century was a growing awareness, thanks to the microscope, of the complexity of very small animals. The multifaceted eye of an insect as revealed by the microscope did not seem something that could spontaneously emerge from rotting matter. Leading microscopists including Jan Swammerdam and Antoni van Leeuwenhoek opposed belief in spontaneous generation. The strongest evidence and arguments against it was provided by another microscopist, Francesco Redi (1626–1697), an Italian physician, poet, courtier, and experimentalist. Redi sealed away pieces of meat in airtight containers and then noted that they did not spontaneously produce maggots or flies. In *Experiments on the Generation of Insects* (1668), Redi claimed that insects must arise from seeds originating in parent insects.

Redi's evidence was widely accepted, but it did not completely destroy the doctrine of spontaneous generation. Even Redi himself was at a loss to explain insects' appearance in oak galls, and he speculated that the insects emerge from a perversion of the life force of the tree. (Marcello Malpighi later traced the connection between the eggs of an insect and the later appearance of both an oak gall and a mature insect.) The doctrine of spontaneous generation persisted, mainly on the level of even smaller creatures. In the eighteenth century, it was claimed that not insects but microscopic *animalcules* are spontaneously generated. This doctrine was only finally disproved in the nineteenth century.

See also Embryology.
Reference
Jacob, Francois. *The Logic of Life: A History of Heredity.* Translated by Betty E. Spillman. New York: Pantheon Books, 1973.

Starkey, George (1627–1665)

The alchemist George Starkey was the most important natural philosopher born and educated in America before Benjamin Franklin (1706–1790). Born in Bermuda and the son of a Scottish minister, Starkey was educated at Harvard, where he learned a version of Aristotelian natural philosophy that emphasized "corpuscles," the smallest particles into which matter could be divided. There he also began his alchemical studies. Although Starkey lacked a medical degree, in physician-poor New England he was able to practice medicine upon his graduation with a B.A. in 1646. Frustrated with the lack of laboratory equipment, he moved to England in 1650. There Starkey became a member of the Hartlib circle and a friend of Robert Boyle, whom he instructed in alchemy and who supported some of his experiments.

Starkey wrote a number of alchemical tracts, some published during his lifetime and others posthumously. Some were published under his own name and some under the name Eirenaeus Philalethes, or "peaceful lover of truth." (This should not be confused with Eugenius Philalethes, the pseudonym of Starkey's contemporary and fellow alchemist, Thomas Vaughan [1622–1666].) Starkey claimed that Philalethes was an alchemical

adept living in America who performed wonders, such as restoring an old woman's hair and teeth. In fact, Philalethes assumed an existence independent of his creator; Sir Kenelm Digby claimed to have met him, and he was said to be alive as late as the mid-eighteenth century.

Philalethes's (Starkey's) alchemy was principally based on that of Johannes Baptista van Helmont; it was expressed in the traditionally obscure alchemical style and sought both to make gold and to cure diseases. Starkey emphasized the use of mercury in the preparation of the philosopher's stone, in contrast to many seventeenth-century alchemists, including Vaughan, who emphasized the use of salts. Starkey was also involved in several chemically oriented business ventures in England, including the refining of precious metals and soap manufacturing.

Although he got along well with the Puritans who ruled England in the 1650s, Starkey also published monarchical tracts in an attempt to ingratiate himself with the Royalists when Charles II (r. 1660–1685) was restored to the throne in 1660. Any hopes of patronage he had were disappointed, however, and the last years of his life were spent in desperate poverty. He attempted to enrich himself by selling "Starkey's pill," which he claimed made other medicines more effective. But like many Helmontian physicians, he found himself squeezed between the Galenists of the London College of Physicians and the unscrupulous charlatans who promised miracle cures. Along with others, Starkey formed a Society of Chemical Physicians to advance true Helmontian medicine. He died while practicing medicine during the great London plague of 1665.

Starkey's corpuscular alchemy was a major influence on the alchemy of both Boyle and Isaac Newton. Newton's alchemical papers contain many references to Starkey's writings, and he was associated with a shadowy group of alchemists claiming to carry on Philalethes's work. Starkey also influenced Continental alchemists and chemists, including Georg Stahl (1660–1734), the founder of the phlogiston school of chemists. His works were printed into the eighteenth century.

See also Alchemy; Boyle, Robert; Chemistry; Helmont, Johannes Baptista van; Newton, Isaac.
Reference
Newman, William R. *Gehennical Fire: The Lives of George Starkey, an American Alchemist in the Scientific Revolution.* Cambridge: Harvard University Press, 1994.

Statistics

See Political Arithmetic.

Steno, Nicolaus (1638–1686)

Niels Stensen (Latinized to Nicolaus Steno) made fundamental contributions to anatomy, crystallography, and geology, a science he may be said to have founded. Born in Copenhagen of middle-class origin, he studied medicine at the University of Copenhagen and the University of Leiden, where he made the acquaintance of Jan Swammerdam, a lifelong friend. He received an M.D. from Leiden in 1664. The same year, he published his great work on anatomy, *On Muscles and Glands,* and traveled to Paris. Steno, an extremely skilled dissector, is associated with the description of the glandular system and the muscles, as well as the discovery of the Stensen duct, the exit of the parotid gland.

After a year in Paris, Steno continued on to Florence, where from 1666 to 1668 he resided at the court of Grand Duke Ferdinand II de' Medici (r. 1621–1670), serving as Ferdinand's physician and associating with some of the members of the Accademia del Cimento. In Florence he engaged in a number of spectacular dissections, wrote a treatise on the shark's head that pointed out the similarity of sharks' teeth to common fossils, and converted to Catholicism. Steno was recalled to Denmark in 1672 to serve as royal anatomist, a position specifically created for him, since as a non-Lutheran he could not hold the anatomy chair at Copenhagen. His stay in his native land was brief and troubled by religious hostility, and he returned to Florence in 1674 and was ordained a priest the year after. After his ordination, Steno largely abandoned science, instead serving the Catholic minority in Lutheran North Germany and Scandinavia as Vicar Apostolic. He took on the formidable task of converting to Catholicism his acquaintance from university days, Baruch Spinoza—a noteworthy sign of Steno's religious sincerity!

Steno's *Prodromus* (1669) was one of the most innovative books of the scientific revolution. Here Steno claimed, correctly, that the angles of the faces of a given crystal are constant regardless of the crystal's size or shape. This idea is the foundation of crystallography. He also described the condensation of rock crystals out of liquid solutions, something he had observed experimentally. Describing the geology of Tuscany, he was the first to analyze the succession of rock strata as a temporal sequence. He argued that each layer had condensed separately out of water. Each stratum was originally parallel to the horizon, and strata in a different positions, for example vertical, had been shifted from their original place. Strata containing things associated with the sea indicate that the sea had at some point been there, and coal and ashes indicate that there had been a volcanic eruption nearby. Steno also argued that fossils originate in once-living things. His geology did not challenge the scriptural story of the flood but incorporated it. He shunned Cartesian rationalism and built his work on careful empirical observation. The *Prodromus* was quickly translated from Latin into English by Henry Oldenburg, and it was published in London in 1671.

See also Anatomy; Fossils; Geology.
Reference
Wolf, A., with the cooperation of F. Dannemann and A. Armitage. *A History of Science, Technology, and Philosophy in the Sixteenth and Seventeenth Centuries.* 2d ed., prepared by Douglas McKie. London: Allen and Unwin, 1950.

Stevin, Simon (1548–1620)

The illegitimate son of a wealthy citizen of Bruges in the Spanish Netherlands, Simon Stevin was a leading mathematician of late-sixteenth-century Holland, best known for his invention of decimals. He enrolled in the University of Leiden in 1583, although he never took a degree, and he set forth his invention in his 1585 *The Tenth,* published in Leiden, where he lived. Stevin was a leading figure in mechanics, hydraulics, and hydrostatics, both as a practicing engineer and a theorist, the first to evaluate the force a liquid exerts on the walls and bottom of a container and to publish a theoretical treatise on mills. He believed that natural and mathematical knowledge is useful and should be accessible to a wide range of intelligent people, whatever their academic training. For this reason, he published all of his work after 1586 in Dutch, a language he believed to be ideally suited to science, and he invented much of the Dutch scientific vocabulary.

Stevin derived his income from working as both a civil and military engineer, receiving a number of patents from the province of Holland and the United Provinces for technical inventions, mostly having to do with hydraulics, always a matter of great concern in the Netherlands. His expertise was internationally known, and he journeyed to Gdansk in 1591 to advise the city government on harbor improvements. By moving from the Spanish Netherlands in the south to the United Provinces in the north, Stevin had transferred his allegiance from the king of Spain to the rebellious rule of Calvinists, although there is no sign of deep religious feeling on his part. Around 1590, he acquired the patronage of the military leader of the United Provinces, Prince Maurice of Nassau (1567–1625), in whose employ he spent the rest of his career.

Stevin tutored Maurice in mathematics and mechanics, and he sat on numerous committees evaluating matters of defense and engineering. In 1600, he helped to establish an engineering school connected with the University of Leiden, using Dutch as the language of instruction. Stevin published treatises and textbooks on navigation, fortification, optics, and astronomy, among other subjects, and collected the course of instruction he had given the prince in *Mathematical Memoirs* (1605–1608). His *De Hemelloop* (1608) was one of the first Copernican works by a Dutchman.

See also Decimals.
Reference
Dijksterhuis, E. J. *Simon Stevin: Science in the Netherlands around 1600.* The Hague: Nijhoff, 1970.

Stoicism

One aspect of the revival of ancient learning in the Renaissance humanist movement was an interest in the ancient Stoic philosophers, most notably the Greek slave Epictetus (c. 55–c. 135) and the Roman senator Seneca (c. 4 B.C.–A.D. 65), author of the influential scientific text *Natural Questions.* Stoic philosophy, to which the medieval Christian philosophers had paid little attention, had the reputation of being close to Christianity in its monotheism and its ethics, lacking the association with impiety that tarnished Epicureanism. In the early modern period, Stoicism was most influential in the areas of politics and ethics, but its natural philosophy also provided an alternative to both Aristotelianism and Epicurean atomism.

The most important Stoic physical concept was that of *pneuma,* or breath, an ethereal force compounded from fire and air. Stoics attacked atomism, identified with their great rivals the Epicureans, and accepted the four elements theory of matter. In the Stoic view, *pneumas* permeate the cosmos, their tensions and activities responsible for the organization of matter into distinct forms. In living things, most importantly human beings, the *pneuma* is identified with the soul. Different sorts of *pneuma* explain different sorts of natural phenomena, and the highest *pneuma* of all, the intellectual pneuma, permeates the entire

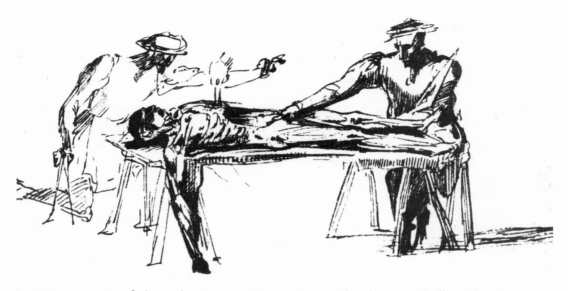

An artistic representation of a human dissection, an activity sometimes carried out by surgeons. (Bodleian Library)

universe. This omnipresent soul of the universe is God, who functions as an active and providential designer of a rationally ordered cosmos.

Stoic emphasis on divine power and providence, which had already influenced the early Christians, was also congenial to the Christian thinkers of the scientific revolution. As mediated through the philosophically eclectic ancient physician Galen, the Stoic *pneumas* would enter medical and alchemical thinking as the "vital spirits." The Stoic notion of *pneumas* also influenced the natural philosophy of such early modern nonphysicians as Francis Bacon, who believed that inert matter is organized by vital spirits.

Thanks to Stoicism's fluid cosmos, it had no use for the massive crystalline spheres containing the planets of traditional Aristotelian-Ptolemaic cosmology. Stoics also denied that absolute difference between the imperfect Earth and the perfect heavens that Aristotelians upheld. Heavenly *pneumas* were thought to be finer and more fiery than earthly ones, but not radically different. Stoics believed that the planets swim through an ethereal medium, a cosmological theory gaining increasing publicity in the late sixteenth century. Stoicism was also more compatible with ideas about the infinity of the universe than was Aristotelianism, providing ancient precedent for new cosmogonical theories such as those of the Neoplatonist university professor Francesco Patrizi (1529–1567) and the accused heretic Giordano Bruno. Stoicism may have influenced Isaac Newton in formulating the doctrine of universal gravitation. Gravitation, like the Stoic *pneuma*, shapes matter and permeates the universe. Newton's library contained most of the ancient writings relevant to Stoicism, and his writings make use of the Stoic distinction between active spirits and passive matter. The Stoic idea of a God who permeates nature also influenced the pantheism of Baruch Spinoza.

See also Aristotelianism; Bruno, Giordano; Epicureanism; Humanism; Newton, Isaac.
Reference
Osler, Margaret J., ed. *Atoms, Pneuma, and Tranquility: Epicurean and Stoic Themes in European Thought.* Cambridge: Cambridge University Press, 1991.

Surgeons and Surgery

From the mundane tasks of lancing boils and setting bones to the removal of an anal fistula

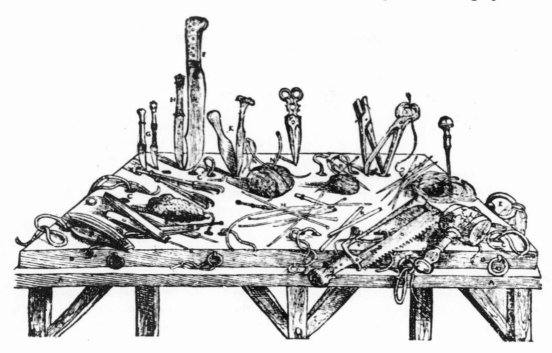

Surgery was not work for the squeamish, as can be seen from this collection of instruments. (National Library of Medicine)

from the backside of the king of France, surgeons performed a variety of medical services in early modern Europe. Unlike physicians, surgeons dealt with the body's surface, not its interior, and they were considered craftsmen, not members of a learned profession. Only in Italy were there chairs of surgery at university medical schools. Surgeons who received some university medical training were often considered the elite of the profession, but most were trained in apprenticeships or in their family's practice; indeed, there were surgical "dynasties" that kept the knowledge of profitable operations to themselves. Less learned surgeons were often lumped together with barbers or combined the two crafts. Although surgery was not revolutionized during the scientific revolution, it did see a number of innovations and a rise in the status and educational levels of its practitioners. Some classical surgical texts were published by humanist physicians and surgeons in the fifteenth century, but these had little impact on the average surgeon, who might not even be literate in Latin. The most

influential surgical literature was that produced in vernacular languages. The most influential surgeon of the sixteenth century, Ambroise Paré (1510–1590), surgeon to the king of France, had little Latin and published in French.

The brutal wars and long voyages of early modern Europe increased the demand for surgeons and led to a voluminous vernacular literature on surgery. Because of the belief that gunpowder poisons wounds, many surgeons increased the patient's agony by cauterizing bullet wounds with boiling oil. Paré denounced the use of boiling oil, describing his own successful use of a poultice made of egg white, rose oil, and turpentine. He also promoted the use of ligatures. So successful was Paré that the physicians of the Paris medical faculty denounced him for encroaching on their discipline.

Other surgical innovations in the early modern period included new techniques for removing bladder stones. By the late seventeenth century, surgery was rising in prestige, particularly in France, where the successful removal of Louis XIV's (r. 1643–1715) fistula by the surgeon C. F. Felix (1650–1703) in 1687 gained him an estate and set off a craze for "the king's operation," often demanded by those without fistulas at all.

See also Medicine.
Reference
Conrad, Lawrence I., Michael Neve, Vivian Nutton, Roy Porter, and Andrew Wear. *The Western Medical Tradition: 800 B.C. to A.D. 1800.* Cambridge: Cambridge University Press, 1995.

Swammerdam, Jan (1637–1680)

The Dutchman Jan Swammerdam was the most noted entomologist of the seventeenth century. The son of an apothecary with a noted natural history collection, Swammerdam studied medicine at the University of Leiden from 1661 to 1663 and again in 1666–1667. He spent the intervening years in France, experimenting and studying Cartesian theories. After he received his M.D. from Leiden in 1667, Swammerdam devoted himself principally to the study of insects.

Inspired by the example of Marcello Malpighi, Swammerdam was the first to systematically dissect a broad range of insects within the category of "small bloodless animals," which at the time included amphibians and spiders among other creatures. He developed incredibly fine dissecting instruments and new techniques for handling specimens. An early microscopist, Swammerdam pioneered the study of the reproductive system of insects, and his discovery of the eggs within the queen bee settled the question of whether bees are ruled by kings or queens. Swammerdam built a large and internationally known collection of insect specimens, containing over 1,000 items. His *General History of Insects* was published in 1669. Although he participated in the Collegium Privatim Amsterdamolense, a group of physicians who practiced dissections and published their results, Swammerdam did not practice medicine. His scientific career was supported by his father.

Swammerdam claimed that the purpose of his work was to reveal the glory of God through his creation, showing the intricate detail and workmanship of living things. Influenced by Cartesianism and the mechanical philosophy, Swammerdam described insects as tiny machines, claiming that they were perfect for their kind rather than being, as Aristotle had claimed, imperfect creatures lacking internal anatomy. He refuted the idea that insects and other animals emerge from spontaneous generation in decaying matter, demonstrating that they emerge from eggs laid by females. Swammerdam found beauty in the insects he dissected, speaking for example of the prettiness of the purple inside the eye of the dung fly. His religious motivations led to a crisis when he decided that his science was distracting him from God, and in the 1670s, he fell under the influence of the mystic Antoinette Bourignon (1616–1680). Eventually, however, he returned to his scientific studies.

See also Microscopes; Spontaneous Generation.
References
Ruestow, Edward G. *The Microscope in the Dutch Republic: The Shaping of Discovery.* Cambridge and New York: Cambridge University Press, 1996.
Van Berkel, Klaas, Albert Van Helden, and Lodewijk Palm, eds. *A History of Science in the Netherlands: Survey, Themes and Reference.* Leiden, the Netherlands: Brill, 1999.

Sydenham, Thomas (1624–1689)

Thomas Sydenham, from a landed family that supported Parliament against the king in the English Civil War, was an outstanding practitioner of antitheoretical and empirical medicine. Sydenham's belief in judging by results led him to question not only the teaching of Galenic, mechanical, or Paracelsian theories of disease, but the whole idea of medicine as a university subject. Sydenham advocated, not theory, but careful studies of individual cases and diseases in the Hippocratic tradition, also claiming Baconian inspiration.

Although he attended Oxford intermittently for several years, Sydenham did not take an M.D. He attacked both the textually based learning of the traditional university curriculum and the academic study and teaching of anatomy and botany. The devout Sydenham even claimed that microscopic investigation is a blasphemous and useless extension of the senses beyond the limits God has placed on them. This may be why he never joined the Royal Society, of which many of his friends were members.

Sydenham began practicing medicine in London in 1655 or 1656. A brief political career in support of his older brother Colonel William Sydenham (1615–1661) ended with the Restoration of King Charles II (r. 1660–1685) in 1660. Sydenham's first publication, *The Method of Curing Fevers* (1666), was a careful empirical study of London epidemics based on clinical evidence.

He undertook the project partly at the urging of his friend Robert Boyle. John Locke was another close friend, who learned much medicine from Sydenham and consulted him on difficult cases, and who also spread Sydenham's fame in his travels in the Dutch Republic and France. Sydenham's *Medical Observations* (1676), an expansion of *The Method of Curing Fevers,* became a standard textbook. Sydenham became a leading London physician. He discouraged the use of drugs in favor of exercise and fresh air, although he did promote the use of quinine and laudanum. Sydenham believed medicine was best taught by apprenticeship, and his several medical apprentices included Hans Sloane (1660–1753), a future president of the Royal Society.

Reference

Dewhurst, Kenneth. *Dr. Thomas Sydenham (1624–1689): His Life and Original Writings.* London: Wellcome Historical Medical Library, 1966.

T

Technology and Engineering

Technology and science had a complex relationship during the scientific revolution. The period saw little or no direct application of new science to technological problems, but advances in technology were driven by the needs of science. Science benefited from technological innovations, and the majority of leading scientists were involved in some aspect of technology and were associated with engineers and other technologists. Famous technical problems, like the determination of the longitude, occupied the time of eminent scientists like Galileo Galilei and Christiaan Huygens. Some people, such as Huygens or Robert Hooke, made major contributions to both science and technology.

Europe had been a technologically dynamic society since the Middle Ages. Before the scientific revolution, European culture and society were transformed by technological innovations like printing, gunpowder, and the magnetic compass (all pioneered in China). Technological improvements were made in many fields, such as agriculture and textiles, with little input from or effect on the scientific revolution. Technological writings from Greek and particularly Roman antiquity were read during the scientific revolution, but because of the strength of the medieval technological tradition, there was no full-fledged

classical revival in technology as there was in natural philosophy, medicine, and architecture. A variety of technical skills and industries existed in early modern Europe, with technical experts ranging from military engineers working on ballistics and fortifications, to mining engineers, to hydraulic experts working on drainage problems. The Low Countries, with all their complicated drainage problems, produced a steady supply of engineers. Another technological leader was Germany, with its advanced mining and chemicals industries. Technology was not always associated with practical use, as there was a great demand for spectacular machines that could provide entertainment at court functions; for instance, the great Dutch engineer Cornelius Drebbel (1572–1633) produced fireworks for King James I of England (r. 1603–1625), and he marched among the court entertainers at James's funeral.

Before the invention of the printing press, works on technology were difficult to reproduce and were not widely known. The brilliant innovations of Leonardo da Vinci (1452–1519), for example, had little impact because they were not published. Like science and medicine, technology benefited from the invention of printing, which allowed the broader distribution of both classical and modern technological works. The Italian

Vannoccio Biringuccio (1480–1537) published an early technological work, *Pyrotechnics* (1540), on a variety of arts employing fire, from refining and glassmaking to artillery. Georgius Agricola was another major early technical writer, writing primarily on mining and metallurgy. Technological writing continued with increasing volume throughout this period.

Science and technology grew closer from the late sixteenth century on. This happened both on an ideological level, as scientists and others proclaimed the usefulness of science in technological applications, and on a practical level, as more areas of science made greater use of the instruments introduced or radically improved in this period. Francis Bacon, with his ideal of "power over nature," is the best known champion of the application of science to human betterment through technology in this period, but similar ideas were championed by Simon Stevin and many other engineers, scientists, and philosophers throughout Europe. This kind of technological ideology, which identified technical progress with the building of a perfect world of human rule over nature, could take bizarre forms. The German apothecary and chemist Johann Rudolf Glauber (1604–1670) identified a healing salt that he recreated in his laboratory, calling it *Elias artista*, the prophet Elijah returned as a herald of the Second Coming of Christ.

Technology also posed problems for science to solve: William Gilbert's study of magnetism originated in the practical use of the compass by navigators, and Galileo was fascinated by problems concerning the strength of building materials. His theories on the subject, set forth in *Discourses on Two New Sciences* (1638), were partly inspired by visits to the famous Venetian Arsenal, where ships were built for the Venetian Republic. The arsenal was the largest single industrial enterprise in Europe.

The period saw not only the rise of the ideal of applied science, but also the belief that technology offers a valuable intellectual approach for scientists. Alchemists and natural magicians such as Giambattista della Porta exhibited a fundamentally technological approach to knowledge, being more concerned with knowing how to do things than with natural-philosophical theory. Some natural philosophers found the more pragmatic technological approach appealing; they began to focus more on understanding how a thing works and less on studying the Aristotelian universal principles.

The most famous example of the application of a new instrument to science in the seventeenth century was of course Galileo's use of the telescope in astronomy. Reliance on instruments was nothing new to astronomy, and such instruments as the astrolabe and quadrant were improved during this period. But other new instruments, such as the microscope, thermometer, barometer, and air pump, and improved instruments, such as clocks, all changed not only the methods but the content of other sciences and medicine. For example, Galileo's colleague at the University of Padua, Santorio Santorio (1561–1636) pioneered the application of new technological devices, such as the thermometer, to medicine. In these fields reliance on instruments marked a fundamental change from Aristotelian natural philosophy. Aristotelian natural philosophers made little use of instruments, and they and others had grave difficulties incorporating knowledge that could only be perceived through instruments into natural philosophy.

This was one of the issues at stake in the controversy over Galileo's use of the telescope or Thomas Hobbes's attack on Robert Boyle's air pump. But the general trend was clearly toward greater use of instruments and apparatus. In the late seventeenth century, entire scientific careers were founded on the mastery of a technology, such as Boyle's air pump or Antoni van Leeuwenhoek's microscope. The manufacture of scientific instruments became a commercial business large enough to support whole communities, producing enterprises such as the families of Italian telescope makers or the London instrument makers who catered to the Royal Society. Some scientists themselves became proficient instrument makers; both the astronomer Johannes Hevelius and Leeuwenhoek ground their own lenses.

Both of the great scientific societies founded in the second half of the seventeenth century, the English Royal Society and the French Royal Academy of Sciences, took improvement of technology as one of their goals, and they included technologists alongside natural philosophers in their membership. For example, the great French military engineers Nicolas-Francois Blondel (1618–1686) and Sébastien Vauban (1633–1707) were members of the Royal Academy, and the instrument dealer Joseph Moxon (1627–1700), author of *Mechanick Exercises* (1677), was a member of the Royal Society. The Royal Society had an ambitious but unfulfilled project for compiling the "histories of trades," which would gather the techniques and knowledges of different crafts.

But technological improvement continued to be mostly a matter of the incremental improvement and refinement of existing machines and devices rather than the direct application of scientific knowledge to technical problems. The most dramatic technological inventions of the late seventeenth and early eighteenth century the steam engines of the Englishmen Thomas Savery (c. 1650–1715) and Thomas Newcomen (1664–1729), were the culmination of a long series of efforts by different inventors rather than a result of the application of science. But the emergence of the new engines was also associated with the milieu of English science; Savery's engine was demonstrated at a meeting of the Royal Society in 1699, and Newcomen had discussed engineering problems with Hooke. The eighteenth century saw a coming together of the cultures of engineering and Newtonian science throughout Europe, particularly western Europe.

See also Air Pumps; Barometers; Clocks and Watches; Microscopes; Mining; Navigation; Telescopes; Thermometers.

References
Jardine, Lisa. *Ingenious Pursuits: Building the Scientific Revolution.* New York: Nan A. Talese, 1999.
Wolf, A., with the cooperation of F. Dannemann and A. Armitage. *A History of Science, Technology, and Philosophy in the Sixteenth and Seventeenth Centuries.* 2d ed., prepared by Douglas McKie. London: Allen and Unwin, 1950.

Telescopes

The origins of the telescope are obscure, but arrangements of lenses and mirrors appear to have been used in late-sixteenth-century England and Italy to look at distant objects. The earliest recorded telescope combining a convex and a concave lens in a tube was in the Netherlands, in the patent application of Hans Lippershey (c. 1570–c. 1619) in 1608. The principle of the telescope was so simple that much of its early spread was caused not by the physical introduction of telescopes, but by word of mouth: people heard about telescopes and constructed their own. The most notable example is Galileo Galilei, who built his own telescope after hearing about a Dutch model. Galileo's telescopes, with a magnification of 30 diameters, were the most powerful of their time, so powerful that they made his observations difficult to duplicate. The greatest technical innovator of the early telescope, however, was Johannes Kepler, who devised the so-called astronomical telescope. This instrument, described in his *Dioptrics* (1611), combined two convex lenses and obtained greater accuracy of observation at the price of turning the image upside-down. This is one of the few innovations in telescope design in this period based on optical theory rather than on craft technique. Despite its greater accuracy, the upside-down image kept it from being widely adopted until the 1640s, when telescopes for astronomy were diverging from the more common "terrestrial" models devised for spying on enemy armies or identifying other vessels at sea.

All these telescopes based on refraction presented difficulties with the obscurity of

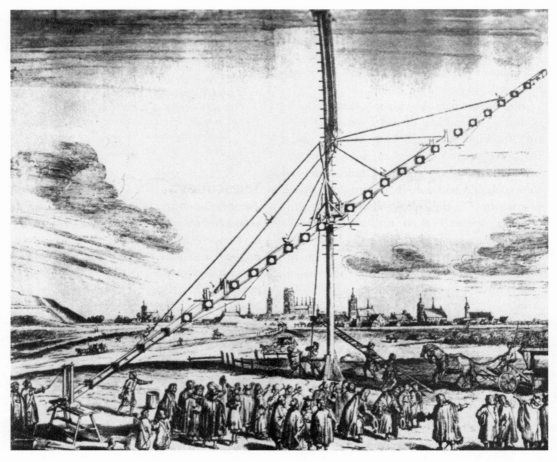

One of Johannes Hevelius's larger telescopes at Danzig. Such enormous telescopes were rare because they were so difficult to use. (Hevelius, Dissertatio de nativa Saturnifacie, *Danzig, 1656)*

images and the scattering of light. René Descartes, in his *La Dioptrique* (1637), suggested that grinding lenses with hyperbolic rather than spherical curves would solve the problem of clarity, but this was beyond the technical capacities of seventeenth-century lens grinders. Isaac Newton devised the reflecting telescope in 1668 to solve these problems, but producing effective reflecting telescopes was also beyond the technological capacities of the time, and the instrument had little impact in the seventeenth century.

The telescope did make a great impression on European culture, as it was the first radical extension of human senses. It made the hitherto neglected study of lenses a vital part of optics, and it brought home to astronomers the relative closeness of the

planets in space and their relative similarity to the Earth, in contrast to the unimaginable distance and difference of the stars. It enabled Galileo to discover the largest satellites of Jupiter and the phases of Venus, and it allowed astronomers to identify sunspots and investigate the geography of the Moon. By the late seventeenth century, good quality telescopes were widely available, the best being Italian. These telescopes could be as long as 50 feet and involved more complicated arrangements of lenses. The so-called aerial telescope even eliminated the tube connecting the lenses. In the same period, telescopic astronomy was booming in the hands of such practitioners as Gian Domenico Cassini, not least because telescopic observers in Catholic countries did not have to commit themselves on the question of the Copernican system. The invention of the micrometer and the addition of telescopic sights to standard astronomical instruments such as quadrants made possible the mapping of the heavens with unprecedented accuracy.

See also Astronomy; Cassini, Gian Domenico; Descartes, René; Galilei, Galileo; Harriot, Thomas; Kepler, Johannes; Observatories; Optics.

References

Van Helden, Albert. "The Telescope in the Seventeenth Century." *Isis* 65 (1974): 38–58.

————. *The Invention of the Telescope.* Transactions of the American Philosophical Society. 67 (special issue) (1977).

Telesio, Bernardino (1509–1588)

From the nobility of Calabria in southern Italy, Bernardino Telesio was an innovative, anti-Aristotelian natural philosopher, called by Francis Bacon "the first of the true philosophers." He was educated by his uncle, the humanist Antonio Telesio, and then at the University of Padua, from which he received a Ph.D. in 1535. He gained a strong background in Aristotle and the ancient philosophers and physicians, along with a dislike for Aristotelian natural philosophy. Following graduation, he spent several years in a Benedictine monastery.

Telesio's *On Nature According to Its Own Principles* was published in 1565, with expanded additions in 1570 and 1586. This work expounded a new natural philosophy combining two active principles, cold and heat, with passive matter. The title is a reference to the epic of the ancient Roman atomist Lucretius (c. 96–c. 55 B.C.), *On the Nature of Things,* but Telesio was not an atomist. Although he thought the heavens were hot and Earth cold, he argued against Aristotle's absolute distinction between the perfect heavens and the corruptible Earth. He also argued against Aristotle's definitions of space and time as defined by relations between bodies, in favor of absolute definitions.

Unlike many Italian anti-Aristotelians, Telesio was uninterested in magic.

Because of his poor management, Telesio's estate near his hometown of Cosenza went to ruin after the death of his wife in 1561. He spent much of the following decades in Naples or Rome. Telesio was friendly with several popes, and Pius IV (pope, 1559–1565) even offered to make him archbishop of Consenza. But after his death, his anti-Aristotelianism and the idea that the natural order he described was independent of God became suspect, and *On Nature* was put on the Index of Forbidden Books in 1593. However, Telesio's philosophy influenced subsequent natural philosophers such as Francis Bacon, William Gilbert, and Tommaso Campanella.

Reference

Kristeller, Paul Oskar. *Eight Philosophers of the Italian Renaissance.* Stanford: Stanford University Press, 1964.

Thermometers

The earliest device developed during the scientific revolution to measure temperature was the thermoscope, a water-filled tube that indicates temperature with heated air that moves the water upward. Marking off the thermoscope so that the movement of the column of water measures the heat makes it a crude thermometer. This device may have been invented by Galileo Galilei around 1600, although the evidence for Galileo's priority is not overwhelming. The modified thermoscope-thermometer was applied to medicine by Galileo's friend and colleague Santorio Santorio (1561–1636), professor of medicine at Padua, whose *Commentary on the Medical Art of Galen* (1612) contains the first printed mention of it. But the thermoscope could not be used for precise quantitative measurement because of the variability of air pressure, nor could measurements be compared without a standard temperature scale.

Various improvements were made in the thermoscope by Otto von Guericke (1602–1686) and others, but the future belonged to the fluid thermometer, which used the expansion of liquid rather than air to measure temperature. The Grand Duke Ferdinand II of Tuscany (r. 1621–1670), a founder of the Accademia del Cimento in Florence, originated a fluid thermometer using colored alcohol in a hermetically sealed tube. The so-called Florentine thermometer was the most commonly used in the late seventeenth century. There were various suggestions about standardizing thermometric measurements around a fixed point, such as Robert Boyle's plan to use the freezing point of aniseed oil, and experiments were also made with alternative liquids. Around 1714, D. G. Fahrenheit (1686–1736), a German living in Amsterdam and a fellow of the Royal Society, introduced a mercury thermometer and a standard scale based on the difference between the melting point of ice and blood temperature, the ancestor of the scale that bears his name.

> *See also* Technology and Engineering; Weather.
> *Reference*
> Middleton, W. E. Knowles. *Invention of the Meteorological Instruments.* Baltimore: The Johns Hopkins University Press, 1969.

Tides

The question of the cause and nature of the tides produced many answers during the scientific revolution, culminating in Isaac Newton's treatment in *Mathematical Principles of Natural Philosophy* (1687). The Middle Ages had combined empirically derived tide tables with several theories, usually based on the Moon's influence over the waters. For example, it was thought that the Moon might cause the waters to heat up and expand. The first major theory put forth during the scientific revolution was William Gilbert's explanation of the tides by the magnetic attraction between Earth and Moon.

The most important early theory, though, was Galileo Galilei's. Galileo wanted to use the tides to prove beyond a doubt the

Copernican motion of the Earth. Developed in 1616, his theory was not published until 1632, in his *Dialogues on the Two Chief Systems of the World.* Galileo claimed the tides are caused by the changing accelerations of the seas, which are due to the combination of the Earth's rotation with its revolution around the Sun. Essentially, the seas slosh over. Galileo was immensely proud of this theory, which unlike most tidal theories eliminated the influence of the Moon. René Descartes, by contrast, explained the tides by proposing that the Moon exerts a pressure on the Earth through the ethereal matter that fills the space between the Earth and the lunar vortex.

Most of this theorizing had been done without precise tidal data. The English Royal Society set forth the first project for gathering tidal data following a controversy over John Wallis's modification of Galileo's theory to include lunar influences in 1666. The most important of the early tide observers was the Reverend Joshua Childrey (1623–1670), whose observation of the Dorset high tides led to him to associate unusually high tides with times when the Moon is closest to the Earth. Later, John Flamsteed made tidal observations at the Royal Observatory at Greenwich, and Edmond Halley observed tides at different places in the English Channel during a voyage undertaken for the English Navy in 1701.

Newton transformed tidal theory by successfully explaining tides as the result of gravitational attractions between the Earth and the Moon and Sun. Newton did not merely explain the tides in general, but also explained specific tidal phenomena such as the cycle of spring and neap tides. He also attempted to describe tides not merely qualitatively but quantitatively. Newton was not above the use of "fudge factors" to increase the fit of his theory to reality, and Descartes's theory found supporters well into the next century. But Newton's theory, however imperfect, was clearly much better than the competition. The eighteenth century saw improvement and refinement of Newton's basic theory, mostly by French scientists.

See also Descartes, René; Galilei, Galileo; Gilbert, William; Navigation; Newton, Isaac.
Reference
Cartwright, David Edgar. *Tides: A Scientific History.* New York: Cambridge University Press, 1999.

Trial of Galileo

The trial and conviction of Galileo Galilei remains a mysterious event, and it has been interpreted in many ways by historians. Galileo's troubles began with the publication of his *Dialogue on the Two Chief Systems of the World* in February 1632. There were two dangers in this book. One is its bold statement of support for the Copernican system, banned by the Catholic Church in 1616. The other is that Pope Urban VIII (pope, 1623–1644) became convinced after the dialogue's publication, which in all probability he himself had licensed, that the dull-witted Aristotelian straw man of Galileo's *Dialogue,* Simplicio, was a satire of himself. The always hot-tempered Urban was under particular stress at this time because of his conflicting diplomatic interests in the Thirty Years War (1618–1648). The conviction that he was being mocked was the last straw, and it shattered what had been a very strong relationship between Urban as patron and Galileo as client that dated to the time before Urban's pontificate. To make things worse, this relationship was ruined at a time when Galileo had lost the support of other Roman patrons. Federico Cesi (1585–1630), founder of the Accademia dei Lincei and Galileo's patron, had died in 1630. Galileo's friend, Papal Secretary Giovanni Ciampoli (1589–1643), had lost Urban's favor and had been banished from Rome. The grand duke of Tuscany was warned by Roman authorities to avoid exerting himself in Galileo's defense. Without the support of powerful men, Galileo was exposed to the attacks of his enemies, notably the Inquisition and the Jesuits.

Urban reacted to Galileo's alleged ridicule by suppressing the *Dialogue* and establishing a commission to investigate the matter. After reading the commission's report, Urban referred the matter to the Inquisition. The Inquisition summoned Galileo to Rome in the winter of 1632–1633, a savage requirement to impose on an old man in ill health during a plague. On his arrival in February, he was imprisoned under fairly humane conditions. The Inquisition charged him with violating an injunction given him in 1616 by Cardinal Robert Bellarmine (1542–1621) not to assert or defend Copernicanism in any way. Galileo responded by claiming that Bellarmine had allowed him to discuss Copernicanism as a hypothesis. Negotiations between Galileo and the Inquisitors involving the threat of torture produced a public confession. According to legend, after publicly abjuring Copernicanism, Galileo audibly muttered, "And yet it moves!," referring to the Earth. But this is a myth. On June 22, 1633, Galileo was condemned to house arrest and the recitation of penitential psalms. He spent the first months of his arrest in Rome, and from the end of 1633 to his death in 1642 he resided at his own house outside Florence.

Galileo's condemnation had a chilling effect on science in the Catholic world, particularly Italy. Descartes abandoned plans to publish his Copernican work, *The World,* and turned from natural philosophy to metaphysics. The trial became the first piece of evidence used by Protestants to claim that Catholicism was particularly hostile to science, and has also been used to argue that science and religion are generally hostile.

See also Bible; Copernicanism; Galilei, Galileo; Jesuits; Papacy; Religion and Science.
References
Biagioli, Mario. *Galileo, Courtier: The Practice of Science in the Culture of Absolutism.* Chicago: University of Chicago Press, 1993.
Westfall, Richard S. *Essays on the Trial of Galileo.* Vatican City: Vatican Observatory, 1989.

U

Universal Languages

In practice, the universal language of early modern scientists was Latin, but its vocabulary was inadequate to handle the flood of new information and new ideas pouring into Europe. So during the seventeenth century a number of schemes were put forward for artificial languages that would directly express the structure of the universe. An artificial language would also be free of the ambiguities of existing languages. In its written form, such a language would be a "real character" language, in which signs would refer not to spoken words, but to ideas. As precedents, universal language proponents pointed to Egyptian hieroglyphics, Chinese characters, and musical and mathematical notation.

Francis Bacon hoped that a universal language would advance natural knowledge. After Bacon's, the earliest proposals for a universal language were French. Commenting on one of these projects, René Descartes argued in a letter to Marin Mersenne that a universal language would have to be based on the "simple notions" that grounded Cartesian philosophy. Mersenne himself later published a plan for a universal language. The most active movements for universal language, however, were associated with the Hartlib circle and John Wilkins in England in the mid-seventeenth century. Samuel Hartlib's associates, most notably the Czech educator John Amos Comenius (1592–1670), put forth a number of universal language schemes, mostly based on existing languages. Comenius's *The World of the Senses Pictured* (1658) combined a radically simplified Latin vocabulary with drawings of what he claimed to be all visible things.

Wilkins and his associates such as George Dalgarno (c. 1626–1687), author of *The Art of Signs* (1661), were more interested in a language that would correspond to the real universe than one that would be based on existing languages. This type of project implied a natural philosophy on which to structure the language and a systematic way of associating the words that expressed simple, or basic, notions into complex ideas. The most elaborate version was Wilkins's *Essay Towards a Real Character and Philosophical Language* (1668), which combined a universal language with a universal system of classification. This latter element would prove particularly important: Wilkins engaged John Ray to work out the classifications of plants and animals, and Ray also worked on an unpublished Latin translation of Wilkins's book. This work led to Ray's new system for the classification of plants in *Botanical Method* (1682).

Although John Wallis and Robert Hooke

learned Wilkins's language well enough to write in it, it never caught on as hoped. Despite the interest of Hooke and Gottfried Wilhelm Leibniz, universal artificial languages in general ceased to be discussed by the end of the seventeenth century. French joined Latin as a universal language of the European scientific community.

See also Ray, John; Wilkins, John.
Reference
Slaughter, M. M. *Universal Languages and Scientific Taxonomy in the Seventeenth Century.* Cambridge: Cambridge University Press, 1982.

Universities

The relationship of universities to the scientific revolution was ambivalent. Science was studied in early modern universities either as part of natural philosophy, or as mathematics, a discipline that included astronomy, or in medical schools. The vast majority of scientific innovators had university educations, and many held university positions, including such luminaries as Galileo Galilei and Isaac Newton. The sixteenth and seventeenth centuries were a time of expansion for the universities—with the foundation of new chairs, such as the Savilian Professorships of Geometry and Astronomy founded at Oxford in 1619, and the founding of new universities, such as the University of Leiden in 1575. Universities provided much of the infrastructure for scientific studies, such as botanical gardens and libraries, adding observatories and chemical laboratories by the end of the seventeenth century.

Science was also studied in various informal gatherings of virtuosi in university towns and communities unaffiliated with the institutions themselves, such as the Experimental Philosophy Club at Oxford during the 1650s, which included John Wilkins and Robert Boyle. Some students could also receive private tutorials in scientific subjects, taught by professors but bypassing the official university curriculum. Indeed, much new science was developed outside the university environment, and in opposition to the university curriculum.

The dominant school of natural philosophy in European universities between the thirteenth and the late seventeenth centuries was Aristotelianism. Thus, those innovative natural philosophers such as Paracelsus and Francis Bacon who denounced Aristotelianism also denounced the university system. University learning was text-based; the professor would read a passage and then comment on it. Most textbooks reduced Aristotelian natural philosophy to a form suitable for teaching, with a systematic and deductive presentation, although some sixteenth-century humanists called for a return to the original Greek texts of Aristotle.

By the late sixteenth century, new scientific findings were being incorporated into university textbooks, even if only to refute them, and in the following century new ideas would be incorporated into teaching, although in a disorganized way. In the late seventeenth century, Cartesianism emerged as an alternative to Aristotelianism—equally logical, deductive, and suitable for teaching, but with a better fit to modern scientific findings. Although Cartesian metaphysics was still thought dubious, Cartesian natural philosophy swept the European university world. However, this Cartesian revolution had little impact on the method of teaching science, which remained textual and deductive.

Mathematics, which included astronomy, was still subordinated to natural philosophy but was an increasingly prominent subject in early modern university curricula. Humanists had urged the study of newly recovered classical Greek mathematical texts in university curricula. Both Protestant universities in the tradition of the University of Wittenberg and Catholic Jesuit institutions emphasized mathematics as necessary for understanding astronomy and Aristotelian natural philosophy. In Protestant universities, the influence of Ramism also favored mathematics. Mathematics was also promoted for its practical importance in navigation, engineering, accounting, and other applications and for its importance in understanding new schools of natural philosophy, such as Cartesianism, that had a more quantitative basis than did the qualitative Aristotelian approach. Newly founded universities such as Leiden were particularly open to mathematics instruction based on practical uses. By contrast, the long-established French universities, particularly the University of Paris, were much more conservative.

Universities dominated medical theory, and medical school professors and graduates researched in a variety of scientific fields. University medical study was beginning to incorporate studies of human health that had been excluded from medieval curricula, such as surgery and pharmacology. The medical schools of the Italian universities, notably the Universities of Padua and Bologna, were centers of innovative medical thinking. Although medical study remained text based, dissections were practiced in some university medical schools, particularly in Italy, and the practice spread widely in the seventeenth century. Many advances in anatomy were associated with university-based physicians, particularly Andreas Vesalius and the anatomists who followed him at Padua.

Universities were teaching institutions, not research institutions, and research-oriented scientists often left universities for other types of appointments, as Galileo Galilei left Padua in 1610 to be the court mathematician of the duke of Tuscany. Universities could also be less welcoming than other institutions for those whose religion differed from that of the state, which is one reason Johannes Kepler, a Lutheran in Catholic Austria, never held a university position. The academies and societies that emerged during the seventeenth century filled the need for religiously diverse research-oriented institutions, and in the eighteenth century, the universities would decline in science as they did in many other fields.

See also Cambridge University; Collegio Romano; Oxford University; University of Leiden; University of Padua; University of Wittenberg.
References
De Ridder-Symoens, Hilde, ed. *Universities in Early Modern Europe.* Cambridge and New York: Cambridge University Press, 1996.
Gascoigne, John. "A Reappraisal of the Role of the Universities in the Scientific Revolution." In Robert S. Westman and David Lindberg, eds., *Reappraisals of the Scientific Revolution.* Cambridge: Cambridge University Press, 1990.

University of Leiden

In the seventeenth century, the University of Leiden displaced the University of Padua as Europe's premier university for science and medicine. Founded in 1575 by the Dutch rebels against the king of Spain, the university attracted leading Protestant scholars such as Charles de l'Ecluse (1526–1609), who was appointed professor of botany in 1593. Leiden acquired an anatomical theater in 1593, and work on a botanical garden, founded in 1577, began in earnest in 1594. A technologically oriented engineering school loosely attached to the university was founded in 1600, and Leiden was the first university to have its own astronomical observatory, founded in 1633. Leiden's recent origin made it less dominated by Aristotelianism than Europe's older universities were. For instance, an early mathematics professor at Leiden, Rudolph Snel van Royen (1546–1613), was a leading Ramist. He was succeeded by his son Willebrord Snel (1580–1626), who is best known for his work in geodesy and optics, being one of the discoverers of Snel's law.

The University of Leiden was particularly important given the state of scientific institutions in the Dutch Republic. Despite its wealth, the republic had less aristocratic or princely patronage to offer than did other European societies, and it lacked a significant scientific academy or society. The universities were more important for Dutch science than

HERMANNI BOERHAAVE
SERMO ACADEMICUS
DE COMPARANDO CERTO
IN PHYSICIS
LUGDUNI BATAVORUM,
Apud Petrum vander Aa, Bibliopolam.
MDCCXV.

University of Leiden professor Hermann Boerhaave was Europe's most famous medical teacher in the early eighteenth century. This is a title page from a collection of his lectures. (National Library of Medicine)

for French or English, and Leiden was the republic's leading university. In addition to educating great Dutch scientists such as Jan Swammerdam, it attracted students from all over Protestant Europe. Leiden was one of the first universities to openly teach Cartesianism; René Descartes himself matriculated there in 1630. Christiaan Huygens, who had studied at Leiden without taking a degree, recognized its unique stature by bequeathing his papers to the university. Experimental philosophy was introduced by Burchardus de Volder (1643–1709), a Leiden professor and correspondent of Gottfried Wilhelm Leibniz who had been influenced by members of the Royal Society during a trip to England in 1674. On his return, experiments were conducted at Leiden in a Theatrum Physicum.

By the early eighteenth century, Leiden scientists were abandoning Cartesianism for Newtonianism. Leiden's two scientific stars at this time were the medical professor Hermann Boerhaave (1668–1738), a student of de Volder, and the astronomy professor Willem s'Gravesande (1688–1742). Boerhaave, appointed a professor in the medical school in 1703, was the leading medical educator of his time. He also received appointments at Leiden as professor of chemistry and botany. His iatromechanical theories applied Newtonian physics to medicine, and these theories, as well as his method of teaching through case histories, were disseminated in popular textbooks and through the education of hundreds of physicians who carried Boerhaavian medicine and pedagogy throughout the university world. S'Gravesande was the leading Continental Newtonian physicist of the time. His Latin Newtonian textbook, *Mathematical Elements of Natural Philosophy* (1720–1721), was reprinted and translated into French and English.

See also Cartesianism; Huygens, Christiaan; Newtonianism.
Reference
Van Berkel, Klaas, Albert Van Helden, and Lodewijk Palm, eds. *A History of Science in the Netherlands: Survey, Themes, and Reference.* Leiden, the Netherlands: Brill, 1999.

University of Padua

The University of Padua was the most important university for science in the sixteenth and early seventeenth centuries, and its medical school was the dominant force in academic medicine. The university forbade natives of Padua from holding the top chairs, and it actively recruited outstanding scholars from throughout Europe and Italy. Andreas Vesalius, a Fleming, was professor of surgery from 1537 to 1544, founding a great academic dynasty of Padua surgeons and anatomists including Gabriele Falloppio (1523–1562) and Girolamo Fabrici, also known as Fabricius of Acquapendente (1533–1619). The first permanent anatomical theater was founded at Padua in 1594, and one of the first chairs of anatomy as distinct from surgery was founded there in 1609. Padua had one of the earliest university botanical gardens, founded in 1546 for the use of the medical school.

Outside medicine, the University of Padua was also home to some of the most innovative Aristotelian natural philosophers of the sixteenth century—from Pietro Pomponazzi (1462–1525), who denied on philosophical grounds the immortality of the soul, to Jacopo Zabarella (1533–1589), an empiricist logician and student of scientific method. Padua also attracted students from all over Europe, one illustrious alumnus being Nicolaus Copernicus, a Padua medical student from 1501 to 1503.

Padua was situated in the territory of the Venetian Republic in northern Italy, and the republic, not the Catholic Church, determined its policies. Theology was less important than at other European universities, and Galileo Galilei had no difficulties as professor of mathematics at Padua from 1592 to 1610. Unlike other Italian universities, Padua remained open to Protestant students after the Reformation, and it was also more open to Jewish medical students than were other European universities, although they were charged triple fees. The most influential Protestant Padua student was the Englishman William Harvey, who studied under Fabrici and received an M.D. in 1602. Into the seventeenth century, the University of Padua remained the most important center for the diffusion of Italian science northwards, and Padua medical students founded anatomical theaters and botanical gardens on Italian models as far north as Denmark.

> See also Galilei, Galileo; Harvey, William; Vesalius, Andreas.
> *References*
> Randall, John Herman. *The School of Padua and the Emergence of Modern Science.* Padua, Italy: Antenore, 1961.
> Woolfson, Jonathan. *Padua and the Tudors: English Students in Italy, 1485–1603.* Toronto: University of Toronto Press, 1998.

University of Wittenberg

The home university of the Protestant Reformers Martin Luther (1483–1546) and Philip Melanchthon (1497–1560), Wittenberg was the first place where Melanchthon, a humanist, tested his ideas for educational reform. Under his leadership, the University of Wittenberg provided a model for other Lutheran universities in Germany and Scandinavia such as Copenhagen and Tübingen. Melanchthon's reforms included an emphasis on natural history, astronomy, and mathematics, as these subjects would help future Lutheran ministers, the largest constituency among the students, to understand the harmony of God's creation and to perform simple astrological calculations. Melanchthon believed that astrology demonstrates God's providential care for creation. Although Luther himself was hostile to Aristotle, and Melanchthon rejected the medieval Aristotelian tradition, Melanchthon thought the natural philosophy of Aristotle and Galen useful in teaching. Melanchthon founded two mathematical chairs at Wittenberg, one in geometry and algebra and the other in astronomy, and he contributed prefaces to many mathematical and astronomical texts published at the university, in addition to writing two science textbooks himself. Melanchthon also encouraged anatomical study as a foundation for understanding the soul, and he was interested in the work of Andreas Vesalius.

Wittenberg is associated with some of the earliest responses to Copernicanism. It was the eccentric Wittenberg professor Georg Iserin (Rheticus) (1514–1574) who first urged Nicolaus Copernicus to publish *On the Revolutions of the Heavenly Spheres.* Rheticus was one of the earliest believers in Copernicanism as a true picture of the universe. However, the so-called Wittenberg interpretation of Copernicanism treated it as a means for making mathematical calculations rather than an accurate picture of the universe. One Wittenberg professor, Erasmus Reinhold (1511–1553), wrote the *Prutenic Tables* (1551), based on Copernican calculations, and another, Kaspar Peucer (1525–1602), attempted to transform the Copernican system into an Earth-centered system. The sixteenth-century Wittenberg medical faculty was also a center for botany.

> See also Copernicanism.
> *References*
> Kusukawa, Sachiko. *The Transformation of Natural Philosophy: The Case of Philip Melanchthon.* Cambridge and New York: Cambridge University Press, 1995.
> Westman, Robert S. "The Melanchthon Circle, Rheticus, and the Wittenberg Interpretation of the Copernican Theory." *Isis* 65 (1974): 165–193.

Uraniborg

See Brahe, Tycho; Observatories.

Utopias

Several early modern writers used stories of ideal societies located in far-off lands to make points about the organization and mission of natural philosophy, and above all its relationship to political power. The original early modern utopia, Sir Thomas More's (1478–1535) *Utopia* (1516), was more practically oriented than its ancient models, such as Plato's *Republic.* Natural philosophy and technology were not More's major concerns, but he did ascribe some new inventions to the

Utopians, including an artificial incubation system for eggs.

Natural philosophy and technology were central to three influential early-seventeenth-century utopian texts. Tommaso Campanella's *The City of the Sun* was written in Italian in 1602 and published in a Latin translation in 1623. The Lutheran minister and alchemist Johann Valentin Andreae's (1586–1684) *Christianopolis* (1619) was associated with the Rosicrucian tracts. Francis Bacon's *New Atlantis* was published posthumously in 1627, and then in Latin translation in 1638. Campanella's utopia has a top official called Wisdom, with a number of subordinates representing different intellectual disciplines, including an astrologer, a cosmographer, and a physician. As Campanella was careful to point out, the wise philosophers of his ideal society are enemies of Aristotle. The structure of the city itself embodies knowledge, and its rulers use natural philosophy and astrology in their decision making, for example employing astrology to determine the most propitious times for planting crops. Yet Campanella's city is not oriented toward advancing knowledge, as Andreae's and Bacon's are. Andreae's *Christianopolis* includes ideal laboratories and other facilities for gathering natural knowledge, and its inhabitants are well ahead of Europe in adopting Copernicanism. They put this knowledge to practical, industrial use.

Both Campanella and Andreae influenced the most celebrated utopia of the scientific revolution, Bacon's *New Atlantis*. Bacon's description of a hierarchical institute for the gathering and expansion of natural knowledge, Solomon's House, is the classic statement of the cooperative nature of scientific endeavor. His portrayal of natural science closely integrated with the state would later be imperfectly embodied in the Royal Society and the Royal Academy of Sciences. Some founders of the Royal Society claimed Solomon's House as an inspiration, although the society vastly differed from Bacon's authoritarian, tightly organized, and well-funded institute. Bacon and Andreae both influenced the Hartlib circle, and one member of the circle, Gabriel Plattes, produced a short utopian text, *A Description of the Famous Kingdom of Macaria* (1641), which emphasizes technology and economic development rather than theoretical natural philosophy.

Margaret Cavendish's imaginative utopia, *A Discovery of a New World, Called the Blazing World* (1666), reverses Bacon's position on the relation between science and power. *The Blazing World,* the only female-authored seventeenth-century utopian text, treats science not as a support to ruling authority, but as a danger. Cavendish's female-ruled society is populated by creatures partly human and partly animal, whose labors are divided according to their natures; bear-men are experimental philosophers, bird-men are astronomers, ape-men are chemists, and so on. Although their intellectual endeavors are presented in a mostly positive light, at the end of the story the empress, advised by the character of Cavendish herself, dissolves the societies of natural inquirers. Her reason is that their inability to agree was endangering the state by breeding factions. The danger of faction would have been very important to Cavendish, as she was on the losing Royalist side in the English Civil War. She was less optimistic than Bacon about the potential for political control of knowledge.

The great anti-utopia of the scientific revolution was Jonathan Swift's (1667–1745) Laputa, which appears in the third book of his *Gulliver's Travels* (1726). Swift was familiar with *Philosophical Transactions* and other sources of scientific ideas at the time, and he satirized them in his description of the floating island of Laputa and the Academy of Lagado in the country of Balnibarbi. The Laputans are so mad about mathematics and natural philosophy that they cut their meat into precise geometrical shapes, and tailors work (very badly) by taking the customer's height with a quadrant. The Academy of Lagado, meant to parody the Royal Society and Baconian ideology, claims it can improve human ability to the point where a palace could be built in a week or agricultural production could improve a hundredfold. Among the many laughable projects of the academicians is the scheme for reducing human excrement back to the original food. Meanwhile, the people of the country live in rags and starve, unaided by the "projects."

The fact that so many major utopian writers of the seventeenth century included discussion of the rule of science in their ideal societies shows its growing cultural importance.

See also Bacon, Francis; Campanella, Tommaso; Cavendish, Margaret, Duchess of Newcastle; Literature; Millenarianism.

References

Eliav-Feldon, Miriam. *Realistic Utopias: The Ideal Imaginary Societies of the Renaissance.* Oxford: Clarendon Press, 1982.

Eurich, Nell. *Science in Utopia: A Mighty Design.* Cambridge: Harvard University Press, 1967.

V

Vacuum

One of the classic scientific questions of the early modern period was that of the existence or nonexistence of empty space, sometimes phrased, Can "nothing" have actual existence? For Aristotle and subsequent Aristotelians, the answer was no. The Aristotelian universe is a "plenum," where every bit of space is occupied by matter, and space itself is defined by the matter it contains. Thus, nature "abhors a vacuum." The classical Greek "atomistic" philosophers, Democritus (c. 460–c. 370 B.C.) and Epicurus (341–270 B.C.), had upheld the existence of the void, but they were intellectually marginalized during the Middle Ages.

The question of the void was reintroduced to European science in the early seventeenth century through two routes. One was the revival of Epicurean atomism by Pierre Gassendi, and the other was a series of experiments and devices originating with Galileo Galilei and culminating with Robert Boyle and his air pump. Galileo had noticed that suction pumps raise water only to a height of about 30 feet, and wondered if other substances had limiting heights. Galileo's disciple Evangelista Torricelli (1608–1647) inverted a tall flask containing mercury into a bowl of mercury, and observed that the column of mercury in the flask sank to a height of around 29 inches. The space between the end of the flask and the top of the column was called the "Torricellian vacuum." Further experiments with barometers that were adapted from Torricelli's, such as the famous trials carried out by Blaise Pascal, demolished alternative explanations put forward by Aristotelians, such as the claim that the Torricellian vacuum was actually an air bubble. This demonstrated that the column itself was upheld by air pressure on the fluid in the bowl, not by nature's abhorrence of a vacuum.

Spaces seemingly emptied of all matter were also produced by air pumps, first invented by the German brewer Otto von Guericke (1602–1686) and then identified with Boyle's highly publicized experiments. The vacuum of the air pump became known in England as the *vacuum Boylianum*. Boyle defined a vacuum not as an existing nothing but as a given space deprived of matter, circumventing the logical objections to the existence of vacuum.

René Descartes had denied the possibility of the vacuum, and later seventeenth-century Cartesians dealt with vacuums by arguing that they are in reality filled with "subtle matter." However, the belief in the existence of vacuums was incorporated into the Newtonian physics that eventually displaced Cartesianism.

See also Air Pumps; Atomism; Barometers; Boyle, Robert.
Reference
Westfall, Richard S. The Construction of Modern Science: Mechanisms and Mechanics. New York: Wiley, 1971.

Vesalius, Andreas (1514–1564)

Andreas Vesalius was the most notable anatomist in sixteenth-century Europe. Originally from Flanders, where his father was an apothecary to the Holy Roman Emperor Charles V (r. 1519–1556), he studied medicine at the University of Paris, then the center of a Galenic revival, and the University of Louvain. In Paris, Vesalius had contributed to a Latin edition of Galen, editing some of Galen's treatises, and although he has often been portrayed as an anti-Galenist, he had a deep knowledge of Galen. In 1537, Vesalius arrived at the University of Padua, where he was quickly given an M.D. There he accepted a newly created position as lecturer in surgery and anatomy.

In 1538, Vesalius published a set of six elaborate anatomical charts, basically following Galenic anatomy, and in 1543 he published his masterpiece, the lavishly illustrated *Of the Fabric of the Human Body*. The illustrations were produced by artists associated with the workshop of the Venetian painter Titian (c. 1476–1576). A masterpiece of printing as well as anatomy, *Of the Fabric of the Human Body* required Vesalius to go to Basel to work very closely with the printer. Although it contained no earthshaking innovations, *Of the Fabric of the Human Body* provided a description of the body unprecedented in its detail. It was the first anatomy book to present the body in logical order, from the skeleton outward, as opposed to medieval anatomy texts that presented the parts of the body in the order they were encountered by dissectors, from the viscera inward.

Whereas Galen had lived in a culture where human dissection was forbidden and so worked from animals, mainly monkeys, projecting from them to humans, Vesalius worked on human bodies, and he taught by cutting up the body himself rather than using the medieval and early Renaissance method of reading from a classic medical text while a servant cut up the body and displayed the various organs. Based on this experience, Vesalius denounced a number of traditional and Galenic errors, such as the belief that men have one more rib than women do or that the human liver has five lobes. Vesalius's approach remained basically Galenic, but by claiming that all of Galen's human anatomy was based on illegitimate extrapolation from animal bodies, he made all of Galen's anatomy potentially wrong. This was the true significance of his work.

Of the Fabric of the Human Body was frequently reprinted and became a standard anatomical text. The illustrations in particular were influential, and cheap reproductions were produced for medical education; students could cut out the different organs and paste them into the appropriate cavities. The same year he published *Of the Fabric of the Human Body,* Vesalius left Padua to become physician to the household of Charles V. He died at sea returning from a pilgrimage to Jerusalem. Vesalius's successors at Padua, such as Gabriele Fallopio (1523–1562), discoverer of the fallopian tubes, dominated sixteenth-century European anatomy.

See also Anatomy; Dissection and Vivisection; Medicine; University of Padua.
Reference
O'Malley, Charles Donald. Andreas Vesalius of Brussels, 1514–1564. Berkeley and Los Angeles: University of California Press, 1964.

Viète, Francois (1540–1603)

The founder of modern symbolic algebra, Francois Viète (also known by the Latin form of his name, Vieta) was from France's class of nobles and royal administrators. He received a law degree from the University of Poitiers in 1560 and practiced law successfully. A moderate Catholic who associated with

ANDREAE VESALII
BRVXELLENSIS, SCHOLAE
medicorum Patauinæ profeſſoris, de
Humani corporis fabrica
Libri ſeptem.

CVM CAESAREAE
Maieſt. Galliarum Regis, ac Senatus Veneti gra-
tia & priuilegio, ut in diplomatis eorundem continetur.

This illustration from Of the Fabric of the Human Body *(1543) shows Andreas Vesalius performing a dissection. The surrounding crowd emphasizes Vesalius's status as a "star" of the medical world. (National Library of Medicine)*

Protestants, Viète enjoyed an outstanding career as an administrator and courtier during the turmoil of the French Wars of Religion. Although he practiced mathematics only in his spare time, he was able to employ his skills to assist the French government in its war with Spain by decoding captured Spanish messages.

Viète's earliest mathematical publication, *Canon Mathematicus* (1571–1579), covered trigonometry and was meant to be the introduction of a work on astronomy, which was never published. The *Canon Mathematicus* was followed by a long silence, probably due to Viète's pressing official duties. His next mathematical publication was the short work *Introduction to the Art of Analysis* (1591), which employed an innovative symbolic form, breaking with the tradition of expressing problems verbally. Viète used the previously devised plus (+) and minus (−) signs for addition and subtraction, and he used letters to denote unknown quantities. This enabled algebraic problems to be solved in general rather than specific forms.

Viète claimed that algebra was an area of investigation independent of geometry, although he did not always carry this through. For example, he did not add or subtract expressions of different degree, since adding a second power to a third power would be the algebraic equivalent of adding the area of a square to the volume of a cube, a geometrically meaningless operation. His work contained significant innovations in the handling of equations and an elegant expression of pi (π) as a continued fraction. Influenced by Ramus, Viète was the first to identify algebra with the ancient art of analysis. He did not present his algebra as an entirely new phenomenon, but as an attempt to reconstruct an ancient art of analysis that the classical mathematicians had kept secret. Viète also applied algebra to trigonometry in *Geometrical Supplement,* which was published the same year as *Introduction to the Art of Analysis* and provided solutions to the problems of doubling the cube and trisecting the angle. Viète founded the branch of trigonometry known later as goniometry, the study of angles. He opposed the Gregorian reform of the calendar, disputing with its champion, Christoph Clavius.

Viète's immediate influence was strongest in England and France. The English mathematicians Thomas Harriot and William Oughtred (1575–1660) worked in the tradition of Viète's algebra, and his greatest disciple was another French lawyer-mathematician, Pierre de Fermat.

See also Mathematics.
Reference
Grattan-Guiness, Ivor. *The Norton History of the Mathematical Sciences: The Rainbow of Mathematics.* New York: W. W. Norton, 1998.

Void
See Vacuum.

W

Wallis, John (1616–1703)

After Isaac Newton, the pugnacious John Wallis was England's leading mathematician in the late seventeenth century. Educated at Oxford and Cambridge, Wallis was ordained in the Church of England in 1640. He combined his mathematical studies with service to every English regime from the Civil War of the 1640s, in which he supported Parliament and used his mathematical skills to decipher Royalist coded messages, to that of William III (r. 1689–1702) at the end of the century. Wallis was a member of John Wilkins's circle at Oxford in the 1650s and a founding member of the Royal Society, of which he became president in 1680. He served as Savilian Professor of Geometry at Oxford for over 50 years, from 1649 to his death in 1703. He also held the position of keeper of the university archives from 1658 to his death.

Wallis's most important mathematical work was *Algebra* (1685), which discussed negative and complex roots of algebraic equations. His *On Conic Sections* (1655) was the first major treatment of conics in terms of analytic geometry. Wallis introduced the modern symbol for infinity (∞), and his *Arithmetic of Infinites* (1656), a major work on integration and infinite series, contained a particularly elegant expression of the value of pi (π) in the form of a continued fraction. Wallis's mathematical works inspired Isaac Newton's discovery of the binomial theorem and his early work on the calculus, and the two became friends. The prolific Wallis also published in mechanics, gravitational theory, grammar, logic, theology, and music, and he prepared editions of several ancient Greek mathematical and scientific texts.

Wallis was involved in a particularly bitter controversy in which he got the better of Thomas Hobbes, who claimed that he had discovered how to square the circle by using a compass and straightedge. Wallis correctly denied that this was possible. He also quarreled with Pierre de Fermat over number theory and with minor figures over accusations of plagiarism or Royal Society politics. Partially motivated by nationalism, Wallis originated the false charge that René Descartes's analytic geometry was plagiarized from the Englishman Thomas Harriot, whose work he did much to publicize.

See also Fermat, Pierre de; Hobbes, Thomas; Mathematics; Oxford University.

Reference

Grattan-Guiness, Ivor. *The Norton History of the Mathematical Sciences: The Rainbow of Mathematics*. New York: W. W. Norton, 1998.

War

During the scientific revolution, Europe saw vast changes in warfare, changes that have been described by some military historians as a military revolution. The relation between the two revolutions was complex. The direct application of scientific knowledge for the creation of more effective weapons was still in the future. However, war did set the agenda for some areas of science, particularly with the founding of the first military schools, which did much to professionalize the military and to educate the officer corps.

The drive to improve navigation, which had so much influence on scientific endeavor, particularly astronomy, had obvious military as well as commercial relevance. Naval architecture also attracted the interest of scientists. Fortification, which grew increasingly complex in this period, demanded mathematical skills, and the great siege engineers were on an equal footing with the scientists of their day. Indeed the two groups overlapped. The greatest military engineers, the Frenchmen Nicolas-Francois Blondel (1618–1686) and Sébastien Vauban (1633–1707), were members of the French Royal Academy of Sciences, and Vauban in particular contributed to a range of scientific questions beyond his military profession. War also presented a variety of problems for the surgeon. Ambroise Paré (1510–1590), surgeon to the king of France and the most influential surgical writer of the entire period, made his reputation in part by denying the necessity of cauterizing bullet wounds with boiling oil.

The most important application of pure science to a military problem, and conversely of military data to science, was in ballistics. Many mathematicians and mechanical philosophers were fascinated by gunnery, particularly as longer cannon designed to be fired at some distance from the target were just being introduced in the late fifteenth century. As crossbows had earlier, gunnery posed in dramatic form a classic problem for Aristotelian physics—that of projectile motion, or motion imparted to the mover and continuing. Mathematicians were also eager to work out the angles at which to place a gun in order to reach a desired target. Ballistics as an applied science was founded by the Italian mathematician Niccolò Tartaglia (c. 1499–1557), who determined that a cannon at a 45-degree angle to the Earth would hurl its ammunition the farthest. Ballistics were further developed by Galileo Galilei, who demonstrated that the path of an ideal projectile was a parabola, and by his disciple Evangelista Torricelli (1608–1647). The impact of their theories on the actual practice of gunnery was minimal, however, particularly given the Galilean propensity to consider problems as abstracted from reality and to ignore things like the resistance of the air and the imperfection of seventeenth-century guns.

Problems of gunnery remained interesting to scientists, some of whom tried to take the air resistance into account with ever more sophisticated mathematics. Others took an experimental approach; for example, the Royal Society, one of whose members was Sir Jonas Moore (1617–1679), mathematician and surveyor-general of the King's Ordnance, held trials on Blackheath to observe the trajectory of missiles.

See also Navigation; Politics and Science.
Reference
Hall, A. Rupert. *Science and Society: Historical Essays on the Relations of Science, Technology and Medicine*. Aldershot, England: Variorum, 1994.

Weapon Salve

The weapon salve was an ointment, described by Paracelsus, that could supposedly cure wounds by application not to the wound, but to the instrument that had caused the wound. The salve itself was plainly magical in its origin, involving ingredients such as the moss from the skull of a hanged thief and warm human blood, and it involved the disputed question of occult action, working from a distance. The weapon-salve debate involved some of the leading natural philosophers of the early seventeenth century, including

Johannes Baptista van Helmont and Robert Fludd. The weapon salve was attacked by anti-Paracelsians like Andreas Libavius, and it was defended by Paracelsians and sympathizers such as Oswald Croll and the German physician Daniel Sennert (1572–1637).

This controversy's most prominent combatants in the early seventeenth century were two university professors, a Protestant believer in natural magic at the University of Marburg, Rudolphus Goclenius the Younger (1572–1621) and a Jesuit from Douai who hated magicians, Jean Roberti. Goclenius published tracts extolling the weapon salve in 1608 and 1613, and Roberti responded with an attack on the weapon salve and Goclenius in 1617. The period from 1617 to 1625 was the most intense phase of the weapon-salve controversy on the European continent.

Van Helmont became involved in 1621, claiming in *The Magnetic Cure of Wounds* that, contrary to Goclenius's view, the success of the cure depended on the sympathy between the blood in the body of the wounded person and the blood remaining on the weapon. Van Helmont tried to make the process more a matter of natural magic than witchcraft by claiming that the moss from any skull would work in the salve. In the early phase of the Thirty Years War, with the shadowy Rosicrucians identified with the Protestant cause, Catholics like van Helmont who supported Paracelsianism or anything magical were viewed with great suspicion by Catholic Church authorities, and it was his weapon-salve tract that led to van Helmont's troubles with the Inquisition.

The English offshoot of the Continental weapon-salve controversy began with the appearance of *Hoplocrisma-spongus, or A Sponge to Wipe Away the Weapon-Salve* (1631) by a Protestant minister named William Foster (1591–1643). Foster's main purpose was to defend the good name of Protestantism, which he thought Goclenius had injured by endorsing a cure that was magical and quite possibly satanic. Foster employed Aristotelian natural philosophy to deny the possibility of a magical cure, attacking a number of magicians but principally Fludd, who was already involved in a heated controversy with Marin Mersenne. Fludd replied in *Doctor Fludd's Answer to M. Foster, or The Squeezing of Parson Foster's Sponge, Ordained by Him for the Wiping Away of the Weapon-Salve* (1631), where he defended the weapon salve and sympathetic action between separated entities in general by employing the example of the magnet. Although Foster never replied, other English anti-Paracelsians continued to use the weapon salve to attack Fludd. The weapon salve, now called the powder of sympathy, would be later championed by Sir Kenelm Digby, the last major natural philosopher to do so.

See also Digby, Sir Kenelm; Fludd, Robert; Helmont, Johannes Baptista van; Magic; Paracelsus.
Reference
Debus, Allen G. *The Chemical Philosophy: Paracelsian Science and Medicine in the Sixteenth and Seventeenth Centuries.* 2 vols. New York: Science History Publications, 1977.

Weather

In a predominantly agricultural society, understanding and predicting the weather was a matter of great urgency. It was also vital for travelers, merchants, and sailors. Advances in science during the scientific revolution were applied to problems of weather, and although there was little success in weather prediction, there were some advances in measuring and understanding weather. Most basic among these was sorting out what belonged to weather and what belonged to astronomy. Meteors and comets, thought by Aristotle and his successors to be gaseous exhalations from the Earth, and therefore part of weather, were shown to exist above the Moon (although the term "meteorology" for the study of the weather persists as a reminder of the earlier idea).

Some natural philosophers were inspired by problems in the study of weather. René

Descartes's system of natural philosophy originated as an attempt to explain the appearance of multiple images of the Sun and other puzzling meteorological phenomena, such as the rainbow. His *Meteors* (1637) insists on purely mechanical explanations for weather. Galileo's disciple Evangelista Torricelli (1608–1647) applied his discovery of the weight of the air to describe winds as caused by differing densities of air in different places. But most of the progress made in understanding the weather was due not to the application of theory but to improved instruments. In addition to the barometer and thermometer invented in this period, the scientific revolution also saw the introduction of the hygroscope for the measuring of the moisture of the air, the wind gauge for measuring the strength of the wind, the rain gauge for measuring the amount of rain, and the weather clock, which combined several of these instruments with a clock to record weather conditions automatically by punching holes in a roll of paper. This last device was described by its inventor Robert Hooke, who improved several devices for weather measurement in 1679. Given its complexity, however, the weather clock was never very effective, and no more seem to have been built after Hooke's prototype.

Francis Bacon hoped that the diligent gathering of facts would lead to a science of the weather, but subsequent efforts showed the limits of Baconian fact gathering. The systematic recording of weather data began in the mid-seventeenth century. Hundreds of people throughout Europe kept weather diaries, and scientific societies such as the Accademia del Cimento and the Royal Society attempted to organize networks of people to observe and record weather (Thomas Sprat's *History of the Royal Society* [1667] included a form drawn up by Hooke for recording weather information in a standardized way). Although diligently keeping their weather diaries may have given people the sense that they were participating in an exciting scientific endeavor, this experiment made little advance in understanding of the weather and still less in the ability to predict it. Some progress was made in mapping the winds. Edmond Halley created a global map of the prevailing winds of the oceans for the benefit of navigators, and he also worked on problems of evaporation, the height of the atmosphere, and the distribution of the Sun's heat.

See also Baconianism; Barometers; Hooke, Robert.

Reference
Wolf, A., with the cooperation of F. Dannemann and A. Armitage. *A History of Science, Technology, and Philosophy in the Sixteenth and Seventeenth Centuries.* 2d ed., prepared by Douglas McKie. London: Allen and Unwin, 1950.

Wilkins, John (1614–1672)

John Wilkins was the foremost popularizer and organizer of the new science in mid-seventeenth century England. He received a B.A. from Oxford in 1631 and an M.A. in 1634. Making a career in the Church of England, Wilkins served as a chaplain in the household of a Puritan noble family, the Fiennes family, and he sided with the Parliament in the English Civil War. Historians have debated whether Wilkins was a Puritan, but if so, he was a very moderate one. Before the English Civil War Wilkins wrote two important popularizations, *The Discovery of a New World* (1638) and *A Discourse Concerning a New Planet* (1640). *The Discovery* demonstrates that the celestial bodies are not qualitatively different from the Earth, as Aristotelian physics would have it, and that they do not revolve in crystalline spheres, as Ptolemaic astronomy would have it. It also contains one of the earliest speculations on travel to other planets. *A Discourse,* drawing heavily on the work of Galileo Galilei, defended the Copernican system. Both books were reprinted, and *The Discovery* was translated into French in 1656. Wilkins popularized mechanics in *Mathematical Magick, or the Wonders That May Be Proved by Mechanical Geometry* (1648).

While working in London as chaplain to

King Charles I's (r. 1625–1649) nephew Charles Louis in 1645, Wilkins joined a group of scientific investigators, including the mathematician John Wallis. When he returned to Oxford as the master of Wadham College in 1649, many members of the London group followed him there. Wilkins led scientific activities at Oxford for the next ten years. Wallis, along with Seth Ward (1617–1689), the Savilian Professor of Astronomy, Christopher Wren, and, by 1656, Robert Boyle, as well as many others held regular meetings in Wilkins's rooms at Wadham to discuss natural philosophy and technological improvements. There was some overlap between this group and the Hartlib circle, notably in the person of Boyle. But Wilkins's group differed from its contemporary in that it was less oriented to millenarian schemes of universal and social reformation.

Although he had married Oliver Cromwell's sister Robina, Wilkins adjusted easily to the changed political and religious situation following the Restoration of the monarchy and Church of England in 1660. Back in London, he was the acknowledged leader of the latitudinarian faction, favoring tolerance for different beliefs and practices within the Church of England. His ecclesiastical career culminated in his appointment as bishop of Chester in 1666. He was a founding member of the Royal Society, helping draw up its charter and serving as one of its secretaries and vice president. Wilkins also supervised Thomas Sprat in writing *The History of the Royal Society* (1667). He remained an active experimenter and continued to tinker with new technology. Wilkins worked on a scheme for a universal, logically structured language, publishing *An Essay Towards a Real Character and Philosophical Language* in 1668.

See also Oxford University; Popularization of Science; Puritanism and Science; Royal Society; Universal Languages.
Reference
Shapiro, Barbara. *John Wilkins, 1614–1672: An Intellectual Biography.* Berkeley and Los Angeles: University of California Press, 1969.

Witchcraft and Demonology

The age of the scientific revolution was also the age of the great European witch-hunts, in which approximately forty to fifty thousand people, about 80 percent of them women, were executed for being witches, and countless others were accused and tortured. The relation between the two phenomena cannot be described as science simply driving out witchcraft and demonological belief, but should rather be understood as a series of attempts to employ science and natural magic to understand demonic action. It was generally agreed that whatever a witch did by virtue of witchcraft was not performed by her, but by Satan or other demons, who were seen as spirits interacting with the material world. Demons, in this view, have to work through natural causes, as true miracles are reserved for God. Therefore, demons and witches and what they could accomplish were a legitimate topic of inquiry for natural philosophers. For instance, natural philosophers studied how witches created storms and whether copulation between witches and demons could produce offspring. Demons, particularly Satan himself, were often portrayed as complete masters of all scientific knowledge, which they employed to carry out their evil deeds. Francis Bacon included a call for a history of witchcraft in his program for the systematic study of nature, and Andreas Libavius discussed witches' alleged power of flight in his study of difficult problems in natural philosophy, *Singularia* (1599–1601).

Large-scale witch persecution was over in most places in Europe by the late seventeenth century. The reasons for this decline are unclear, but the scientific revolution had little to do with it. Much more likely causes are fear of the social disorder that witch-hunts unleashed, a growing reluctance among elite magistrates to believe accusations by plebeians, and the cooling of religious passions by the end of the Thirty Years War, the last war of religion, in 1648. Whatever its cause, the decline in persecution did not entail a decline in the belief in

This sixteenth-century engraving (Hans Weiditz, 1532) of a witch burdened with the weight of a sphere containing the moon and stars emphasizes the interaction between witches and the natural world. (National Library of Medicine)

witches, demons, and their power to interact with the material world, among either Europe's elite or its common people. Many believed that denying the possibility of witchcraft would lead to materialism and atheism, as these, too, were based on a common denial of the power of spirits. Strict versions of the mechanical philosophy and insistence on mechanical causation might have led to a denial of witchcraft, but in practice they were seldom applied to the question. One exception was the work of the Dutch Cartesian minister Baltasar Bekker (1634–1698), whose *The Enchanted World* (1691) argues that the Devil is bound in Hell and unable to affect the natural world. Bekker's work came out decades after the Dutch Republic had actually ceased persecuting witches, so its impact on practice was minimal.

The debate on witchcraft that most closely involved early modern science took place in late-seventeenth-century England, a time

and place that saw few actual witch trials (the last English execution of a witch took place in 1685). One important English advocate of science was the Reverend Joseph Glanvill (1636–1680), a fellow of the Royal Society. He wrote an influential study of witchcraft that was first published in 1666 but was best known in the 1681 expanded edition, where it was titled *Saducismus Triumphatus,* or *Sadducism Conquered.* Glanvill, one of the many English thinkers who rejected mechanical Cartesianism for its materialism, linked those who denied the possibility of witchcraft with the ancient Jewish Sadducees, who had denied the immortality of the soul. The work includes a number of narratives about the actions of witches, including the author's own investigations, and Glanvill clearly placed his argument in the context of the debate over spirit and matter. Employing the principle of the slippery slope, he claimed that any denial of spiritual causation leads to the denial of God. "Atheism is begun in

Sadducism," he wrote (Glanvill 1966). Glanvill collected witch stories, attaching specific names, dates, and places, in order to empirically demonstrate the existence and powers of witches. The 1681 edition, appearing after Glanvill's death, included contributions from the eminent Cambridge Platonist and fellow of the Royal Society, Henry More (1614–1687), another supporter of witchcraft belief.

Other collaborators in the investigation of the powers of witches included distinguished scientists and Royal Society fellows like the natural historian Robert Plot (1640–1696), the anatomist Thomas Willis (1621–1675), and Robert Boyle. But the relation of the new science to witchcraft could be interpreted in more than one way. The Puritan and natural magician John Webster (1610–1682) was the author of the skeptical reply to Glanvill entitled *The Displaying of Supposed Witchcraft* (1677). In addition to explaining the supposed deeds of witches as natural rather than as demonic magic, Webster also credited the experimental philosophy of the Royal Society with combating witchcraft belief; in fact, the Royal Society actually licensed Webster's book to appear in print. Another skeptic, John Wagstaffe (1633–1677), asserted in *The Question of Witchcraft Debated* (1669) that ignorance about nature was so great that attributing events to demonic rather than natural causation could never be intellectually justified. Despite the radicalism of his antidemonological position, Wagstaffe's natural philosophy was conservative and Aristotelian. This more skeptical view of witchcraft would become more common by the eighteenth century, when the Enlightenment belief in science would come to be seen as incompatible with belief in the power of demons and witches.

See also Magic; Women.
References
Clark, Stuart. *Thinking with Demons: The Idea of Witchcraft in Early Modern Europe.* Oxford: Clarendon Press, 1997.
Glanvill, Joseph. *Saducismus Triumphatus; or, Full and Plain Evidence Concerning Witches and Appar[i]tions (1689).* A facsimile reproduction, with an introduction by Coleman O. Parsons. Gainesville, FL: Scholar's Facsimiles & Reprints, 1966.
Hunter, Michael. *Science and the Shape of Orthodoxy: Intellectual Change in Late Seventeenth-Century Britain.* Woodbridge, England: Boydell, 1995.

Women

Although early modern science, like other areas of early modern culture, was strongly male dominated, some women did manage to participate. They faced formidable handicaps. Fewer women were literate or had the necessary education to participate in science at any level. Lack of fluency in Latin was a particular source of difficulty, although one that diminished over the course of the scientific revolution as vernacular languages were more widely used. Women were also barred from most of the institutions of early modern science; indeed, the general trend was for women to be less active as science became more institutionalized. European universities, with a few rare exceptions in Italy, did not admit women, nor did the new scientific societies and academies. For a woman, even a natural philosopher such as the Duchess of Newcastle, Margaret Cavendish, to attend a meeting of the Royal Society as a guest provoked great controversy. Early modern science inherited the stereotype of women as intellectually inferior, although the full development of scientific sexism lay in the future. Explanations of women's purported intellectual inferiority were mainly couched in the idiom of religion or of Aristotelianism, rather than that of innovative science.

Women of the upper classes in Europe could function as patrons of the sciences, as did Queen Christina of Sweden (1626–1689), unfortunately with disastrous consequences, when she invited René Descartes to Stockholm, where he died of the cold in 1650. In her later exile in Rome after her conversion to Catholicism, Christina was the patron of a small scientific academy, the Physical-Mathematical Academy. Another area where women could play a scientific role was in support of male scientists as wives or daughters. This pattern was particularly marked in astronomy, as many seventeenth-century astronomers worked from their homes with female relatives working as assistants. Sophie Brahe, Tycho Brahe's sister, was the one member of his family to share his interests in astronomy and alchemy. This tradition of familial astronomy was particularly strong in central Europe, where it was represented in the seventeenth century by Maria Cunitz, Caterina Elisabeth Hevelius (1647–1693), Maria Eimmart (1676–1707), and Maria Kirch née Winkelmann. Yet the refusal of the Berlin Academy of Sciences to employ Kirch as assistant astronomer on the death of her astronomer husband exemplifies the difficulties women had in participating in scientific institutions. The academy did not question her competence but believed that it was inappropriate for a woman to hold the post, which might possibly open the recently founded academy to ridicule. Further, the growth of specialized and government-sponsored observatories in general relegated the astronomical household to the sidelines and limited female participation, although Margaret Flamsteed (c. 1670–1730) did assist her husband John Flamsteed, the first royal astronomer, at the Greenwich Observatory.

Knowledge handed down in a family could also lead women into science through scientific illustration. Eimmart's principal scientific accomplishment was a series of 250 illustrations of the surface of the Moon, and Anna Maria Sibylla Merian, like Eimmart the daughter of an engraver, expanded her interests from illustration to original natural history. Other areas where women could possess scientific authority were those associated with traditionally feminine functions. The most obvious example is midwifery, but some technological areas were also associated with household management. Manuals of so-called

domestic chemistry, often female-authored, contained practical procedures and recipes, but as chemistry developed, these works were increasingly scorned by male chemists. Women of the upper classes also collected and published medical recipes, as tending simple medical problems of their servants was one of their traditional responsibilities.

Some hoped to incorporate science into women's educations. The French alchemist Marie Le Jars de Gournay (1565–1645), author of *The Equality of Men and Women* (1622), and the Englishwoman Bathsua Makin (c. 1612–c. 1674) both put forth proposals for educating women in the sciences. However, like many proposals for educational reform in the seventeenth century, these had little practical effect.

Cartesianism seemed to legitimize women's scientific and intellectual activity and was especially appealing to women. It was an alternative to the Aristotelianism of the completely male-dominated universities, and many of the canonical works of the Cartesian tradition, starting with those of René Descartes himself, were written in a clear and elegant French intended to be accessible to people without university educations, including women. Descartes was comfortable with intellectual women, carrying on a rich philosophical correspondence with Princess Elizabeth of Bohemia (1618–1680), to whom he dedicated *Principles of Philosophy* (1644) and who promoted Cartesianism in Germany. The Cartesian emphasis on the separation of the soul from the body was attractive to women intellectuals, as it meant their minds were not limited by their allegedly inferior female bodies. One of Descartes's followers, Francois Poullain de La Barre (1647–1723), used Cartesianism to argue for women's equality with men in *The Equality of the Sexes* (1673), which was translated into English in 1677. Female and feminist Cartesianism was particularly characteristic of the Parisian salon culture of the late seventeenth century.

If women were viewed with suspicion or

hostility as creators of science, the situation was quite different when they assumed the role of the audience for male displays of scientific knowledge. Women were frequently addressed in works of scientific popularization, most notably Bernard Le Bouvier de Fontenelle's *Conversations on the Plurality of Worlds* (1686), which presented Cartesian science in the form of a dialogue between a male scientist and a young noblewoman. In France and to a lesser extent in England, women attended public lectures and demonstrations, and the annual *Ladies' Diary, or The Woman's Almanack,* first appearing in 1704, appealed to the scientific and mathematical interests of educated Englishwomen.

Women interested in the sciences were the targets of satirical ridicule in the late seventeenth century, notably in the great French comic playwright Molière's (1622–1673) *The Learned Ladies* (1672), satirizing Cartesian salon women, and more affectionately, in the English playwright Susannah Centlivre's (c. 1677–1723) *The Basset Table* (1705), which included a female experimentalist. Male virtuosi were not immune from satire, of course, but women were often attacked more viciously, sometimes drawing on the traditional prejudice that learned women must be unchaste, unattractive, or bad housekeepers and mothers.

Between the position of scientific creator and audience lay the translator, and women were surprisingly active in translating works of natural philosophy. The poet and dramatist Aphra Behn (1640–1689) translated Fontenelle's *Conversations* into English, including a commentary of her own. The most significant female translator of the period was Gabrielle-Émilie, Marquise du Chatelet (1706–1749), whose translation of Newton's *Mathematical Principles of Natural Philosophy* from Latin to French was posthumously published in 1759, contributing to the spread of Newtonianism in France. It remains the only French version of Newton's masterpiece. Du Chatelet also published or contributed to other works in French on Newtonian physics.

See also Cavendish, Margaret, Duchess of Newcastle; Conway, Anne; Cunitz, Maria; Kirch née Winkelmann, Maria; Merian, Anna Maria Sibylla; Midwives; Salons; Sexual Difference.
References
Hunter, Lynette, and Sarah Hutton, eds. *Women, Science, and Medicine, 1500–1700: Mothers and Sisters of the Royal Society.* Stroud, England: Sutton Publishing, 1997.
Phillips, Patricia. *The Scientific Lady: A Social History of Women's Scientific Interests, 1520–1918.* London: Weidenfeld and Nicolson, 1990.
Schiebinger, Londa. *The Mind Has No Sex? Women in the Origins of Modern Science.* Cambridge: Harvard University Press, 1989.

Wren, Christopher (1632–1723)

Chiefly known today as one of England's greatest architects, Christopher Wren was one of the most versatile scientists and engineers of Restoration England, although he published little. Born into a distinguished family of Church of England clergymen, Wren attended Wadham College, Oxford, from 1649 to 1653, receiving a B.A. and an M.A. While at Oxford, Wren was a member of John Wilkins's scientific circle. His earliest interests were in mathematics and astronomy, and he made a reputation as a brilliant young mathematician by rectifying the cycloid.

Possibly due to Wilkins's influence with the Lord Protector Oliver Cromwell (1599–1658), Wren was appointed Gresham College Professor of Astronomy in 1657. Wren's teaching of astronomy, both as Gresham Professor until 1661 and as Savilian Professor of Astronomy at Oxford from 1661 to 1673, contributed to the dissemination of Keplerian astronomy in England, and his discussions on cosmogony with Robert Hooke contributed to the pre-Newtonian development of gravitational theory. In 1660, after one of Wren's lectures at Gresham, a group was founded in London that would become the Royal Society, of which Wren was an original member.

Throughout his career, Wren was a fertile deviser of instruments and gadgets in a huge variety of fields, including a micrometer for telescopes, a double telescope to be operated by two observers, engines, and meteorological instruments. He created a pasteboard globe of the Moon that reflected his observations on its surface, which he gave to Charles II (r. 1660–1685), who kept it as a curiosity. Like many astronomers of the time, Wren devoted much of his efforts over the course of his life to solving the problem of determining the longitude (and like theirs, his efforts did not succeed). He performed vivisections and medical experiments and made the illustrations for Thomas Willis's (1621–

1675) *Brain Anatomy* (1664). In physics, he did important experimental work on the laws of impact. Although the focus of his interests shifted to architecture after the Great Fire of London in 1666, Wren remained involved with natural philosophy, serving as president of the Royal Society from 1681 to 1682.

See also Cycloid; Royal Society.
Reference

Bennet, J. A. *The Mathematical Science of Christopher Wren*. Cambridge: Cambridge University Press, 1982.

Z

Zoology

During the scientific revolution, knowledge about animals both expanded rapidly and was redefined, becoming closer than ever to modern zoology. Zoology, like other subfields of natural history, was a basically textual discipline in the sixteenth century. Its heritage included the biological works of Aristotle, Pliny's *Natural History,* and the work of the medieval encyclopedists, most notably Albert the Great (c. 1200–1280). There was also the practical knowledge of hunters and herders, and physicians as well, although medical uses of animals were nowhere near as extensive as those of plants. Some traditional myths concerning various animals were still matters of scientific concern during this period. These included both mythical animals, such as the unicorn, whose reality was very widely accepted, and fabulous stories about real animals, such as the claim that Irish geese grow from barnacles on driftwood.

The major encyclopedic zoologists of the sixteenth century were both physicians, the Swiss polymath Konrad Gesner (1516–1565) and the Italian Ulisse Aldrovandi. In addition to ancient and medieval sources, Gesner and Aldrovandi drew on their own firsthand observations and on contemporary writers on animals. Specialized works of zoology in this period included the French physician Guillaume Rondelet's (1507–1566) enormous work, *Books of Fish* (1554–1555), which set the standard for ichthyological research. The Italian papal physician Ippolito Salviani (1514–1572) produced another ichthyological work, concentrating on the fish of the Mediterranean. Another Frenchman, Pierre Belon (1517–1564), wrote about birds and has some claim to be the founder of comparative anatomy, comparing the skeleton of a chicken with that of a human.

Both Gesner, whose four-volume *History of Animals* appeared from 1551 to 1558, and Aldrovandi, most of whose books appeared posthumously, wanted to produce complete repositories of information about animals. The role of animals in history, proverbs, mythology, and other areas of human culture was as important to them as what we would now deem biological information. (This approach was not universal; Rondelet and Belon omitted such information.) Their cultural emphasis required Gesner and Aldrovandi to devote far more space to animals with a prominent cultural role, such as horses, eagles, and bees, than to more obscure ones such as the newly discovered animals of the New World, who had few associations for Europeans. Whether or not the animals actually existed was a question of secondary importance, and Gesner included in his compendium such legendary creatures as the unicorn, the phoenix, and the sea serpent. Aldrovandi's works were larger and more complete than Gesner's, but their bulk was truly intimidating, and Gesner's profusely illustrated books had more influence, partly because they were published earlier and were more widely translated. Edward Topsell's (1572–1638) *Historie of Four Footed Beasts* (1607) was essentially an English partial translation and expansion of Gesner's work.

The New World natural histories that appeared did discuss animals, although zoology took a definite second place to medical botany. Their authors had to take a different tack than the encyclopedists, given the lack of Old World cultural associations of the animals they described. Thus a New World book like Georg Markgraf's (1610–1644) *Natural History of Brazil* (1648) looks much more like a modern zoology book than does anything by Gesner or Aldrovandi. The approach of the encyclopedists, whose ideal was to gather everything said about an animal whether or not it was true, also came under fire from the opponents of "vulgar errors," who subjected their claims to empirical validation. In his *Pseudoxia Epidemica* (1646), Sir Thomas Browne (1605–1682) used experiment to attack such traditional beliefs as the innate antipathy between toads and spiders. After putting a toad and some spiders in a jar, he noted that the spiders crawled around undisturbed, while the toad ate them, one by one.

Various classification schemes for animals, mostly based on obvious physical resemblances or on Aristotelian categories, were employed during this period. The principle of the great chain of being—that all life forms are linked—was one premise in classification. Gesner, for example, included bats among birds, as intermediary forms between birds and mice. Whales and porpoises could not be classified among other animals that we would now call "mammals," a term invented in the eighteenth century, because most mammals were then referred to as "quad-

rupeds," four-legged animals, and animals that lacked legs entirely could not be included in this category. Even after the major groups of animals had been sorted out, it was still often difficult to arrange individual species within a group; Gesner, for instance, listed animals alphabetically within broad categories, which meant that his arrangement changed every time the book was translated.

The encyclopedic approach of Gesner and Aldrovandi was on the wane by the seventeenth century. The appearance of a new encyclopedia, Joannes Jonston's (1603–1675) *Natural History* (1650), marks the era when zoological works focused narrowly on the description, characteristics, and anatomies of specific animals rather than their role in the larger cultural system. The most distinguished zoologist of the second half of the seventeenth century, John Ray, had as his greatest legacy a new approach to classification based on internal physical characteristics. In his initial plan for organizing a system of nature, Ray divided the labor between his friend and patron Francis Willughby (1635–1672), who would handle animals, and Ray himself, who would handle plants. Willughby's early death forced Ray to handle animals as well, bringing out two posthumous volumes of Willughby's work on birds and fishes as well as works of his own, one on "serpents and quadrupeds" and one on insects. Ray's classifications made greater use of internal anatomy than of external similarities, although he still divided birds into land and waterfowl and with some misgivings included whales and dolphins as a special group among the fishes; he suggested the abandonment of the term "quadruped." Ray was also responsible for the introduction of the concept of species.

Comparative anatomy also advanced in the late seventeenth century, practiced by Edward Tyson (1650–1708) in England and by a school of anatomists associated with the Royal Academy of Sciences in Paris, including Joseph-Guichard Duverney (1648–1730) and Claude Perrault (1613–1688). Tyson's works include some of the first monographs

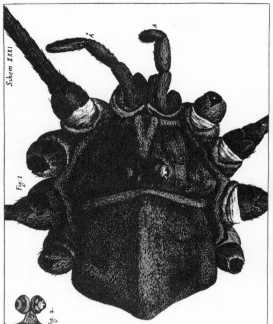

These Shepherd spiders were among the many creatures examined and illustrated in Robert Hooke's Micrographia *(1665).*

devoted to the dissection of particular animals. These included a dolphin, which he demonstrated was not a fish, and an "Orang-Outang" (actually a chimpanzee), whose resemblance to humanity Tyson emphasized, arguing that the ape was more like a man than a monkey—another intermediate on the great chain of being. Entomology was also booming; the insect life cycle received close attention and the microscope was applied to the study of insects and other small creatures by Jan Swammerdam and Antoni van Leeuwenhoek, among others. Martin Lister (1639–1712), an English physician and friend of Ray's, published important works on mollusks and British spiders. In all, the work of the seventeenth-century zoologists set the stage for the achievements of Carl Linnaeus (1707–1778) and Georges-Louis Leclerc, Comte de Buffon (1707–1788), in the eighteenth century.

See also Aldrovandi, Ulisse; Natural History; Ray, John.

References
Ashworth, William. "Natural History and the Emblematic World View." In David C. Lindberg and Robert S. Westman, eds., *Reappraisals of the Scientific Revolution.* New York: Cambridge University Press, 1990.
Ley, Willy. *Dawn of Zoology.* Englewood Cliffs, NJ: Prentice-Hall, 1968.

Chronology

1503   Spanish King Ferdinand II founds the Casa de la Contratación, repository of navigational and cartographical knowledge.

1512   Nicolaus Copernicus is appointed canon of the cathedral at Frauenburg, where he sets up a small observatory.

1514   The last arithmetic book using Roman numerals is published. Copernicus declines invitation to Rome to work on calendar reform.

1518   The London College of Physicians is chartered by Henry VIII.

1525   A massive complete edition of Galen's writings, compiled using techniques of humanist scholarship, is published.

1528   Paracelsus is expelled from his chair at the University of Basel.

1530   Otto Brunfels's innovative herbal, *Living Images of Herbs,* and Girolamo Fracastoro's *Syphilis, or the French Disease,* which gives the disease its name, are published.

1531   The publication of a Latin translation of Galen's *On Anatomical Procedures* sparks interest in anatomy.

1537   The papal bull *Sublimi Deus* declares that the natives of the Americas are human beings with souls.

1539   Georg Rheticus visits Nicolaus Copernicus and urges him to publish his manuscript *On the Revolutions of the Celestial Spheres,* which Copernicus then entrusts to him.

1543   Nicolaus Copernicus's *On the Revolutions of the Celestial Spheres* and Andreas Vesalius's *Of the Fabric of the Human Body* are published.

1544   The first edition of Pier Andrea Mattioli's commentary on the ancient Greek botanist Dioscorides is published, to become the most popular and frequently reprinted herbal of the sixteenth century.

1545   The first university botanical garden is founded at the University of Pisa. The first modern medical description of the clitoris is given by Charles Estienne in *Dissection of the Parts of the Human Body.*

1546   The botanical garden at the University of Padua is founded.

1550   Bartoloméo de Las Casas and Juan Ginés de Sepúlveda debate at Valladolid on the question of whether the inhabitants of the New World are natural slaves.

1551   The Collegio Romano is founded in Rome by Ignatius of Loyola. Erasmus Reinhold issues the *Prutenic Tables,* a set of astronomical tables based on Copernican assumptions.

1551–   Konrad Gesner's *History of Animals* is
1558   published.

1554   The mercury amalgamation process for refining silver is developed by Bartolomé de Medina in Peru.

1555   Ignatius of Loyola receives Christoph Clavius into the Jesuit order.

1556   The body of Ignatius of Loyola is dissected by Realdo Colombo.

1558   Padua-educated English physician John Caius re-founds Gonville and Caius College of Cambridge University, introducing new anatomical techniques to England.

1559   Realdo Colombo identifies the clitoris as the seat of female sexual pleasure.

1562   The works of the ancient skeptical philosopher Sextus Empiricus are published in Latin translation, contributing to the spread of skeptical ideas.

1563   García d'Orta's *Colloquies on the Herbs and Drugs of India* becomes the first scientific book printed in India.

1568   The botanical garden of Bologna is founded under the leadership of Ulisse Aldrovandi.

1570   Abraham Ortelius's cartographic collection, *Theater of the World,* is published. Girolamo Cardano is arrested by the Inquisition for publishing the horoscope of Jesus Christ.

1571–   Francisco Hernández undertakes his
1577   scientific exploration of Mexico.

1572   The appearance of a new star inspires astronomical and astrological debate. Ugo Buoncompagni is elected pope, taking the name Pope Gregory XIII.

1575   The University of Leiden is founded.

1576   King of Denmark Frederick II grants Tycho Brahe the island of Hven on which to found his observatory, Uraniborg. Rudolf II is elected Holy Roman Emperor.

1577   A comet appears; its parallax will be used by Tycho Brahe and others to show that it is above the Moon and that, contrary to Aristotle, the heavens are subject to change. The Leiden Botanical Garden is founded.

1579   Englishman Christopher Saxton produces the first volume of maps covering a single nation. The Paris medical faculty succeeds in having a Paracelsian, Roch Le Baillif, Sieur de la Rivière, condemned and expelled from France by the Parlement of Paris.

1582   Pope Gregory XIII announces the Gregorian reform of the calendar.

The Lumleian Lectures are founded at the London College of Physicians.

1585   Simon Stevin publishes *The Tenth,* setting forth the system of decimal fractions. Thomas Harriot leaves England for Sir Walter Ralegh's colony in Virginia.

1588   Tycho Brahe's *Of More Recent Phenomena of the Ethereal World* sets forth the Tychonic system describing the motions of the Sun and planets.

1589   Galileo Galilei wins the mathematics chair at University of Pisa.

1589–   The standard ten-volume edition of
1591   the works of Paracelsus, edited by the physician Johannes Huser, is published.

1591   Francois Viète's *Introduction to the Art of Analysis,* the foundation of modern algebra, is published.

1592   Galileo Galilei becomes professor of mathematics at the University of Padua.

1594   The first permanent university anatomy theater is founded at the University of Padua.

1596   Gresham College is founded in London.

1597   Tycho Brahe abandons Uraniborg and leaves Denmark for the court of the Holy Roman Emperor, Rudolf II.

1598   The failure of a Neapolitan revolt leads to the imprisonment of Tommaso Campanella.

1600   Giordano Bruno is burned at the stake in Rome. William Gilbert's *On the Magnet* is published.

1601   On the death of Tycho Brahe in Vienna, Johannes Kepler inherits his astronomical data and succeeds him as imperial astronomer.

1603   The Accademia dei Lincei is founded.

1604   The appearance of a new star inspires Johannes Kepler to write *Of the New Star,* setting forth what becomes the widely accepted theory that new stars are caused by the burning of celestial waste.

1608   Hans Lippershey applies for a patent on the telescope in the Netherlands.

1609   The first university chair in anatomy, separate from surgery, is founded at Padua. A chair of chemical medicine is founded at the University of Marburg, the first chair requiring laboratory work.

1610   Galileo Galilei's *The Starry Messenger* is published, announcing his telescopic discoveries.

1611   The Oratorian order is introduced into France by Cardinal Pierre de Berulle. Galileo Galilei joins the Accademia dei Lincei. A solemn convocation at the Jesuit Collegio Romano honors Galileo's telescopic discoveries. The Jesuit Superior General Claudio Aquaviva requires Jesuits to defend the authority of Aristotle in philosophy. Rudolf II abdicates as Holy Roman Emperor. Johannes Kepler publishes *Dioptrics,* the first examination of the optics of telescopes.

1612   *Commentary on the Medical Art of Galen* by Galileo Galilei's friend Santorio Santorio is published; it contains the first printed mention of the thermometer.

1613   The publication of Galileo Galilei's *Letters on Sunspots* leads to a feud with the Jesuit astronomer Christoph Scheiner, eventually poisoning Galileo's heretofore good relations with the order.

1614   Isaac Casaubon demonstrates that the Hermetic writings originated in the early Christian era. John Napier's *Description of the Marvelous Canon of Logarithms* is published, setting forth his invention of logarithms. The first Rosicrucian tract, *The Discovery of the Fraternity of the Most Noble Order of the Rosy Cross* (Fama Fraternitatis) is published.

1615   William Harvey is appointed Lumleian Lecturer in Surgery and Anatomy to the London College of Physicians.

1616   Nicolaus Copernicus's *On the Revolutions* is put on the Index of Forbidden Books. Galileo Galilei is admonished not to teach or publicly hold Copernican doctrine.

1617   The Society of Apothecaries of London is chartered.

1617 – 1625   Johannes Roberti's attack on Rudolphus Goclenius the Younger inaugurates the most intense phase of the Continental weapon-salve controversy.

1618   Comets are observed through the telescope for the first time. René Descartes meets Dutch mechanical philosopher Isaac Beeckman.

1619   The Savilian Professorships of Geometry and Astronomy at Oxford are founded. Kepler's *Harmonies of the World* is published, setting forth his third law and inaugurating a dispute with Robert Fludd. René Descartes is inspired to found a new philosophy.

1621   The botanical garden at Oxford is founded. Johannes Baptista van Helmont becomes involved in the weapon-salve controversy.

1625   The first work of microscopic science, a broadsheet on the bee by Francesco Stelluti, is published. The Spanish Inquisition denounces Johannes Baptista van Helmont, partly for his role in the weapon-salve controversy; he will spend most of the next two decades under house arrest.

1626   The French Royal Botanical Garden is founded, although it will not be in operation until 1640.

1627   Johannes Kepler's *Rudolphine Tables,* astronomical tables based on his theories and the most accurate to date, are published. Domenicus Petavius's *On the Doctrine of Time* sets forth the modern B.C./A.D. dating system.

1628   The publication of William Harvey's *On the Motion of the Heart* sets forth the theory of the circulation of blood. A quarrel between René Descartes and Isaac Beeckman ends their personal relationship. Unable to find the ratio between the area of a cycloid and the area of the generating circle, Marin Mersenne passes the problem to Gilles Personne de Roberval, who solves it.

1629   Tommaso Campanella is released from prison.

1631   The transit of Mercury across the face of the Sun is observed by Pierre Gassendi. The English phase of the weapon-salve controversy begins with an exchange of pamphlets between William Foster and Robert Fludd.

1632   Galileo Galilei's *Dialogue on the Two Chief Systems of the World,* espousing Copernicanism, is published. The first published description of the slide rule appears.

1633   Galileo Galilei is tried and convicted. Athanasius Kircher arrives in Rome. The first university observatory is founded at Leiden. Théophraste Renaudot begins his weekly open meetings to discuss various topics, including natural philosophy, in Paris.

1634   Jesuit missionaries present a telescope to the Chinese emperor. Gilles Personne de Roberval wins the Ramus Chair of Mathematics at the Royal College in Paris.

1635   This is the approximate date of the beginning of Marin Mersenne's weekly meetings of natural philosophers at his cell in the Minim monastery in Paris. Through French diplomats, Nicolas-Claude Fabri de Peiresc coordinates widely scattered observations of a solar eclipse.

1636   Pierre de Carcavi introduces the work of Pierre de Fermat to Marin Mersenne, who begins circulating it in Europe's mathematical community.

1637   The controversy between Pierre de Fermat and René Descartes begins with Fermat's criticism of Descartes's *Optics.*

1638   Galileo Galilei's *Discourse on Two New Sciences* is published in Leiden. René Descartes publishes *Discourse on Method* and associated scientific works.

1639   The transit of Venus across the face of the Sun is observed by Jeremiah Horrocks.

1642   Théophraste Renaudot's conferences end.

1643   The Torricellian experiment, leading to the development of the barometer, is conducted for the first time by Evangelista Torricelli.

1645   A Western Jesuit, Johann Adam Schall von Bell, is appointed head of the Chinese Imperial Astronomical Bureau.

1647   Blaise Pascal's *New Experiments on the Void* demonstrates the existence of a vacuum.

1649   John Wilkins arrives at Oxford University as master of Wadham College, founding an active scientific circle there. Queen Christina of Sweden convinces René Descartes to come to Stockholm, where he dies the next year.

1651   The German College of the Curiosities of Nature is founded.

1653   Henry Oldenburg arrives in England.

1654   The first public demonstration of an air pump is performed by Otto von Guericke. Correspondence between

1654
(cont.)
Blaise Pascal and Pierre de Fermat leads to the foundation of mathematical probability theory. Walter Charleton's *Physiologia Epicuro-Gassendo-Charletoniana* introduces Gassendist atomism to England.

1655
A dispute begins between Thomas Hobbes and John Wallis over Hobbes's claim to have squared the circle.

1657
The Florentine Accademia del Cimento is founded. Olof Rudbeck establishes a botanical garden at the University of Uppsala. Otto von Guericke demonstrates the vacuum by evacuating the space between two brass hemispheres and then showing that teams of horses cannot pull the hemispheres apart.

1658
The complete works of Pierre Gassendi are published. Christiaan Huygens's paper setting forth his discovery of Saturn's rings is read at a meeting of the Montmor Academy.

1660
The first meetings of what would become the Royal Society are held in London.

1661
Marcello Malpighi's *On the Lungs* describes the circulation of blood through the capillaries.

1662
The Royal Society receives a charter from King Charles II.

1663
Works of René Descartes are placed on the Catholic Church's Index of Forbidden Books. The Lucasian Chair of Mathematics at Cambridge is founded, with the first incumbent being Isaac Barrow.

1665
The first issues of *Journal des scavans* and *Philosophical Transactions* appear, the latter edited by Henry Oldenburg. Robert Hooke's *Micrographia* is published. Richard Lower performs the first blood transfusion experiments, between dogs.

1665–
1666
The plague of London kills George Starkey and drives Isaac Newton to the country, where he experiences his *annus mirabilis*.

1666
The Royal Academy of Sciences is founded in Paris. Controversy over John Wallis's modification of Galileo's tide theory leads the Royal Society to set up a project to gather tidal information. The Great Fire of London drives the society from Gresham House and destroys many of William Harvey's manuscripts at the London College of Physicians.

1667
The Accademia del Cimento disbands. Margaret Cavendish attends a Royal Society meeting, the only woman to do so. The Paris Observatory is founded. The first human blood transfusion is performed in Paris. The body of René Descartes is returned to France and given a public burial. Henry Howard donates the Arundel Library to the Royal Society. Thomas Sprat's *History of the Royal Society* is published.

1668
Isaac Newton devises the reflecting telescope. Francesco Redi's *Experiments on the Generation of Insects* is published, giving experimental evidence against spontaneous generation. Louis XIV invites Gian Domenico Cassini to come to Paris and run the Paris Observatory.

1669   Gian Domenico Cassini arrives in Paris, never to return to Italy. Isaac Barrow resigns the Lucasian Chair of Mathematics in favor of Isaac Newton. Oxford University appoints Robert Morison as its first professor of botany with the task of supervising the botanical garden. Hennig Brand separates phosphorus from urine.

1670   The first issue of *Miscellanea Curiousa,* a periodical sponsored by the College of the Curiosities of Nature, is published; it concentrates on medical issues.

1672   Reinier de Graaf mistakenly identifies the Graafian follicles as female human eggs. A letter from Isaac Newton in *Philosophical Transactions* recounts his prism experiments, demonstrating that white light is composed of colored light. Molière's French comedy *The Learned Ladies* ridicules Cartesian salon women.

1673   The first letter of Antoni van Leeuwenhoek to be published in *Philosophical Transactions* appears. Francois Poullain de La Barre's tract arguing for the equality of the sexes on Cartesian grounds is published.

1675   Ole Rømer concludes from astronomical observation that the speed of light is finite. Greenwich Observatory is founded. Thomas Shadwell's play *The Virtuoso* ridicules virtuosi. Christiaan Huygens and Robert Hooke both successfully design and supervise the creation of spring watches, provoking a bitter priority dispute.

1677   Antoni van Leeuwenhoek's description of the male sperm is published. The death of Henry Oldenburg leads to the collapse of most of the Royal Society's foreign correspondence and eventually the suspension of *Philosophical Transactions.*

1679   Edmond Halley visits Johannes Hevelius's observatory in Danzig to evaluate the quality of Hevelius's observations made with instruments not fitted with telescopic sights; shortly after Halley's departure, Hevelius's observatory burns down.

1680   Gottfried Kirch becomes the first astronomer to discover a comet through the telescope. Robert Hooke sends a letter to Isaac Newton formulating an inverse-square law of attraction. Antoni van Leeuwenhoek is unanimously elected to the Royal Society.

1681   The increasingly harsh anti-Protestant policies of Louis XIV cause Christiaan Huygens to leave Paris and return to the Netherlands.

1682   Halley's comet appears.

1683   The first issue of the German journal *Acta Eruditorum* is published. *Philosophical Transactions* resumes publication under the editorship of Robert Plot. The death of Jean-Baptiste Colbert leads to cuts in the funding of the Royal Academy of Sciences.

1684   Edmond Halley's visit to Isaac Newton leads Newton to formulate his theory of universal gravitation. *New Division of the World among the Different Species or Races of Men That Inhabit It,* by French Gassendist and

1684 (cont.)   traveler Francois Bernier, expound early scientific racism. Shibukawa Harumi employs Chinese-language Jesuit works in creating Japan's first native calendar.

1685   Goverd Bidloo publishes *Anatomy of the Human Body,* the first atlas of the human body to incorporate microscopic data.

1686   Elias Ashmole founds the Ashmolean Museum in Oxford, England's first public museum, to house the Tradescant Collection. In an article in *Acta Eruditorum,* Gottfried Wilhelm Leibniz argues that force should be defined as mass × velocity^2 rather than mass × velocity.

1687   Newton's *Mathematical Principles of Natural Philosophy* is published. The French surgeon C. F. Felix successfully operates on Louis XIV for an anal fistula, contributing to a rise in the prestige of surgery. The College of the Curiosities of Nature receives patronage from the Holy Roman Emperor Leopold I, changing its name to the Leopoldine Academy.

1691   Dutch Cartesian Baltazar Bekker publishes *The Enchanted World,* arguing on philosophical grounds that witches and the devil have no supernatural powers. The Royal Academy of Sciences enjoys a revival under the patronage of Louis Phelypeaux de Ponchartrain. *Philosophical Transactions* resumes publication after a four-year hiatus. Robert Boyle dies; his will establishes the Boyle Lectures. John Dunton founds his periodical, the *Athenian Mercury,* which includes

scientific popularization and runs until 1697.

1692   The first Boyle Lectures are delivered by Richard Bentley, with help from Isaac Newton.

1694   Rudolph Camerarius gives the first detailed explanation of plant sexuality.

1695   Hans Sloane takes over and revitalizes *Philosophical Transactions.*

1696   Johann Bernoulli announces a mathematical contest to find the brachistochronous curve.

1697   Bernard Le Bouvier de Fontenelle becomes secretary to the Royal Academy of Sciences, a position he will hold until 1740.

1698–1701   Edmond Halley undertakes his Atlantic voyages in search of astronomical and geomagnetic information.

1699   The Royal Academy of Sciences is reorganized and allows the admission of Cartesians, including the Oratorian Nicholas Malebranche. Thomas Savery's steam engine is demonstrated at a meeting of the Royal Society. The Gregorian Calendar is adopted by Denmark and the Protestant states of the Holy Roman Empire.

1699–1701   Anna Maria Sibylla Merian and her daughter Dorothea gather natural-historical knowledge and specimens in the Dutch colony of Surinam.

1700   The Berlin Academy of Sciences is founded, with Gottfried Wilhelm Leibniz as its first president. The first Spanish scientific society, the Royal Society of Medicine and Other Sciences, is founded in Seville.

1702 The French Royal Academy of Sciences begins to publish annual reports. William Whiston succeeds Isaac Newton as Lucasian Professor of Mathematics.

1703 Isaac Newton becomes president of the Royal Society. The English House of Lords gives apothecaries the right to practice medicine. Edmond Halley succeeds John Wallis as Savilian Professor of Geometry after Wallis's death. Hermann Boerhaave is appointed professor of medicine at the University of Leiden.

1704 Isaac Newton's *Opticks* is published. The first issue of the English periodical *The Ladies Diary* appears, introducing women to mathematics and science.

1705 Edmond Halley's *Synopsis of Cometary Astronomy* predicts the return of the comet of 1682 in 1758.

1707 The publication of Tobias Cohn's Hebrew textbook of medicine and natural philosophy spreads knowledge of modern science among the Jewish community.

1709 On Isaac Newton's recommendation, John Flamsteed is discharged from the Royal Society for nonpayment of dues.

1710 Maria Kirch petitions the Berlin Academy of Sciences for an appointment as astronomer following the death of her husband,

Gottfried Kirch, and is denied on account of gender. William Whiston is expelled from Cambridge and forfeits his chair for heresy.

1712 Edmond Halley publishes an abbreviated version of John Flamsteed's astronomical data, igniting a feud between Flamsteed on one side and Halley and Isaac Newton on the other. A committee of the Royal Society, chosen by Newton, publishes a report that backs Newton against Gottfried Wilhelm Leibniz in the controversy over the origin of the calculus.

1714 Edmond Halley takes over editorship of *Philosophical Transactions* from Hans Sloane, reorienting it from natural history to mathematics, astronomy, and physics. This is the approximate date of D. G. Fahrenheit's introduction of the temperature scale.

1715 Two issues of *Philosophical Transactions* are devoted to attacks on Gottfried Wilhelm Leibniz.

1715– The Leibniz-Clarke controversy
1716 rages.

1719 John Flamsteed dies; Edmond Halley succeeds him the next year as royal astronomer.

1720 Japanese shogunate liberalizes the law banning foreign books, exposing Japanese intellectuals to Western science.

1723 Antoni van Leeuwenhoek dies, and his work appears for the last time in *Philosophical Transactions*.

1724   The St. Petersburg Academy of
Sciences is founded.

1725   John Flamsteed's star catalog is
published posthumously in a version
that he had authorized.

1727   Isaac Newton dies and is succeeded
as president of the Royal Society by
Sir Hans Sloane.

Bibliography and Web Sites

Bibliography

Adamson, Donald. *Blaise Pascal: Mathematician, Physicist, and Thinker about God.* New York: St. Martin's, 1995.

Adamson, J. R. "The Administration of Gresham College and Its Fluctuating Fortunes as a Scientific Institution in the Seventeenth Century." *History of Education* 9 (no. 1, March 1980): 13–25.

Adelmann, Howard B. *Marcello Malpighi and the Evolution of Embryology.* 5 vols. Ithaca: Cornell University Press, 1966.

Akerman, Susanna. *Rose Cross over the Baltic: The Spread of Rosicrucianism in Northern Europe.* Leiden, the Netherlands: Brill, 1998.

Ambrosoli, Mauro. *The Wild and the Sown: Botany and Agriculture in Western Europe: 1350–1850.* Translated by Mary McCann Salvatori. Cambridge and New York: Cambridge University Press, 1997.

Arber, Agnes. *Herbals, Their Origin and Evolution: A Chapter in the History of Botany, 1470–1670.* 2d ed. Cambridge: Cambridge University Press, 1938.

Armitage, Angus. *Copernicus: The Founder of Modern Astronomy.* London: Allen and Unwin, 1938.

Ashworth, William. "Natural History and the Emblematic World View." In David C. Lindberg and Robert S. Westman, eds., *Reappraisals of the Scientific Revolution.* New York: Cambridge University Press, 1990.

Atkinson, Dwight. *Scientific Discourse in Historical Context: The Philosophical Transactions of the Royal Society of London, 1675–1975.* Mahwah, NJ: L. Erlbaum Associates, 1999.

Barrett, C. R. B. *The History of the Society of Apothecaries of London.* London: E. Stock, 1905.

Beekman, E. M. Introduction to *The Ambonese Curiosity Cabinet,* by Georgius Eberhardus Rumphius, trans. by E. M. Beekman. New Haven, CT: Yale University Press, 1999.

Bell, A. E. *Christian Huygens and the Development of Science in the Seventeenth Century.* London: Arnold, 1947.

Bennet, J. A. *The Mathematical Science of Christopher Wren.* Cambridge: Cambridge University Press, 1982.

Beretta, Marco. *The Enlightenment of Matter: The Definition of Chemistry from Agricola to Lavoisier.* Canton, MA: Science History Publications, 1993.

Biagioli, Mario. "Scientific Revolution, Social Bricolage, and Etiquette." In Roy Porter and Mikulas Teich, eds., *The Scientific Revolution in National Context.* Cambridge and New York: Cambridge University Press, 1992.

———. *Galileo, Courtier: The Practice of Science in the Culture of Absolutism.* Chicago: University of Chicago Press, 1993.

Blackwell, Richard J. *Galileo, Bellarmine, and the Bible: Including a Translation of Foscarini's Letter on the Motion of the Earth.* Notre Dame, IN: University of Notre Dame Press, 1991.

Bloch, Marc. *The Royal Touch: Sacred Monarchy and Scrofula in England and France.* Translated by J. E. Anderson. London: Routledge and Kegan Paul, 1973.

Boorstin, Daniel. *The Discoverers: A History of Man's Search to Know His World and Himself.* New York: Random House, 1983.

Bowen, James. *A History of Western Education.* Vol. 3, *The Modern West: Europe and the New World.* London: Methuen, 1981.

Bowen, Margarita. *Empiricism and Geographical Thought: From Francis Bacon to Alexander von Humboldt.* Cambridge and New York: Cambridge University Press, 1981.

Briggs, John C. *Francis Bacon and the Rhetoric of Nature.* Cambridge: Harvard University Press, 1989.

Brock, William H. *The Norton History of Chemistry.* New York: W. W. Norton, 1992.

Brooke, John Hedley. *Science and Religion: Some Historical Perspectives.* Cambridge: Cambridge University Press, 1991.

Brown, Harcourt. *Scientific Organization in Seventeenth-Century France (1620–1680).* Baltimore: Johns Hopkins University Press, 1934.

Brundell, Barry. *Pierre Gassendi: From Aristotelianism to a New Natural Philosophy.* Dordrecht, the Netherlands: Reidel, 1987.

Burgaleta, Claudio M., S.J. *José de Acosta, S.J. (1540–1600): His Life and Thought.* Chicago: Jesuit Way, 1999.

Burns, William E. "'Our Lot Is Fallen into an Age of Wonders': John Spencer and the Controversy over Prodigies in the Early Restoration." *Albion* 27 (1995): 237–252.

Burtt, E. A. *The Metaphysical Foundations of Modern Physical Science.* 2d ed., rev. Garden City, NY: Doubleday, 1954.

Bylebyl, Jerome J., ed. *William Harvey and His Age: The Professional and Social Context of the Discovery of the Circulation.* Baltimore: Johns Hopkins University Press, 1979.

Cartwright, David Edgar. *Tides: A Scientific History.* New York: Cambridge University Press, 1999.

Caspar, Max. *Kepler.* Translated from the German by C. Doris Hellman. London: Abelard-Schuman, 1959.

Christianson, J. R. *On Tycho's Island: Tycho Brahe and His Assistants, 1570–1601.* Cambridge and New York: Cambridge University Press, 2000.

Clark, Sir George. *A History of the Royal College of Physicians of London.* 3 vols. Oxford: Clarendon Press for the Royal College of Physicians, 1966.

Clark, Stuart. *Thinking with Demons: The Idea of Witchcraft in Early Modern Europe.* Oxford: Clarendon Press, 1997.

Clatterbaugh, Kenneth. *The Causation Debate in Modern Philosophy, 1637–1739.* New York: Routledge, 1999.

Clulee, Nicholas H. *John Dee's Natural Philosophy: Between Science and Religion.* London: Routledge, 1988.

Cohen, H. Floris. *Quantifying Music: The Science of Music at the First Stage of the Scientific Revolution.* Dordrecht, the Netherlands: Reidel, 1984.

————. *The Scientific Revolution: A Historiographical Enquiry.* Chicago: University of Chicago Press, 1994.

Cohen, I. Bernard. *The Birth of a New Physics.* New York: W. W. Norton, 1985.

————, ed. *Puritanism and the Rise of Modern Science: The Merton Thesis.* New Brunswick, NJ: Rutgers University Press, 1990.

Conrad, Lawrence I., Michael Neve, Vivian Nutton, Roy Porter, and Andrew Wear. *The Western Medical Tradition: 800 B.C. to A.D. 1800.* Cambridge: Cambridge University Press, 1995.

Cook, Alan. *Edmond Halley: Charting the Heavens and the Seas.* Oxford: Clarendon Press, 1998.

Cook, Harold. *The Decline of the Old Medical Regime in Stuart London.* Ithaca: Cornell University Press, 1986.

Cooke, Roger. *The History of Mathematics: A Brief Course.* New York: Wiley, 1997.

Cottingham, John, ed. *The Cambridge Companion to Descartes.* Cambridge: Cambridge University Press, 1992.

Coudert, Allison P. *The Impact of the Kabbalah in the Seventeenth Century: The Life and Thought of Francis Mercury Van Helmont (1614–1698).* Leiden, the Netherlands: Brill, 1999.

Coyne, G. V., S.J., M. A. Hoskin, and O. Pedersen, eds. *Gregorian Reform of the Calendar: Proceedings of the Vatican Conference to Commemorate Its 400th Anniversary, 1582–1982.* Vatican City: Specolo Vaticana, 1983.

Cranston, Maurice. *John Locke: A Biography.* New York: Macmillan, 1957.

Crosby, Alfred W. *The Measure of Reality: Quantification and Western Society, 1250–1600.* Cambridge: Cambridge University Press, 1997.

Cunningham, Andrew. *The Anatomical Renaissance: The Resurrection of the Anatomical Projects of the Ancients.* Aldershot, England: Scolar Press, 1997.

Curry, Patrick. *Prophecy and Power: Astrology in Early Modern England.* Princeton: Princeton University Press, 1989.

Daston, Lorraine, and Katharine Park. *Wonders and the Order of Nature, 1150–1750.* New York: Zone Books, 1998.

Davis, Natalie Zemon. *Women on the Margins: Three Seventeenth-Century Lives.* Cambridge: Harvard University Press, 1995.

Dear, Peter. *Discipline and Experience: The Mathematical Way in the Scientific Revolution.* Chicago: University of Chicago Press, 1995.

Debus, Allen G. *The Chemical Philosophy: Paracelsian Science and Medicine in the Sixteenth and Seventeenth Centuries.* 2 vols. New York: Science History Publications, 1977.

————, ed. *Science and Education in the Seventeenth Century: The Webster-Ward Debate.* New York: American Elsevier, 1970.

Debus, Allen G., and Michael T. Walton, eds. *Reading the Book of Nature: The Other Side of the Scientific Revolution.* Kirksville, MO: Sixteenth Century Journal Publishers, 1998.

De Figuerido, John M. "Ayurvedic Medicine in Goa According to the European Sources in the Sixteenth and Seventeenth Centuries." *Bulletin of the History of Medicine* 58 (1984): 225–235.

de Leon-Jones, Karen Silvia. *Giordano Bruno and the Kabbalah: Prophets, Magicians, and Rabbis.* New Haven, CT: Yale University Press, 1997.

De Ridder-Symoens, Hilde, ed. *Universities in Early Modern Europe.* Cambridge and New York: Cambridge University Press, 1996.

Dewhurst, Kenneth. *Dr. Thomas Sydenham (1624–1689): His Life and Original Writings.* London: Wellcome Historical Medical Library, 1966.

Dickenson, Victoria. *Drawn from Life: Science and Art in the Portrayal of the New World.* Toronto: University of Toronto Press, 1998.

Dijksterhuis, E. J. *Simon Stevin: Science in the Netherlands around 1600.* The Hague: Nijhoff, 1970.

DiLiscia, Daniel, Eckhard Kessler, and Charlotte Methuen, eds. *Method and Order in Renaissance Philosophy of Nature: The Aristotle Commentary Tradition.* Aldershot, England: Ashgate, 1997.

Dobbs, B. J. T. *The Foundations of Newton's Alchemy: or "The Hunting of the Greene Lyon."* Cambridge: Cambridge University Press, 1975.

————. *The Janus Faces of Genius: The Role of Alchemy in Newton's Thought.* Cambridge: Cambridge University Press, 1991.

Dobbs, B. J. T., and Margaret Jacob. *Newton and the Culture of Newtonianism.* Atlantic Highlands, NJ: Humanities Press, 1994.

Dobell, Clifford. *Antoni van Leeuwenhoek and His "Little Animals."* New York: Russell and Russell, 1958.

Dobryzyicki, Jerzy, ed. *The Reception of Copernicus' Heliocentric Theory: Proceedings of a Symposium Organised by the Nicholas Copernicus Committee of the International Union of the History and Philosophy of Science.* Dordrecht, the Netherlands, and Boston: Reidel, 1972.

Donnely, Marian Card. *A Short History of Observatories.* Eugene: University of Oregon Books, 1973.

Dreyer, J. L. E. *Tycho Brahe: A Picture of Scientific Life and Work in the Sixteenth Century.* Edinburgh: A. & C. Black, 1890.

Dugas, Rene. *A History of Mechanics.* Translated by J. R. Maddox. New York: Dover Publications, 1988.

Duval, Marguerite. *The King's Garden.* Translated by Annette Tomarken and Claudine Cowen. Charlottesville: University Press of Virginia, 1982.

Eamon, William. *Science and the Secrets of Nature: Books of Secrets in Medieval and Early Modern Culture.* Princeton: Princeton University Press, 1994.

Eisenstein, Elizabeth. *The Printing Press as an Agent of Change: Communications and Cultural Transformations in Early Modern Europe.* 2 vols. Cambridge: Cambridge University Press, 1979.

Eliav-Feldon, Miriam. *Realistic Utopias: The Ideal Imaginary Societies of the Renaissance.* Oxford: Clarendon Press, 1982.

Emerton, Norma E. *The Scientific Reinterpretation of Form.* Ithaca: Cornell University Press, 1984.

Eriksson, Gunnar. *The Atlantic Vision: Olaus Rudbeck and Baroque Science.* Canton, MA: Science History Publications, 1994.

Espinasse, Margaret. *Robert Hooke.* London: William Heinemann, 1956.

Eurich, Nell. *Science in Utopia: A Mighty Design.* Cambridge: Harvard University Press, 1967.

Evans, R. J. W. *Rudolf II and His World: A Study in Intellectual History, 1576–1612.* Oxford: Clarendon Press, 1973.

Feingold, Mordechai. *The Mathematicians' Apprenticeship: Science, Universities, and Society in England, 1560–1640.* Cambridge: Cambridge University Press, 1984.

Ferrone, Vincenzo. *The Intellectual Roots of the Italian Enlightenment: Newtonian Science, Religion, and Politics in the Early Eighteenth Century.* Translated by Sue Brotherton. Atlantic Highlands, NJ: Humanities Press, 1995.

Field, J. V. *The Invention of Infinity: Mathematics and Art in the Renaissance.* Oxford: Oxford University Press, 1997.

Findlen, Paula. *Possessing Nature: Museums, Collecting, and Scientific Culture in Early Modern Italy.* Berkeley and Los Angeles: University of California Press, 1994.

Foster, Sir Michael. *Lectures on the History of Physiology during the Sixteenth, Seventeenth, and Eighteenth Centuries.* Cambridge: Cambridge University Press, 1901.

Fournier, Marian. *The Fabric of Life: Microscopy in the Seventeenth Century.* Baltimore: Johns Hopkins University Press, 1996.

French, Peter. *John Dee: The World of an Elizabethan Magus.* London: Routledge, 1972.

French, Roger. *William Harvey's Natural Philosophy.* Cambridge and New York: Cambridge University Press, 1994.

———.*Dissection and Vivisection in the European Renaissance.* Aldershot, England: Ashgate, 1999.

Friedenwald, Harry. *The Jews and Medicine: Essays.* 2 vols. Baltimore: The Johns Hopkins University Press, 1944.

Funkenstein, Amos. *Theology and the Scientific Imagination from the Middle Ages to the Seventeenth Century.* Princeton: Princeton University Press, 1986.

Fussell, G. E. *Farms, Farmers, and Society: Systems of Food Production and Population Numbers.* Lawrence, KS: Coronado, 1976.

Garrett, Don, ed. *The Cambridge Companion to Spinoza.* Cambridge: Cambridge University Press, 1996.

Gascoigne, John. *Cambridge in the Age of the Enlightenment: Science, Religion, and Politics from the Restoration to the French Revolution.* Cambridge: Cambridge University Press, 1989.

———. "A Reappraisal of the Role of the Universities in the Scientific Revolution." In Robert S. Westman and David Lindberg, eds., *Reappraisals of the Scientific Revolution.* Cambridge: Cambridge University Press, 1990.

Gatti, Hilary. *Giordano Bruno and Renaissance Science.* Ithaca: Cornell University Press, 1999.

Gaukroger, Stephen. *Descartes: An Intellectual Biography.* Oxford: Oxford University Press, 1995.

———, ed. *The Uses of Antiquity: The Scientific Revolution and the Classical Tradition.* Dordrecht, the Netherlands: Kluwer Academic Publishers, 1991.

Genuth, Sara Schechner. *Comets, Popular Culture, and the Birth of Modern Cosmology.* Princeton: Princeton University Press, 1997.

Geymonat, Ludovico. *Galileo Galilei: A Biography and Inquiry into His Philosophy of Science.* Translated from the Italian with additional notes and appendix by Stillman Drake. New York: McGraw-Hill, 1965.

Gillispie, Charles Coulston. *The Edge of Objectivity: An Essay in the History of Scientific Ideas.* Princeton: Princeton University Press, 1960.

Glanvill, Joseph. *Saducismus Triumphatus; or, Full and Plain Evidence Concerning Witches and Appar[i]tions (1689).* A facsimile reproduction, with an introduction by Coleman O. Parsons. Gainesville, FL: Scholar's Facsimiles & Reprints, 1966.

Gohau, Gabriel. *A History of Geology.* Revised and translated by Albert V. Carozzi and Marguerite Carozzi. New Brunswick, NJ: Rutgers University Press, 1990.

Goldgar, Anne. *Impolite Learning: Conduct and Community in the Republic of Letters, 1680–1750.* New Haven, CT: Yale University Press, 1995.

Goodman, David. "The Scientific Revolution in Spain and Portugal." In Roy Porter and Mikulas Teich, eds., *The Scientific Revolution in National Context.* Cambridge: Cambridge University Press, 1992.

Gouk, Penelope. *Music, Science, and Natural Magic in Seventeenth-Century England.* New Haven, CT: Yale University Press, 1999.

Grafton, Anthony. *Joseph Scaliger: A Study in the History of Classical Scholarship.* Vol. 2, *Historical Chronology.* Oxford: Oxford University Press, 1993.

Grafton, Anthony, with April Shelford and Nancy Siraisi. *New Worlds, Ancient Texts: The Power of Tradition and the Shock of Discovery.* Cambridge: Belknap Press of Harvard University Press, 1992.

Grafton, Anthony, and Nancy Siraisi, eds. *Natural Particulars: Nature and the Disciplines in Renaissance Europe.* Cambridge: MIT Press, 1999.

Grant, Edward. *Planets, Stars, and Orbs: The Medieval Cosmos, 1200–1687.* Cambridge: Cambridge University Press, 1994.

Grattan-Guiness, Ivor. *The Norton History of the Mathematical Sciences: The Rainbow of Mathematics.* New York: W. W. Norton, 1998.

Hacking, Ian. *The Emergence of Probability: A Philosophical Study of Early Ideas about Probability, Induction, and Statistical Inference.* Cambridge: Cambridge University Press, 1975.

Hadden, Richard W. *On the Shoulders of Merchants: Exchange and the Mathematical Conception of Nature in Early Modern Europe.* Albany: State University of New York Press, 1994.

Hahn, Roger. *The Anatomy of a Scientific Institution: The Paris Academy of Sciences, 1666–1803.* Berkeley, Los Angeles, and London: University of California Press, 1971.

Hall, A. Rupert. *The Scientific Revolution, 1500–1800: The Formation of the Modern Scientific Attitude.* London: Longmans, 1954.

————. *Science and Society: Historical Essays on the Relations of Science, Technology, and Medicine.* Aldershot, England: Variorum, 1994.

Hall, Marie B. *Promoting Experimental Learning: Experiment and the Royal Society, 1660–1727.* Cambridge: Cambridge University Press, 1991.

Hall, Thomas S. *Ideas of Life and Matter: Studies in the History of General Physiology, 600 B.C.–1900 A.D.* Vol. 1, *From Pre-Socratic Times to the Enlightenment.* Chicago: University of Chicago Press, 1969.

Hanafi, Zakiya. *The Monster in the Machine: Magic, Medicine, and the Marvelous in the Time of the Scientific Revolution.* Durham, NC, and London: Duke University Press, 2000.

Hannaford, Ivan. *Race: The History of an Idea in the West.* Washington, DC: Woodrow Wilson Center Press, 1996.

Hannaway, Owen. *The Chemists and the Word: The Didactic Origins of Chemistry.* Baltimore: Johns Hopkins University Press, 1975.

————. "Laboratory Design and the Aim of Science." *Isis* 77 (1986): 585–610.

Harrison, Peter. *The Bible, Protestantism, and the Rise of Natural Science.* Cambridge and New York: Cambridge University Press, 1998.

Headley, John. *Tommaso Campanella and the Transformation of the World.* Princeton: Princeton University Press, 1997.

Hobart, Michael E. *Science and Religion in the Thought of Nicholas Malebranche.* Chapel Hill: University of North Carolina Press, 1982.

Hooykaas, Reijer. *Religion and the Rise of Modern Science.* Edinburgh: Scottish Academic, 1972.

Howson, J. B. *A History of the Practice of Navigation.* Glasgow: Brown, Son and Ferguson, 1951.

Hunter, Andrew, ed. *Thornton and Tully's Scientific Books, Libraries, and Collectors: A Study of Bibliography and the Book Trade in Relation to the History of Science.* 4th ed. Brookfield, VT: Ashgate, 2000.

Hunter, Lynette, and Sarah Hutton, eds. *Women, Science, and Medicine, 1500–1700: Mothers and Sisters of the Royal Society.* Stroud, England: Sutton Publishing, 1997.

Hunter, Michael. *The Royal Society and Its Fellows: The Morphology of an Early Scientific Institution.* Chalfont St. Giles, England: British Society for the History of Science, 1982.

————. *Establishing the New Science: The Experience of the Early Royal Society.* Woodbridge, England: Boydell, 1989.

————. *Science and the Shape of Orthodoxy: Intellectual Change in Late Seventeenth-Century Britain.* Woodbridge, England: Boydell, 1995.

————, ed. *Robert Boyle Reconsidered.* Cambridge: Cambridge University Press, 1994.

————, ed. *Archives of the Scientific Revolution: The Formation and Exchange of Ideas in Seventeenth-Century Europe.* Woodbridge, England: Boydell, 1998.

Hunter, Michael, and Simon Schaffer, eds. *Robert Hooke: New Studies.* Woodbridge, England: Boydell, 1989.

Impey, Oliver, and Arthur MacGregor, eds. *The Origins of Museums: The Cabinet of Curiosities in Sixteenth- and Seventeenth-Century Europe.* Oxford: Clarendon, 1985.

Jacob, Francois. *The Logic of Life: A History of Heredity.* Translated by Betty E. Spillman. New York: Pantheon Books, 1973.

Jacob, Margaret. *The Newtonians and the English Revolution, 1689–1720.* Ithaca: Cornell University Press, 1976.

————. *Scientific Culture and the Making of the Industrial West.* New York: Oxford University Press, 1997.

Jardine, Lisa. *Ingenious Pursuits: Building the Scientific Revolution.* New York: Nan A. Talese, 1999.

Jardine, Nicholas, Emma Spary, and James A. Secord, eds. *Cultures of Natural History.* Cambridge: Cambridge University Press, 1996.

Johns, Adrian. *The Nature of the Book: Print and Knowledge in the Making.* Chicago: University of Chicago Press, 1998.

Johnson, Francis R. *Astronomical Thought in Renaissance England: A Study of the English Scientific Writings from 1500 to 1645.* Baltimore: The Johns Hopkins University Press, 1937.

Jolley, Nicholas, ed. *The Cambridge Companion to Leibniz.* Cambridge: Cambridge University Press, 1995.

Jones, Kathleen. *A Glorious Fame: The Life of Margaret Cavendish, Duchess of Newcastle, 1623–1673.* London: Bloomsbury, 1988.

Jones, R. F. *Ancients and Moderns: A Study of the Rise of the Scientific Movement in Seventeenth-Century England.* 2d ed., with an index, new preface, and minor revisions. Gloucester, MA: P. Smith, 1975.

Kearney, Hugh. *Science and Change, 1500–1700.* New York: McGraw-Hill, 1971.

Keynes, Geoffrey. *The Life of William Harvey.* Oxford: Clarendon Press, 1966.

Koyre, Alexander. *From the Closed World to the Infinite Universe.* Baltimore: Johns Hopkins University Press, 1957.

———. *The Astronomical Revolution: Copernicus-Kepler-Borelli.* Translated from the French by Dr. R. E. W. Maddison. Ithaca: Cornell University Press, 1973.

Kremers, Edward, and George Urdang. *History of Pharmacy: A Guide and a Survey.* 2d ed., revised and enlarged. Philadelphia: Lippincott, 1951.

Kristeller, Paul Oskar. *Eight Philosophers of the Italian Renaissance.* Stanford: Stanford University Press, 1964.

Kronick, David A. *A History of Scientific and Technical Periodicals: The Origins and Development of the Scientific and Technical Press, 1665–1790.* 2d ed. Metuchen, NJ: Scarecrow Press, 1976.

Kuhn, Thomas S. *The Copernican Revolution: Planetary Astronomy in the Development of Western Thought.* Cambridge: Harvard University Press, 1957.

Kusukawa, Sachiko. *The Transformation of Natural Philosophy: The Case of Philip Melanchthon.* Cambridge and New York: Cambridge University Press, 1995.

Landes, David. *Revolution in Time: Clocks and the Making of the Modern World.* Cambridge: Harvard University Press, 1983.

Laqueur, Thomas. *Making Sex: Body and Gender from the Greeks to Freud.* Cambridge: Harvard University Press, 1990.

Lattis, James M. *Between Copernicus and Galileo: Christoph Clavius and the Collapse of Ptolemaic Cosmology.* Chicago: University of Chicago Press, 1994.

Lennon, Thomas M. *The Battle of the Gods and Giants: The Legacies of Descartes and Gassendi, 1655–1715.* Princeton: Princeton University Press, 1993.

Ley, Willy. *Dawn of Zoology.* Englewood Cliffs, NJ: Prentice-Hall, 1968.

Lindberg, David C., and Ronald L. Numbers, eds. *God and Nature: Historical Essays on the Encounter between Christianity and Science.* Berkeley and Los Angeles: University of California Press, 1986.

Lindberg, David C., and Robert S. Westman, eds. *Reappraisals of the Scientific Revolution.* Cambridge: Cambridge University Press, 1990.

Livingstone, David N. *The Geographical Tradition: Episodes in the History of a Contested Enterprise.* Oxford: Blackwell, 1993.

Long, Pamela O. "The Openness of Knowledge: An Ideal and Its Context in Sixteenth-Century Writings on Mining and Metallurgy." *Technology and Culture* 32 (1991): 318–355.

Lovejoy, Arthur O. *The Great Chain of Being: A Study of the History of an Idea.* Cambridge: Harvard University Press, 1936.

MacPike, E. F. *Hevelius, Flamsteed, and Halley: Three Contemporary Astronomers and Their Mutual Relations.* London: Taylor and Francis, 1937.

Maddison, R. E. W. *The Life of the Honourable Robert Boyle, F.R.S.* London: Taylor and Francis, 1969.

Mahoney, Michael Sean. *The Mathematical Career of Pierre de Fermat, 1601–1665.* 2d ed. Princeton: Princeton University Press, 1994.

Mandrou, Robert. *From Humanism to Science, 1480–1700.* Translated by Brian Pearce. N.p: Penguin Books, 1978.

Marland, Hilary, ed. *The Art of Midwifery: Early Modern Midwives in Europe.* New York: Routledge, 1993.

Martin, Julian. *Francis Bacon, the State, and the Reform of Natural Philosophy.* Cambridge and New York: Cambridge University Press, 1992.

Mazzolini, Renato G., ed. *Non-Verbal Communication in Science Prior to 1900.* Florence: L. S. Olschki, 1993.

McClellan, James E. *Science Reorganized: Scientific Societies in the Eighteenth Century.* New York: Columbia University Press, 1985.

McEwen, Gilbert D. *The Oracle of the Coffee House: John Dunton's Athenian Mercury.* San Marino, CA: Huntington Library, 1972.

Merchant, Carolyn. *The Death of Nature: Women, Ecology, and the Scientific Revolution.* San Francisco: Harper & Row, 1980.

Merkel, Ingrid, and Allen G. Debus, eds. *Hermeticism and the Renaissance: Intellectual History and the Occult in Early Modern Europe.* Washington, DC: Folger Shakespeare Library, 1988.

Merton, Robert K. *Science, Technology, and Society in Seventeenth-Century England.* New York: Howard Fertig, 1970.

Middleton, W. E. Knowles. *Invention of the Meteorological Instruments.* Baltimore: The Johns Hopkins University Press, 1969.

———. *The Experimenters: A Study of the Accademia del Cimento.* Baltimore: Johns Hopkins University Press, 1971.

Moran, Bruce, ed. *Patronage and Institutions: Science, Technology, and Medicine at the European Court, 1500–1750.* Woodbridge, England: Boydell, 1991.

Morton, A. G. *History of Botanical Science: An Account of the Development of Botany from Ancient Times to the Present Day.* London: Academic Press, 1981.

Needham, Joseph. *A History of Embryology.* 2d ed., revised with the assistance of Arthur Hughes. Cambridge: Cambridge University Press, 1959.

Newman, William R. *Gehennical Fire: The Lives of George Starkey, an American Alchemist in the Scientific Revolution.* Cambridge: Harvard University Press, 1994.

Nicolson, Marjorie Hope. *The Breaking of the Circle: Studies in the Effect of the "New Science" upon Seventeenth-Century Poetry.* Evanston, IL: Northwestern University Press, 1950.

———. *Science and Imagination.* Ithaca: Great Seal Books, 1956.

———. *Pepys' Diary and the New Science.* Charlottesville: University Press of Virginia, 1965.

———, ed. *The Conway Letters: The Correspondence of Anne, Viscountess Conway, Henry More, and Their Friends.* Rev. ed. with an introduction and new material, edited by Sarah Hutton. Oxford: Clarendon Press, 1992.

North, John. *The Norton History of Astronomy and Cosmology.* New York: Norton, 1995.

Oakley, Francis. *Omnipotence, Covenant, and Order: An Excursion in the History of Ideas from Abelard to Leibniz.* Ithaca: Cornell University Press, 1984.

Oldroyd, D. R. *Thinking about the Earth: A History of Ideas in Geology.* Cambridge: Harvard University Press, 1996.

Olson, Richard S. *Science Deified and Science Defied: The Historical Significance of Science in Western Culture.* Vol. 1, *From the Bronze Age to the Beginnings of the Modern Era, ca. 3500 B.C. to ca. A.D. 1640.* Berkeley, Los Angeles, and London: University of California Press, 1982.

————. *Science Deified and Science Defied: The Historical Significance of Science in Western Culture.* Vol. 2, *From the Early Modern Age through the Early Romantic Era, ca. 1640 to ca. 1820.* Berkeley, Los Angeles, and Oxford: University of California Press, 1990.

O'Malley, Charles Donald. *Andreas Vesalius of Brussels, 1514–1564.* Berkeley and Los Angeles: University of California Press, 1964.

O'Malley, John W., S.J., and Garvin Alexander Bailey, Steven J. Harris, and J. Frank Kennedy, S.J., eds. *The Jesuits: Cultures, Sciences and the Arts, 1540–1773.* Toronto: University of Toronto Press, 1999.

Ong, Walter J. *Ramus: Method and the Decay of Dialogue; from the Art of Discourse to the Art of Reason.* Cambridge: Harvard University Press, 1958.

Ore, Oystein. *Cardano: The Gambling Scholar.* Princeton: Princeton University Press, 1953.

Osler, Margaret J. *Divine Will and the Mechanical Philosophy: Gassendi and Descartes on Contingency and Necessity in the Created World.* Cambridge: Cambridge University Press, 1994.

————, ed. *Atoms, Pneuma, and Tranquility: Epicurean and Stoic Themes in European Thought.* Cambridge: Cambridge University Press, 1991.

————, ed. *Rethinking the Scientific Revolution.* Cambridge: Cambridge University Press, 2000.

Pagden, Anthony. *The Fall of Natural Man: The American Indian and the Origins of Comparative Ethnology.* Cambridge: Cambridge University Press, 1982.

Pagel, Walter. *Paracelsus: An Introduction to Philosophical Medicine in the Era of the Renaissance.* Basel, Switzerland, and New York: S. Karger, 1958.

————. *Joan Baptista van Helmont: Reformer of Science and Medicine.* Cambridge: Cambridge University Press, 1982.

Park, Katherine. "The Rediscovery of the Clitoris: French Medicine and the Tribade, 1570–1620." In David Hillman and Carla Mazzio, eds., *The Body in Parts: Fantasies of Corporeality in Early Modern Europe.* New York: Routledge, 1997: 170–193.

Patai, Raphael. *The Jewish Alchemists: A History and Source Book.* Princeton: Princeton University Press, 1994.

Peck, Linda Levy. "Uncovering the Arundel Library at the Royal Society: Changing Meanings of Science and the Norfolk Donation." *Notes and Records of the Royal Society of London* 52 (1998): 3–24.

Petersson, R.T. *Sir Kenelm Digby: The Ornament of England, 1603–1665.* London: Jonathan Cape, 1956.

Phillips, Patricia. *The Scientific Lady: A Social History of Women's Scientific Interests, 1520–1918.* London: Weidenfeld and Nicolson, 1990.

Pinto-Correia, Clara. *The Ovary of Eve: Egg and Sperm and Preformation.* Chicago and London: University of Chicago Press, 1997.

Popkin, Richard H. *The History of Scepticism from Erasmus to Spinoza.* Rev. ed. Berkeley and Los Angeles: University of California Press, 1979.

Porter, Roy. *The Greatest Benefit to Mankind: A Medical History of Humanity.* New York: W. W. Norton, 1998.

Porter, Roy, and Mikulas Teich, eds. *The Scientific Revolution in National Context.* Cambridge: Cambridge University Press, 1992.

Prest, John. *The Garden of Eden: The Botanic Garden and the Re-Creation of Paradise.* New Haven, CT: Yale University Press, 1981.

Principe, Lawrence. *The Aspiring Adept: Robert Boyle and His Alchemical Quest, Including Boyle's "Lost" Dialogue on the Transmutation of Metals.* Princeton: Princeton University Press, 1998.

Pullman, Bernard. *The Atom in the History of Human Thought.* Translated by Axel Reisinger. New York and Oxford: Oxford University Press, 1998.

Qaisar, Ahsan Jan. *The Indian Response to European Culture and Technology,* A.D. *1498–1707.* Delhi: Oxford University Press, 1982.

Quinton, Anthony. *Francis Bacon.* Oxford: Oxford University Press, 1980.

Randall, John Herman. *The School of Padua and the Emergence of Modern Science.* Padua, Italy: Antenore, 1961.

Raven, Charles E. *John Ray, Naturalist: His Life and Works.* 2d ed., with an introduction by S. M. Walters. Cambridge and New York: Cambridge University Press, 1986.

Reeves, Eileen. *Painting the Heavens: Art and Science in the Age of Galileo.* Princeton: Princeton University Press, 1997.

Roller, Duane H. D. *The De Magnete of William Gilbert.* Amsterdam: Hertzberger, 1959.

Rosen, Edward. *Three Imperial Mathematicians: Kepler Trapped between Tycho Brahe and Ursus.* New York: Abaris Books, 1986.

Ruderman, David B. *Kabbalah, Magic, and Science: The Cultural Universe of a Sixteenth-Century Jewish Physician.* Cambridge: Harvard University Press, 1988.

———. *Jewish Thought and Scientific Discovery in Early Modern Europe.* New Haven, CT: Yale University Press, 1995.

Ruestow, Edward G. *The Microscope in the Dutch Republic: The Shaping of Discovery.* Cambridge and New York: Cambridge University Press, 1996.

Russell, G. A., ed. *The 'Arabick' Interest of the Natural Philosophers in Seventeenth-Century England.* Leiden, the Netherlands: E. J. Brill, 1994.

Sarasohn, Lisa. "Nicolas-Claude Fabri de Peiresc and the Patronage of the New Science in the Seventeenth Century." *Isis* 84 (1993): 70–90.

Schiebinger, Londa. *The Mind Has No Sex? Women in the Origins of Modern Science.* Cambridge: Harvard University Press, 1989.

Schieerbeek, A., and Maria Rosenboom. *Measuring the Invisible World: The Life and Works of Antoni Van Leeuwenhoek, FRS.* London: Abelard-Schuman, 1959.

Schleiner, Winfried. "Early Modern Controversies about the One-Sex Model." *Renaissance Quarterly* 53 (spring, 2000): 180–191.

Schmitt, Charles B. *Aristotle and the Renaissance.* Cambridge: Harvard University Press, 1983.

Shackelford, Joel. "Tycho Brahe, Laboratory Design, and the Aim of Science: Reading Plans in Context." *Isis* 84 (1993): 211–230.

Shaffer, Simon, and Steven Shapin. *Leviathan and the Air-Pump: Hobbes, Boyle, and the Experimental Life.* Princeton: Princeton University Press, 1985.

Shapin, Steven. "The House of Experiment in Seventeenth-Century England." *Isis* 79 (1988): 373–404.

———. *A Social History of Truth: Civility and Science in Seventeenth-Century England.* Chicago: University of Chicago Press, 1994.

Shapiro, Barbara. *John Wilkins, 1614–1672: An Intellectual Biography.* Berkeley and Los Angeles: University of California Press, 1969.

———. *Probability and Certainty in Seventeenth-Century England: A Study of the Relationships between Natural Science, Religion, History, Law, and Literature.* Princeton: Princeton University Press, 1983.

Shea, William R. *Galileo's Intellectual Revolution.* London: Macmillan, 1972.

Shirley, John W. *Thomas Harriot: A Biography.* Oxford: Clarendon Press, 1983.

Singh, Simon. *Fermat's Enigma: The Epic Quest to Solve the World's Greatest Mathematical Problem.* New York: Walker, 1997.

Siraisi, Nancy G. *Avicenna in Renaissance Italy: The Canon and Medical Teaching in Italian Universities after 1500.* Princeton: Princeton University Press, 1987.

———. *The Clock and the Mirror: Girolamo Cardano and Renaissance Medicine.* Princeton: Princeton University Press, 1997.

Sivin, Nathan. *Science in Ancient China: Researches and Reflections.* Aldershot, England: Variorum, 1995.

Slaughter, M. M. *Universal Languages and Scientific Taxonomy in the Seventeenth Century.* Cambridge: Cambridge University Press, 1982.

Slawinski, Maurice. "Rhetoric and Science/Rhetoric of Science/Rhetoric as Science." In Stephen Pumphrey, Paolo L. Rossi, and Maurice Slawinski, eds., *Science, Culture, and Popular Belief in Renaissance Europe.* Manchester, England: Manchester University Press, 1991: 71–99.

Smith, Pamela H. *The Business of Alchemy: Science and Culture in the Holy Roman Empire.* Princeton: Princeton University Press, 1994.

Sobel, Dava. *Longitude: The True Story of a Lone Genius Who Solved the Greatest Scientific Problem of His Time.* New York: Walker, 1995.

Solomon, Howard M. *Public Welfare, Science, and Propaganda in Seventeenth-Century France: The Innovation of Théophraste Renaudot.* Princeton: Princeton University Press, 1972.

Sorell, Tom, ed. *The Cambridge Companion to Hobbes.* Cambridge: Cambridge University Press, 1996.

Spencer, John. *Discourse Concerning Prodigies.* Cambridge: N.p., 1663.

Stephenson, Bruce. *Kepler's Physical Astronomy.* Princeton: Princeton University Press, 1987.

Stewart, Larry. *The Rise of Public Science: Rhetoric, Technology, and Natural Philosophy in Newtonian Britain, 1660–1750.* Cambridge: Cambridge University Press, 1992.

Stone, Richard. *Some British Empiricists in the Social Sciences, 1650–1900.* Cambridge: Cambridge University Press, 1997.

Storey, William K., ed. *Scientific Aspects of European Expansion.* Vol. 6, *An Expanding World.* Aldershot, England: Variorum, 1996.

Stroup, Alice. *A Company of Scientists: Botany, Patronage, and Community at the Seventeenth-Century Parisian Royal Academy of Sciences.* Berkeley and Los Angeles: University of California Press, 1990.

Sugimoto, Masayoshi, and David L. Swain. *Science and Culture in Traditional Japan, A.D. 600–1854.* Cambridge: MIT Press, 1978.

Sutton, Geoffrey V. *Science for a Polite Society: Gender, Culture, and the Demonstration of Enlightenment.* Boulder, CO: Westview Press, 1995.

Taton, Rene, and Curtis Wilson, eds. *Planetary Astronomy from the Renaissance to the Rise of Astrophysics. Part A: Tycho Brahe to Newton. The General History of Astronomy,* ed. M. Hoskin, vol. 2. Cambridge: Cambridge University Press, 1984.

Taylor, F. Sherwood. *The Alchemists: Founders of Modern Chemistry.* London: W. Heinemann, 1951.

Temkin, Owsei. *Galenism: Rise and Decline of a Medical Philosophy.* Ithaca and London: Cornell University Press, 1973.

Tester, S. J. *A History of Western Astrology.* Woodbridge, England, and Wolfeboro, NH: Boydell, 1987.

Thomas, Keith. *Religion and the Decline of Magic.* New York: Charles Scribner's Sons, 1971.

Thoren, Victor, with contributions by John R. Christianson. *The Lord of Uraniborg: A Biography of Tycho Brahe.* Cambridge: Cambridge University Press, 1990.

Thorndike, Lynn. *A History of Magic and Experimental Science.* 8 vols. New York: Macmillan, 1923–1958.

Tillyard, E. M. W. *The Elizabethan World Picture.* New York: Macmillan, 1944.

Toulmin, Stephen, and June Goodfield. *The Fabric of the Heavens: The Development of Astronomy and Dynamics.* New York: Harper, 1961.

———. *The Architecture of Matter.* New York: Harper and Row, 1962.

Turnbull, H. W., ed. *The Correspondence of Isaac Newton*. 3 vols. Cambridge: For the Royal Society at the University Press, 1959–1977.

Tyacke, Nicholas, ed. *The History of the University of Oxford*. Vol. 4, *Seventeenth-Century Oxford*. Oxford: Oxford University Press, 1997.

Vailati, Ezio. *Leibniz and Clarke: A Study of Their Correspondence*. New York: Oxford University Press, 1997.

Van Berkel, Klaas, Albert Van Helden, and Lodewijk Palm, eds. *A History of Science in the Netherlands: Survey, Themes, and Reference*. Leiden, the Netherlands: Brill, 1999.

Van Helden, Albert. "The Telescope in the Seventeenth Century." *Isis* 65 (1974): 38–58.

———. "The Invention of the Telescope." *Transactions of the American Philosophical Society* 67 (special issue) (1977).

Van Nouhuys, Tabitta. *The Age of Two-Faced Janus: The Comets of 1577 and 1618 and the Decline of the Aristotelian World View in the Netherlands*. Leiden, the Netherlands: Brill, 1998.

Vickers, Brian, ed. *English Science, Bacon to Newton*. New York: Cambridge University Press, 1987.

Wallace, William A. *Causality and Scientific Explanation*. 2 vols. Ann Arbor: University of Michigan Press, 1972 and 1974.

Wear, Andrew, R. K. French, and I. M. Lonie, eds. *The Medical Renaissance of the Sixteenth Century*. Cambridge: Cambridge University Press, 1985.

Webster, Charles. *The Great Instauration: Science, Medicine, and Reform, 1626–1670*. New York: Holmes and Meier, 1976.

———, ed. *Samuel Hartlib and the Advancement of Learning*. London: Cambridge University Press, 1970.

Weiner, Friedel, ed. *Laws of Nature: Essays on the Philosophical, Scientific, and Historical Dimensions*. Berlin: Walter de Gruyter, 1995.

Wertheim, Margaret. *Pythagoras' Trousers: God, Physics, and the Gender Wars*. New York: Random House, 1995.

Westfall, Richard S. *Science and Religion in Seventeenth-Century England*. New Haven, CT: Yale University Press, 1958.

———. *The Construction of Modern Science: Mechanisms and Mechanics*. New York: Wiley, 1971.

———. *Force in Newton's Physics: The Science of Dynamics in the Seventeenth Century*. London: Macdonald, 1971.

———. *Never at Rest: A Biography of Isaac Newton*. New York: Cambridge University Press, 1980.

———. *Essays on the Trial of Galileo*. Vatican City: Vatican Observatory, 1989.

Westman, Robert S. "The Melanchthon Circle, Rheticus, and the Wittenberg Interpretation of the Copernican Theory." *Isis* 65 (1974): 165–193.

———. "Proof, Poetics, and Patronage: Copernicus's Preface to *De Revolutionibus*." In Robert S. Westman and David Lindberg, eds., *Reappraisals of the Scientific Revolution*. New York: Cambridge University Press, 1990.

Wilcox, Donald J. *The Measure of Times Past: Pre-Newtonian Chronologies and Relative Time*. Chicago and London: University of Chicago Press, 1987.

Willmoth, Frances, ed. *Flamsteed's Stars: New Perspectives on the Life and Work of the First Astronomer Royal (1646–1719)*. Woodbridge, England: Boydell and Brewer, in association with the National Maritime Museum, 1997.

Wilson, Dudley. *Signs and Portents: Monstrous Births from the Middle Ages to the Enlightenment*. London and New York: Routledge, 1993.

Wolf, A., with the cooperation of F. Dannemann and A. Armitage. *A History of Science, Technology, and Philosophy in the Sixteenth and Seventeenth Centuries*. 2d ed., prepared by Douglas McKie. London: Allen and Unwin, 1950.

Woolfson, Jonathan. *Padua and the Tudors: English Students in Italy, 1485–1603.* Toronto: University of Toronto Press, 1998.

Yates, Frances. *Giordano Bruno and the Hermetic Tradition.* London: Routledge and Kegan Paul, 1964.

———. *The Rosicrucian Enlightenment.* London: Routledge and Kegan Paul, 1972.

Yeomans, Donald K. *Comets: A Chronological History of Observation, Science, Myth, and Folklore.* New York: Wiley, 1991.

Web Sites

The Alchemy Web Site and Virtual Library: http://www.levity.com/alchemy/home.html

Adam Maclean organized this elaborate database, sponsored by Levity Press. It includes alchemical texts and images, with a bibliography of modern articles on alchemy.

The Galileo Project: http://es.rice.edu/ES/humsoc/Galileo/

In addition to a wealth of information on Galileo and his contemporaries, this Rice University site contains Richard Westfall's "Catalog of the Scientific Community," a superb database on over 600 early modern scientists.

Images from the History of Medicine: http://wwwihm.nlm.nih.gov/

From the National Library of Medicine, this site contains thousands of images relevant to early modern medicine, including portraits, frontispieces, and diagrams.

Jesuit Scientists and the Jesuit Tradition: http://www.faculty.fairfield.edu/jmac/sjscient.htm

This site, created by Joseph MacDonald, S.J., includes biographical information, portraits, and illustrations relating to a number of early modern Jesuit scientists.

Jesuits and the Sciences, 1540–1995: http://www.luc.edu/libraries/science/jesuits/index.html

This resource from the Loyola University Chicago Web site contains a number of reproductions from early modern Jesuit scientific books, as well as discussion of Jesuit scientists.

The MacTutor History of Mathematics Archive: http://www-groups.dcs.st-andrews.ac.uk/~history/

From the School of Mathematics at the University of St. Andrews, this is an exhaustive database of mathematical biographies and treatments of specific problems in the history of mathematics.

Project Boulliau: http://web.clas.ufl.edu/users/rhatch/pages/11-ResearchProjects/boulliau/index.htm

Professor Robert Hatch, a historian of early modern French science at Cambridge University and the University of Florida, developed this site devoted to astronomer Ismael Boulliau (1605–1694) and his contemporaries.

Index

About the Author

William Burns is a historian specializing in Early Modern England. His study of prodigies in the late seventeenth century, *An Age of Wonders,* is scheduled to appear in 2002, and he is currently working on a reference book on eighteenth-century science for ABC-Clio. Dr. Burns lives in Washington, D.C.